Springer Theses

Recognizing Outstanding Ph.D. Research

Aims and Scope

The series "Springer Theses" brings together a selection of the very best Ph.D. theses from around the world and across the physical sciences. Each volume has been selected for its scientific excellence by an internationally renowned institute or department. For greater accessibility to non-specialists, the published versions include an extended introduction, as well as a foreword by the student's supervisor explaining the special relevance of the work for the field. The series as a whole provides a valuable resource both for newcomers to the research fields, and for other scientists seeking detailed background information on special questions. Finally, it serves as an accredited documentation of the valuable contributions made by today's younger generation of scientists.

How to Apply

Theses prepared at internationally leading universities and institutes may be nominated for publication by supervisors with the endorsement of the head of department. Each such department is entitled to nominate one thesis per year. Enquiries and requests for a nomination form should be sent, together with the thesis, to Ms. Nathalie Berg (see link on right).

Eligibility Criteria

To qualify for consideration, theses should fulfil the following criteria:

- They must be written in good English.
- The topic should fall within the confines of Chemistry, Physics, Earth Sciences, Engineering and related interdisciplinary fields such as Materials, Nanoscience, Chemical Engineering, Complex Systems and Biophysics.
- The work reported in the thesis must represent a significant scientific advance.
- A maximum of 30% of the text may be a verbatim reproduction from the author's previous publications.
- They must have been examined and passed during the 12 months prior to nomination.
- The theses should have a clearly defined structure including an introduction accessible to new PhD students and scientists not expert in the relevant field.
- After acceptance, theses should be supplemented by a "Supervisor's Foreword" outlining the wider scientific relevance of the work

Indexed by zbMATH.

Gunda Kipp

Self-Cavity Electrodynamics of Van der Waals Heterostructures

Springer

Gunda Kipp
Condensed Matter Dynamics Department
Max Planck Institute for the Structure
and Dynamics of Matter
Hamburg, Germany

ISSN 2190-5053 ISSN 2190-5061 (electronic)
Springer Theses
ISBN 978-3-032-31978-4 ISBN 978-3-032-31979-1 (eBook)
https://doi.org/10.1007/978-3-032-31979-1

This work was supported by Deutsche Forschungsgemeinschaft (170620586, 390715994). Open Access funding provided by Max Planck Society.

This Springer imprint is published by the registered company Springer Nature Switzerland AG
The registered company address is: Gewerbestrasse 11, 6330 Cham, Switzerland

To my family,
for their love and support,
and their inexhaustible, inspiring creativity.

Foreword

Over the past decade, van der Waals heterostructures—artificial stacks of atomically thin crystals—have become central to the study of emergent quantum phenomena in condensed matter. Their electronic and optical properties are highly sensitive to geometry, dielectric environment, and layer composition, offering an unusual degree of control that is difficult to achieve in conventional bulk materials. Many of the most intriguing effects in these systems occur at energy scales in the gigahertz–terahertz range, where collective electronic modes, plasmonic excitations, and low-energy dynamics shape the behavior of quantum phases. Yet this frequency window has remained experimentally challenging to access, especially for micron-sized devices encapsulated in insulating layers.

In this context, Gunda Kipp's doctoral thesis makes two particularly notable contributions. First, it demonstrates that van der Waals heterostructures naturally form what may be called self-defined terahertz cavities: confined electromagnetic modes that arise solely from the finite geometry and dielectric boundaries within the heterostructure. These cavity modes strongly interact with the electronic excitations of conductive layers such as graphene and graphite gates, leading to hybridized resonances that can enter the ultrastrong coupling regime. That such cavities emerge intrinsically, without externally fabricated mirrors or resonators, provides a new perspective on the electrodynamics of two-dimensional materials and suggests that cavity effects may influence their ground-state properties.

The second major element of the thesis is the development of a chip-scale terahertz spectroscopy platform capable of quantitatively measuring the complex optical conductivity of fully encapsulated, gate-tunable two-dimensional materials. Traditional far-field terahertz techniques are incompatible with micron-scale samples, and previously available near-field probes could not reliably access both real and imaginary conductivity. The method introduced here overcomes these limitations through a carefully engineered on-chip architecture, enabling direct measurements of electrodynamic response across a range of materials and device geometries. This tool is likely to be valuable well beyond the specific systems studied in the thesis, offering a general approach for probing low-energy excitations in quantum materials.

A further strength of the work lies in its theoretical analysis. The thesis develops a clear and instructive model of cavity–material coupling in van der Waals heterostructures, identifying the mechanisms by which plasmonic modes hybridize and how the resulting spectrum depends on geometry, carrier density, and dielectric environment. The analysis is closely tied to experiment: the framework not only interprets observed features such as avoided crossings and mode reshaping, but also provides practical guidance for designing future devices either to minimize cavity influence or to use it deliberately as a control knob.

The clarity with which these ideas are presented reflects the author's careful approach to both experiment and interpretation. The thesis is well structured, beginning with an accessible introduction that situates the problem in the wider landscape of cavity electrodynamics and two-dimensional materials. The experimental chapters describe the fabrication and measurement techniques with a level of detail that will be useful to newcomers, while the later chapters synthesize the results into a coherent picture of how cavity physics emerges in these systems. The figures, many of them based on time-domain spectroscopic data, are especially effective in conveying complex concepts in a visually intuitive way.

Taken together, the work offers a new framework for understanding and controlling electrodynamics in van der Waals heterostructures. By showing that cavity effects are intrinsic to many commonly used device architectures, and by providing the tools needed to probe them quantitatively, the thesis opens promising paths for future research on correlated phases, plasmonic engineering, and cavity-mediated phenomena in two-dimensional systems.

It has been a pleasure to supervise this research and to witness the development of the author into a thoughtful and independent scientist. I believe this thesis will serve as a valuable resource for researchers entering the field as well as for those seeking deeper insight into the interplay of geometry, electrodynamics, and quantum materials.

Assistant Professor of Physics,
Department of Physics,
Columbia University,
New York, NY, USA
2025

James McIver

Publications List

This thesis is the outcome of my research as a graduate student at the Max Planck Institute for the Structure and Dynamics of Matter in Hamburg, Germany, within the Quantum Condensed Matter Dynamics Department, and at the University of Hamburg. The findings presented in this manuscript are largely drawn from the following publications:

[1] **Gunda Kipp***, Hope M Bretscher*, Benedikt Schulte, Dorothee Herrmann, Kateryna Kusyak, Matthew W Day, Sivasruthi Kesavan, Toru Matsuyama, Xinyu Li, Sara Maria Langner, Jesse Hagelstein, Felix Sturm, Alexander M Potts, Christian J Eckhardt, Yunfei Huang, Kenji Watanabe, Takashi Taniguchi, Angel Rubio, Dante M Kennes, Michael A Sentef, Emmanuel Baudin, Guido Meier, Marios H Michael, James W McIver, *Cavity electrodynamics of van der Waals heterostructures*, https://doi.org/10.1038/s41567-025-03064-8, Nature Physics 21, 1926-1933 (2025).

[2] Marios H Michael, **Gunda Kipp**, Alexander M Potts, Matthew W Day, Toru Matsuyama, Guido Meier, Hope M Bretscher, James W McIver, *Resolving self-cavity effects in two-dimensional quantum materials*, arXiv:2505.12799 (2025).

The author was further involved in the following publications:

[3] Alexander M Potts*, Marios H Michael*, **Gunda Kipp**, Sara Maria Langner, Hope M Bretscher, Jonathan Stensberg, Kelson Kaj, Toru Matsuyama, Matthew W Day, Felix Sturm, Abhay K Nayak, Liam A Cohen, Xiaoyang Zhu, Andrea Young, James McIver, *Finite-momentum Cooper plasmons in superconducting terahertz microcavities*, arXiv:2601.10692 (2026).

[4] **Gunda Kipp**, Marios H Michael, Alexander M Potts, Dorothee Herrmann, Toru Matsuyama, Guido Meier, Matthew W Day, Hope M Bretscher, James W McIver, *Response regimes in on-chip THz spectroscopy*, arXiv:2511.23365 (2025).

[5] Xinyu Li, Jesse Hagelstein, **Gunda Kipp**, Felix Sturm, Kateryna Kusyak, Yunfei Huang, Benedikt Schulte, Alexander M Potts, Victoria Q Cordero, Chiara Trovatello, Zhi Hao Peng, Chaowei Hu, Jonathan M DeStefano, Takashi

Taniguchi, Kenji Watanabe, P James Schuck, Xiaodong Xu, Angel Rubio, Marios H Michael, Matthew W Day, Hope M Bretscher, James W McIver, *Purcell enhancement of photogalvanic currents in WTe_2 plasmonic cavities*, arXiv:2507.07987 (2025).

[6] Kateryna Kusyak, Benedikt Schulte, Toru Matsuyama, **Gunda Kipp**, Hope M Bretscher, Matthew W Day, Guido Meier, Alexander M Potts, James McIver, *Monolithic optoelectronic circuit design for on-chip terahertz applications*, APL photonics 10, 076117 (2025).

[7] Xinyu Li, Jesse Hagelstein, Felix Sturm, Kateryna Kusyak, Juan I Aranzadi, **Gunda Kipp**, Yunfei Huang, Benedikt Schulte, Alexander M Potts, Victoria Q Cordero, Chiara Trovatello, Zhi Hao Peng, Chaowei Hu, Jonathan M DeStefano, Takashi Taniguchi, Kenji Watanabe, Michael A Sentef, Dongbin Shin, P James Schuck, Xiaodong Xu, Jiun-Haw Chu, Hope M Bretscher, Matthew W Day, James W McIver, *On-chip terahertz emission from Floquet-Bloch states*, Opt. Mater. Express 15, 1765–1776 (2025).

[8] Matthew W Day, Kateryna Kusyak, Felix Sturm, Juan I Aranzadi, Hope M Bretscher, Michael Fechner, Toru Matsuyama, Marios H Michael, Benedikt F Schulte, Xinyu Li, Jesse Hagelstein, Dorothee Herrmann, **Gunda Kipp**, Alexander M Potts, Jonathan M DeStefano, Chaowei Hu, Yunfei Huang, Kenji Watanabe, Takashi Taniguchi, Guido Meier, Dongbin Shin, Angel Rubio, Jiun-Haw Chu, Dante M Kennes, Michael A Sentef, James W McIver, *Nonperturbative Nonlinear Transport in a Floquet-Weyl Semimetal*, arXiv:2409.04531 (2024).

Role of the Author

This thesis work started in February 2020 and ended in December 2024, and is the result of many great collaborations. James W. McIver conceived the experiment and supervised the project. The author developed the sample fabrication protocol for high-quality vdW heterostructure self-cavities embedded in on-chip THz circuitry, with support from Benedikt Schulte, and built more than 200 devices in total, from which 44 THz devices were successfully measured. From the devices discussed in this thesis, the author built devices TD3, TD4, TD6, BG7, BG8, GG2, GG3 and GG4. Many of the bare graphite devices were built in collaboration with Dorothee Herrmann as part of her master's thesis project (Devices TD1, TD2, TD5, TD7, BG1, BG2, BG3, BG5, BG6, BG9 and BGXL). Device BG4 was built by Sivasruthi Kesavan, and Devices GG1 and GG5 were built together with Sivasruthi Kesavan and Sara Maria Langner as part of their master's and bachelor's theses. The sample fabrication was further assisted through flake exfoliation and clean room support by Kateryna Kusyak, Felix Sturm, Benedikt Schulte, Xinyu Li, and Jesse Hagelstein. The on-chip THz spectroscopy circuitry with symmetric THz antenna and reference arm was developed by Kateryna Kusyak in her master's thesis project with experimental support by Benedikt Schulte and finite element simulations by Toru Matsuyama. Benedikt Schulte built the measurement setup. The author was involved in building a copy of the setup, and in designing and implementing an optical millikelvin cryostat to enable on-chip THz spectroscopy to be performed at a base temperature of 300 mK and under high magnetic fields. The author measured the test devices TD2, TD3, TD4, TD6; graphite devices BG2, BG4, BG7, BG8, and BGXL; and the ultrastrong coupling device GG2. The measurement of a plethora of early test devices that are not included in this thesis and of the device GG3 were performed by the author together with Benedikt Schulte, and the measurements of the devices TD1, TD5, TD7, BG1, BG3, BG5, BG6, BG9, GG1, GG4, and GG5 were performed by the author in collaboration with Hope M. Bretscher. Many of the measurements were further supported by Benedikt Schulte, Matthew W. Day, Guido Meier, Kateryna Kusyak, and Felix Sturm. Hope M. Bretscher developed the data analysis framework. The experimental data analysis was performed by the author together with Hope M. Bretscher and with assistance from Matthew W. Day and Benedikt Schulte. Marios H. Michael developed the analytical theory in close collaboration with the author and with support by Toru Matsuyama, who performed the finite element simulations. The analytical theory simulations were carried out by the author. Custom measurement electronics were provided by Toru Matsuyama and Guido Meier. Kenji Watanabe and Takashi Taniguchi provided high-quality hBN single crystals. Emmanuel Baudin, Dante M. Kennes, Michael A. Sentef, Angel Rubio, Christian J. Eckhardt, Alex M. Potts, Matthew W. Day, Andrea Cavalleri, and Guido Meier provided conceptual insight. The author, Hope M. Bretscher, Marios H. Michael, Toru Matsuyama, Alex M. Potts, Guido Meier, and James W. McIver, together with all coauthors, contributed to the development of the interpretation of the experimental and theoretical findings that resulted in the two main publications of the previous list.

Acknowledgments I am deeply grateful to everyone who supported, encouraged, and inspired me throughout this PhD journey.

A special thank you to Prof. Dr. James McIver for offering me this exciting PhD position in a group made up of truly amazing people. Thanks for your support, ideas, and guidance that continuously challenged and encouraged me to think big and aim higher.

A big thank you to Prof. Dr. Andrea Cavalleri for valuable scientific exchanges and for inspiring me from the very beginning to pursue both my master's and PhD theses in his department.

I would like to thank PD Dr. habil. Guido Meier for his unwavering support since my master's thesis. Thank you for your enthusiasm and collaborative spirit, and especially for your remarkable patience in carefully proofreading our 100-page paper and countless rebuttals time and again.

I am very grateful to Prof. Dr. Henning Moritz for being such a great mentor throughout my PhD. Your fantastic Bachelor lectures were a huge motivation for me early on in my studies, and I'm glad to have you as a part of my examining board. I also want to thank Prof. Dr. Ludwig Mathey for taking on the role of chair of my committee.

I extend my heartfelt thanks to Dr. Hope M. Bretscher, with whom I had the pleasure of working closely. Thank you for creating such an incredibly supportive and fun work environment in Hamburg, being always up for scientific discussions and being quick to offer thoughtful feedback, also for this thesis. I'll never forget our crazy night shifts in the lab, taking turns sleeping (you on a yoga mat in the hallway, me in a tent in the experimental hall). Team work makes the dream work! I'm so excited to see all the great things you'll do in the future.

I'm also deeply grateful to Dr. Marios H. Michael. Without you, we might still be scratching our heads over those resonances in the data. Working through Maxwell's equations together, developing the analytical theory, and finally making sense of the experimental results across so many devices was hands down the most exciting part of my PhD. Thank you for enlightening us, for your supervision, and for your patience with my experimental-physicist mindset.

Huge thanks to Dr. Toru Matsuyama for supporting the understanding of the experimental data and for backing up the analytical theory with countless full 3D electromagnetic simulations. I really appreciated your careful approach, and your endless patience in handling all the extra requests we threw your way. I still can't believe you sketched actual random flake shapes to precisely capture the response of specific heterostructures.

A special thank you to Dr. Benedikt Schulte for introducing me to the experimental workflows of the group and for guiding me through my first on-chip THz spectroscopy measurements. I thoroughly enjoyed our work together on the "Icequeen" (our optical mK cryostat) and was always happy to have you around in the lab, also because of your excellent music taste. I'm especially grateful for your helpful feedback while proofreading this thesis.

I am very grateful to Dorothee Herrmann for all her support in building the series of graphite heterostructures. Thank you for filling the lab with kindness, creativity, and ducklings. Sample fabrication was undoubtedly at its most fun when you were around. I'm especially glad that our collaboration turned into a close friendship, and that we still get to meet up, now not to craft masterpieces from van der Waals materials, but from clay!

Many thanks to Kateryna Kusyak, who joined me early on in my PhD journey. Your professionalism, ambition, and discipline have continually inspired me. I'm especially grateful for the memories we shared at the Gordon Research Conference in Ventura and the March Meeting in Minneapolis, from the excitement of scientific exchange to sticking together through some unsettling moments on the train.

I would like to thank Dr. Alex M. Potts, who immediately proved to be on the same (0.3 mm to 6 mm) wavelength as me. I am thankful for all scientific discussions we had, and the career guidance you provided. You were always someone I could ask about big physics questions or tiny technical details.

A big thank you to Dr. Matthew W. Day, who provided invaluable support across *all* sub-projects in the group while always bringing a cheerful and humorous spirit. Having you around in the lab was a true pleasure (though, I must admit, the laughter from your jokes occasionally created a bit too much noise for my sensitive measurements!).

A heartfelt thank you to Sivasruthi Kesavan and Sara Maria Langner for bringing such joyful and vibrant energy to our team. With both of you, everything became a little brighter (or should I say... pinker?). I'm grateful for the support we shared, especially during the toughest moments, with plenty of comforting food for the soul. Your motivation, openness, and authenticity have been truly inspiring. Thanks also to Sara for your thoughtful proofreading of some chapters of this thesis.

Tack så jättemycket to Felix Sturm, who was always ready to lend a hand whenever needed. I am grateful for your support during some of the most successful measurements on the ultrastrong coupling device, fittingly carried out in the midst of a Hamburg hurricane. As the campus emptied, the signal-to-noise ratio peaked: perfect conditions for great data!

Many thanks to Xinyu Li and Jesse Hagelstein for being fantastic office mates, for always asking the most insightful (and often the most challenging) questions, and for providing very helpful feedback on some of the chapters of this thesis.

I would like to acknowledge the support of the McIver group members in New York, some of whom I haven't yet had the pleasure of meeting in person, but whose valuable scientific exchange have come through Slack and Zoom: Yunfei Huang, Dr. Jonathan Stensberg, and Lucia Rondini.

I am sincerely thankful to Prof. Dr. Angel Rubio, Prof. Dr. Dante M. Kennes, Prof. Dr. Andrew Mackenzie, Prof. Dr. Michael A. Sentef, Prof. Dr. Emmanuel Baudin, Dr. Michael Fechner, Dr. Christian J. Eckhardt, and Alex Gómez Salvador for all the discussions we had and the valuable conceptual insights you provided.

I am also grateful for the great cleanroom support provided by Dr. Miriam Barthelmess, Dr. Eryin Wang, and Dr. Mariana Chavez Cervantes. A special thanks goes to Dr. Elena König for having me introduced to the cleanroom work, for always keeping supplies well-stocked, and for welcoming curious students into the lab during their internships. You are truly helping to inspire the next generation of researchers.

Further thanks go to Dr. Michele Buzzi, Dr. Daniele Nicoletti, and Dr. Carsten Putzke for generously sharing your expertise and for your valuable support in navigating some of the more complex experimental challenges encountered in the lab.

I am grateful to Boris Fiedler, Birger Höhling, Michael Volkmann, and Peter Licht for your excellent support in designing and fabricating custom-built equipment, from sample holders and electronics to an elevator for the optical mK cryostat, which greatly facilitated our experiments.

I would like to express my gratitude to Ulrike Warda and Sergiy Dovgopol for managing the many costly orders of laboratory equipment, particularly the difficult one of the mK system. Thank you for all your support, no matter the challenge.

Special thanks to all the wonderful people at MPSD, past and present, who have created such a unique and supportive environment. While there are too many to name individually, I would especially like to acknowledge: Niloofar Taherian Hosseinabadi, Danica Pavicevic, Dr. Edward Rowe, Dr. Mona Lively-Kalthoff, Dr. Kevin Lively, Dr. Giovanni de Vecchi, Dr. Sebastian Fava, Dr. Ting-Han Chou, Jyotiprakash Satapathy, Dr. Alex von Hoegen, Dr. Tobia Nova, Gregor Praedel, Dr. Michael Först, Dr. Jörg Harms, Jenny Witt, and Renukha Singam.

I am also deeply grateful to Dr. Neda Lotfiomran, Dr. Zrinka Gattin, Björn Witt and to Kathja Schröder, whose unwavering support was invaluable throughout my PhD.

I would also like to express my heartfelt gratitude to the remarkable mentors who have guided me both in life and throughout my PhD journey: Jost Freyer and Jens Schliecker. Jens, who would have thought that the Casimir effect you once explained during one of our piano lessons would eventually find its place in my PhD thesis?

A sound thank you to *WestendJazz*! Making music with you has always been food for the soul, and I'm deeply grateful for your patience during my absences, whether for spontaneous late-night measurements or when too much thesis typing left my hands unusable. A special thanks to Helga König, not only for being a wonderful rehearsal companion but also for your support with firing our many clay creations, born from both PhD frustration and joy.

A huge thank you to my friends, whose patience and support over the past years meant the world to me: from beach walks and knitting sessions to parties in Berlin, fun dinners, board game evenings and delicious meals. A heartfelt thanks to Nina, Maike, Laura, Torben, Oli, Sven, Lars, and the *Nanos*.

Last, but by no means least, I want to express my deepest gratitude to my family: To my mother Inga, for being a constant source of inspiration and encouragement. To my father Lutz—you might insist you had nothing to do with my decision to study physics, but I know it was you who first ignited my passion for science! To my brother Linus, who always reminds me that there is more to life than just physics, for example *Saunli*. To my grandparents Inge, Irene, and Ingo, whose unconditional love (and the world's best jams) have been a source of comfort and support throughout my journey. To Julie and Nige, for your special gift of making everyone around you feel seen, valued, and uplifted. And to Joe—you are my favorite "result" of this PhD. I'm endlessly grateful to have shared this journey with you, for all your support (including bravely proofreading this entire beast of a thesis), and I can't wait to see where life takes us next!

Competing Interests The author has no competing interests to declare that are relevant to the content of this manuscript.

Funding

This work was supported by the German Research Foundation (DFG) through the Cluster of Excellence CUI: Advanced Imaging of Matter (EXC 2056, Project ID 390715994) and through the SFB 925 (Project ID 170620586).

Contents

Chapter 1
Introduction: Van der Waals Materials

Abstract In this chapter, I introduce the material class investigated in this thesis: van der Waals materials. I survey the rich spectrum of physical phenomena they host, highlighting their exceptional tunability and the characteristic energy scales at which these phenomena emerge. I then discuss the strengths of THz spectroscopic techniques for probing the microscopic mechanisms underlying these systems, alongside the experimental challenges inherent to applying THz spectroscopy to van der Waals materials, arising from their low energy scales and typically small sample dimensions. To address these challenges, I introduce the on-chip THz spectroscopy approach developed and employed throughout this thesis. Finally, I outline the overall structure of the thesis.

New technologies are enabled by materials with extraordinary properties, often discovered through the curiosity-driven exploration of physicists rather than intentional design [1, 2]. When seeking to uncover new physical phenomena in materials, this exploration generally follows one of two main paths (see Fig. 1.1):

1. Creation of new materials.
2. Manipulation of existing materials.

In an ideal scenario, one would pursue both paths: developing new materials to uncover their fascinating physical properties, and then manipulating them further to explore additional phenomena or enhance their remarkable properties.

There is one material class, in particular, that is exceptionally well-suited for advancing both of these pathways: quantum materials. Quantum materials are materials that have intriguing macroscopic properties that fundamentally differ from anything predicted by classical pictures [3]. They encompass a broad range of phenomena: for instance, superconductors, in which electrical current flows without resistance or energy loss. If superconductivity could be realized at room temperature, it would offer a revolutionary solution to many of today's energy challenges. Another striking example are strange metals, which defy the conventional description of electron transport and exhibit a linear-in-temperature resistivity that persists down to low temperatures. This behavior suggests a breakdown of

G. Kipp, *Self-Cavity Electrodynamics of Van der Waals Heterostructures*,
Springer Theses, https://doi.org/10.1007/978-3-032-31979-1_1

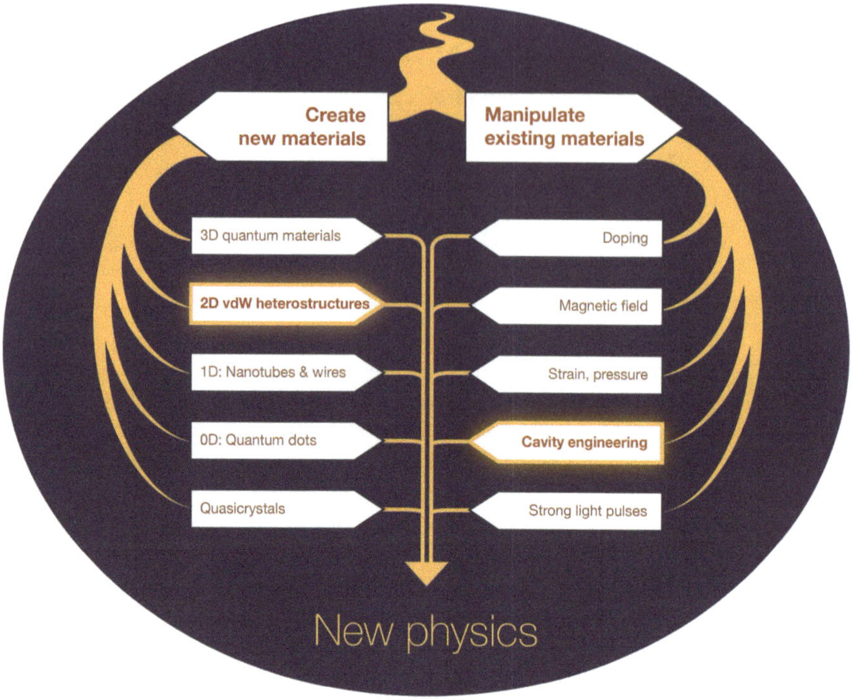

Fig. 1.1 Pathways to discover new physics. Typically, there are two main attractive routes to new physical phenomena: One can create new materials in various different dimensions, or one can manipulate existing, known materials. While three-dimensional (3D) quantum materials [3], one-dimensional (1D) nanotubes and wires [4], zero-dimensional (0D) quantum dots [5], and quasicrystals [6] offer exciting possibilities to create new materials, one particularly appealing approach is to stack various different two-dimensional (2D) crystals to create van der Waals (vdW) heterostructures [7, 8]. Manipulating materials can also be performed in many ways: One can dope a material, can expose it to strong magnetic fields, induce strain to the material or expose the material to enormous pressure. Alternatively, one can manipulate materials due to enhanced light-matter interactions provided by strong light pulses or by placing a material inside a confined space ("cavity") [9]. The advantage of cavity engineering over strong light pulses is that only few or even zero photons (quantum fluctuations) are required to potentially tailor macroscopic properties [10]. This thesis explores 2D vdW materials and discovers that they inherently form plasmonic cavities for THz light. This raises the intriguing possibility that the *self-cavity* electrodynamics of vdW materials enable intrinsic cavity engineering of gate-tunable vdW heterostructures

the traditional Fermi liquid paradigm and points to a deeper, yet unresolved, organizing principle behind their quantum dynamics [11]. Additionally, many quantum materials exhibit strong electronic correlations, where electron–electron interactions become so dominant that conventional band theory fails [12]. This gives rise to emergent phases such as Mott insulators, where materials that should behave as conductors instead exhibit insulating behavior due to these strong correlations.

While quantum materials can be zero-, one-, two-, or three-dimensional, an ideal playground for creating new materials was found with the two-dimensional class

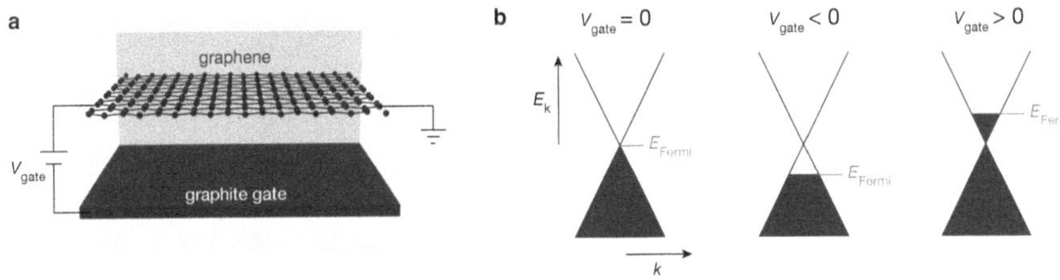

Fig. 1.2 Electrostatic gating of monolayer graphene. (**a**) A graphene monolayer is encapsulated in electrically insulating material, such as hexagonal boron nitride (hBN), and placed on top of a metallic graphite gate (graphite is a stack of multiple graphene layers). By applying a voltage V_{gate} between the graphite gate and graphene, the Fermi energy and with that the carrier density of the graphene layer can be tuned. (**b**) If $V_{gate} = 0$ is applied, the Fermi energy level is located at the Dirac point of graphene, where it is charge neutral. By applying a negative gate voltage, the number of electrons in the graphene is reduced, lowering the Fermi level. When applying a positive gate voltage, the opposite effect occurs and the Fermi energy is increased

of van der Waals (vdW) heterostructures, that can be created by stacking different two-dimensional (2D) crystals. Many of the constituent 2D crystals are already of intrinsic interest, exhibiting properties ranging from insulating and semiconducting to metallic, superconducting, or semimetallic behavior [13]. However, by combining them in specific sequences or at controlled twist angles, it becomes possible to engineer entirely new materials that exhibit emergent phenomena not present in the individual layers. Notable examples include superconductivity in twisted semimetals [12, 14, 15], correlated insulating states [16], polaritonic excitations [17–21], two-dimensional magnetism [22], and both integer and fractional quantum anomalous Hall effects [23, 24].

The most famous vdW material, and the primary focus of this thesis, is graphene, a single layer of carbon atoms arranged in a hexagonal lattice, just one atom thick. Renowned for its exceptional mechanical and electronic properties, graphene is approximately 200 times stronger than steel while remaining extraordinarily lightweight, and it enables electron mobilities up to 100 times higher than those in silicon [25]. Owing to its atomic-scale thickness, graphene is extremely sensitive to its surrounding electromagnetic environment [26], and its carrier density and electronic structure can be precisely tuned in situ via electrostatic gating (see Fig. 1.2). Moreover, graphene's properties are highly tunable through additional external parameters such as temperature, magnetic field, pressure, and strain.

In contrast to 3D quantum materials, where each carrier density typically requires the fabrication of a new sample, 2D devices offer the remarkable advantage of accessing a wide spectrum of low-energy quantum phenomena by simply tuning external parameters [27, 28]. A striking example is magic-angle twisted bilayer graphene, where two graphene sheets are rotated relative to each other by 1.1 degrees. In this system, a rich phase diagram, including superconductivity, correlated insulating states, and strange metallic behavior, can be explored continuously as a function of carrier density and temperature [12, 14, 15].

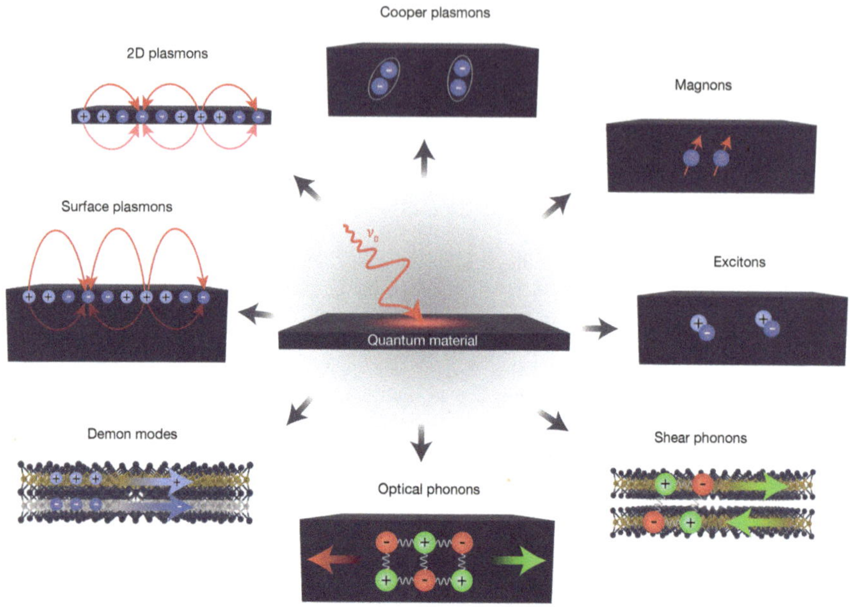

Fig. 1.3 Collective excitations in quantum materials. Quantum materials, especially two-dimensional ones, exhibit a pronounced sensitivity to light-matter interactions, and exhibit a wide range of collective excitations that can emerge within them. *2D plasmons* can form in thin or purely two-dimensional materials, which are collective oscillations of free charges. These charge oscillations can also form in thicker materials, where *surface plasmons* can oscillate at the surface of a material at a metal-dielectric interface. *Demon modes* can form in a material, where electrons and holes oscillate in-phase, leading to an "optically silent" mode, as there is not net dipole. Further, *optical phonons*, where atoms in the lattice move out-of-phase, or *shear phonons*, where atoms in adjacent layers oscillate collectively in opposite directions, can be excited. *Excitons*, bound electron-hole pairs, and *magnons*, collective excitations of spin systems, are other examples. Finally, *Cooper plasmons* can arise in superconductors, which are collective oscillations of Cooper pairs resulting in superconducting plasmons

A further advancement of these 2D quantum materials is their exceptional sensitivity to light–matter interactions [29]. Their strong in-plane coupling and weak interlayer interactions enable the formation of light–matter hybrids such as plasmon, magnon, and phonon polaritons (Fig. 1.3), with enhanced tunability due to reduced screening and high carrier mobility [30]. These collective excitations not only reveal rich physical behavior but also serve as sensitive probes of the underlying electronic order. In quantum materials, where competing microscopic interactions (like electron correlations) occur on similar energy scales [31], understanding which interactions drive emergent quantum phases is crucial for the potential design of novel materials tailored to specific requirements.

While significant progress has been made in understanding the physics behind these emergent phases and collective modes with current experimental techniques, certain aspects remain difficult to explore. Notably, the energy scale of many collec-

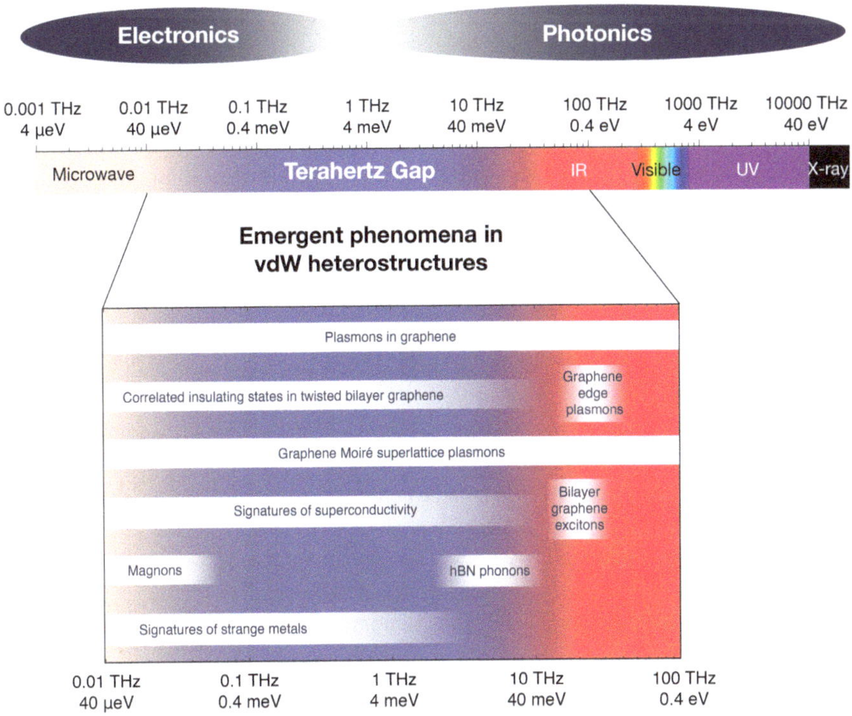

Fig. 1.4 Characteristic energy scale of phenomena in 2D materials. Many emergent phases and collective modes in vdW heterostructures are resonant to frequencies that fall into the so-called "Terahertz gap", that lies energetically between frequencies either accessible with electronics or photonics and is challenging to address experimentally. However, optics resonant to the characteristic energy scale of the phenomena in these materials can provide valuable insights. The graphic was created using the latest experimental findings from the following references: plasmons in graphene [33], correlated insulating states in twisted bilayer graphene [34], graphene edge plasmons [35], graphene Moiré supperlattice plasmons [36], signatures of superconductivity [15, 37], bilayer graphene excitons [38], magnons [39], hBN phonons [40, 41] and signatures of strange metals [42]. *Note*: the depicted phenomena potentially manifest across broader energy scales and in different heterostructures as future experiments will reveal

tive excitations, along with the emergent phenomena in 2D quantum materials, often falls within the 0.1 meV to 10 meV range, corresponding to frequencies between 0.024 THz to 2.4 THz (Fig. 1.4). This range lies within the so-called "Terahertz gap", which is an experimentally challenging region to probe, as it lies beyond the reach of conventional electronics, yet below the accessible domain of photonics [32].

However, optical probes operating at frequencies in the "Terahertz gap" resonant with the characteristic energy scale of the emergent phenomena in vdW heterostructures would directly enable the exploration of the evolution of energy gaps as a function of various tuning parameters, like carrier density, temperature or magnetic field. For example, by tracking how the energy gap of an insulator closes as a function of these tuning parameters, one can determine whether the microscopic origin of the insulating phase is due to electron correlations, indicating a "Mott-like" behavior, or if it has a more trivial cause [43]. In superconductors, probing THz dynamics offers insights into the symmetry of the superconducting gap, whether it is fully opened or contains nodes. This can be further perceived by detecting the temperature dependence of the superfluid density, which is encoded in the imaginary part of the optical conductivity [32, 44]. Thus, optical probes that operate at energy scales in the meV range (0.024 THz to 2.4 THz in frequency) are ideally suited to explore the origin of various different quantum phenomena in vdW heterostructures.

While local THz techniques, such as THz SNOM [45–47] and THz STM [48–54] provide vital local information and can detect spectral changes, extracting the global conductivity of a material, especially in samples with encapsulation and electrostatic gates, remains difficult. While THz spectroscopy addresses these limitations, far-field techniques struggle to probe small samples, which are much smaller than the THz diffraction limit (sub-wavelength).

The gap between free-space THz wavelengths and small sample sizes can be bridged by confining THz light to the near-field through metallic transmission lines using a method called on-chip THz spectroscopy [55–61]. By integrating micron-sized vdW heterostructures into optoelectronic circuitry, THz spectroscopy can be brought to the relevant temperature (0.3 K to 300 K), frequency (0.05 THz to 1 THz), energy (0.2 meV to 4 meV), time (0.1 ps to 20 ps) and length scales (nm – μ m) of the low-energy phenomena in vdW heterostructures [62–68]. On-chip THz spectroscopy does not only enable advanced characterization but also represents a step toward harnessing quantum materials as functional components in emerging technologies, where both efficient sample interfacing and signal propagation are essential.

The recent appearance of an increasing number of on-chip THz spectroscopy related studies demonstrates that this is a growing field, garnering keen interest as a technique to probe the THz electrodynamics of 2D quantum materials [55–61]. The work described in this thesis represents a significant advancement in the ultrafast optoelectronic circuitry interfaced with quantum materials for performing on-chip THz spectroscopy. Previous approaches [55, 57, 60, 61] relied on relative changes in the THz transmission as a function of gate voltage to approximate the conductivity of gate-tunable 2D materials. Such an approach always required the presence of a Dirac point or spectral gap in which the 2D material under investigation is assumed to be insulating to serve as a reference point, and the gate was treated as transparent and non-interacting with the sample. In contrast, the on-chip THz spectroscopy circuitry presented in this thesis introduces a separate transmission line reference arm, enabling the accurate determination of the absolute near-field conductivity of an entire heterostructure for the first time.

The new generation of on-chip THz spectroscopy circuitry with in situ referencing is first applied to characterize a series of graphite heterostructures. Graphite, being metallic and lacking an intrinsic charge neutrality point, could not be characterized in previous experiments without the additional reference arm introduced in this thesis. Intriguingly, the measured spectra reveal a new insight: rather than displaying the expected Drude response in the near-field conductivity spectrum, typical of large-area metallic samples [69, 70] and expected from the results of previous on-chip THz experiments [55, 60, 61], the graphite heterostructures exhibit resonances that are tunable by geometry and align with the Drude-Lorentz model. This raises the question: what is the origin of these resonances?

Structure of the Thesis

In Chap. 2, the principles of on-chip THz spectroscopy are introduced, including the generation and detection of THz frequencies (Sect. 2.1), and how THz light can be confined to the near-field by using transmission lines (Sect. 2.2). The new circuit architecture for on-chip THz spectroscopy used in this thesis is shown in Sect. 2.3.

Chapter 3 outlines the developed fabrication protocol for on-chip THz devices. Starting with the exfoliation of vdW materials (Sect. 3.1) and the method of atomic force microscopy lithography to precisely define the shape of vdW materials (Sect. 3.2), all fabrication steps are detailed in the following sections that lead to a full on-chip THz device with an embedded gate-tunable vdW heterostructure. Section 3.7 provides a summary of the entire sample fabrication protocol, highlighting personal tips that most probably result in high-quality on-chip THz devices.

Chapter 4 discusses the experimental setup (Sect. 4.1) and measurement protocols, including preparations and calibrations (Sect. 4.2), DC transport measurements (Sect. 4.3) and the on-chip THz measurement protocol (Sect. 4.4). How the data is processed is outlined in Sect. 4.5.

Chapter 5 introduces electrical and optical conductivity models (Sect. 5.1), and shows how the optical conductivity can be extracted from on-chip THz spectroscopy data (Sect. 5.3).

Chapter 6 explores the THz electrodynamics of a series of graphite heterostructures. Starting with a THz spectrum obtained from one graphite heterostructure that exhibits a clean Drude-Lorentz-like resonance in both real and imaginary parts of the near-field optical conductivity (Sect. 6.1), a series of different heterostructure designs are built, measured and analysed to characterize the origin of the resonance (Sect. 6.2). Based on the experimental data and full 3D electromagnetic simulations (Sect. 6.3), an analytical theory is developed to describe the THz electrodynamics of graphite heterostructures that are found to form "self-cavities" for THz light (Sect. 6.4). The analytical theory then enables the investigation of the self-cavity tunability through geometry and dielectric environment, as explored in Sect. 6.5. A final section of Chap. 6 discusses in which parameter space cavity responses can be expected with on-chip THz spectroscopy, and in which regime a Drude response can be detected (Sect. 6.6).

Chapter 7 provides the basic concepts of cavity engineering, including the definition of the typical four coupling regimes (weak, strong, ultrastrong and deep strong) in Sect. 7.4. Section 7.5 clarifies the outstanding suitablility of 2D materials for controlling macroscopic material properties via cavity engineering. Section 7.6

discusses how the known cavity engineering principles can be adapted to the plasmonic self-cavities found in this thesis, which led to the question: can plasmonic self cavities formed by electrostatic graphite gates modify the THz electrodynamics of a gated 2D material?

To find an answer to this question, Chap. 8 explores gate-tunable heterostructures consisting of two self-cavities, a graphene and a graphite gate layer. A first heterostructure is designed such that the graphite self-cavity resonance lies outside the spectral sensitivity range, to purely characterize the response of the graphene self-cavity (Sect. 8.1). Surprisingly, the resonance frequencies of the detected graphene plasmonic cavity modes in this device follow a carrier density dependence that is strongly suppressed from the expectation of (uncoupled) graphene plasmons. To understand this deviation, the analytical theory is extended in Sect. 8.2 to capture the near-field conductivity ("cavity conductivity") of a heterostructure consisting of two self-cavities (electrostatic gate and gated vdW material).

This extended theory reveals two distinct coupling mechanisms that enable hybridization between the self-cavities, as discussed in Sect. 8.3. Building on this understanding, a second gate-tunable graphene heterostructure is designed to spectrally resolve the ultrastrong coupling between graphene and graphite plasmonic modes within the detection range (Sect. 8.4). Moreover, the theory provides a foundation for formulating design principles aimed at either probing unperturbed ("cavity sensing") or intentionally tailored ("cavity control") phenomena in gate-tunable quantum materials (Sect. 8.6).

This thesis concludes with Chaps. 9 and 10 that summarize the results and offer an outlook on a selection of experiments inspired by the findings of this thesis, to be conducted in the near future.

References

1. Johnson S (2015) How we got to now: six innovations that made the modern world. Riverhead Books
2. Wellmann PJ (2021) The search for new materials and the role of novel processing routes. Discover Mater 1:14. https://doi.org/10.1007/s43939-021-00014-y10.1007/s43939-021-00014-y.
3. Keimer B, Moore JE (2017) The physics of quantum materials. Nat Phys 13:1045–1055. https://doi.org/10.1038/nphys430210.1038/nphys4302.
4. Xia Y, Yang P, Sun Y, Wu Y, Mayers B, Gates B, Yin Y, Kim F, Yan H (20023) One-dimensional nanostructures: synthesis, characterization, and applications. Adv Mater 15:353–389. https://doi.org/10.1002/adma.200390087
5. Bawendi MG, Steigerwald ML, Brus LE (1990) The Quantum Mechanics of larger semiconductor clusters ('quantum dots'). Annu Rev Phys Chem 41:477–496. https://doi.org/10.1146/annurev.pc.41.100190.002401
6. Levine D, Steinhardt PJ (1984) Quasicrystals: a new class of ordered structures. Phys Rev Lett 53:2477–2480. https://doi.org/10.1103/PhysRevLett.53.2477
7. Geim AK, Grigorieva IV (2013) Van der Waals heterostructures. Nature 499:419–425. https://doi.org/10.1038/nature12385
8. Castellanos-Gomez A, Duan X, Fei Z, Gutierrez HR, Huang Y, Huang X, Quereda J, Qian Q, Sutter E, Sutter P (2022) Van der Waals heterostructures. Nat Rev Methods Primers 2:58. https://doi.org/10.1038/s43586-022-00139-1

9. Luo Y, Zhao J, Fieramosca A, Guo Q, Kang H, Liu X, Liew TC, Sanvitto D, An Z, Ghosh S, Wang Z, Xu H, Xiong Q (2024) Strong light-matter coupling in van der Waals materials. Light: Sci Appl 13:203. https://doi.org/10.1038/s41377-024-01523-0
10. Forn-Díaz P, Lamata L, Rico E, Kono J, Solano E (2019) Ultrastrong coupling regimes of light-matter interaction. Rev Mod Phys 91:025005. https://doi.org/10.1103/RevModPhys.91.025005
11. Phillips PW, Hussey NE, Abbamonte P (2022) Stranger than metals. Science 377. https://doi.org/10.1126/science.abh4273
12. Balents L, Dean CR, Efetov DK, Young AF (2020) Superconductivity and strong correlations in moiré flat bands. Nat Phys 16:725–733. https://doi.org/10.1038/s41567-020-0906-9
13. Ajayan P, Kim P, Banerjee K (2016) Two-dimensional van der Waals materials. Phys Today 69:38–44. https://doi.org/10.1063/PT.3.3297
14. Andrei EY, MacDonald AH (2020) Graphene bilayers with a twist. Nat Mater 19:1265–1275. https://doi.org/10.1038/s41563-020-00840-0
15. Cao Y, Fatemi V, Fang S, Watanabe K, Taniguchi T, Kaxiras E, Jarillo-Herrero P (2018) Unconventional superconductivity in magic-angle graphene superlattices. Nature 556:43–50. https://doi.org/10.1038/nature26160
16. Mak KF, Shan J (2022) Semiconductor moiré materials. Nature Nanotechnol 17:686–695. https://doi.org/10.1038/s41565-022-01165-6
17. Grigorenko AN, Polini M, Novoselov K () Graphene plasmonics. Nat Photon 6:749–758. https://doi.org/10.1038/nphoton.2012.262
18. Basov DN, Asenjo-Garcia A, Schuck PJ, Zhu X, Rubio A (2020) Polariton panorama. Nanophotonics 10:549–577. https://doi.org/10.1515/nanoph-2020-0449
19. Low T, Chaves A, Caldwell JD, Kumar A, Fang NX, Avouris P, Heinz TF, Guinea F, Martin-Moreno L, Koppens F (2017) Polaritons in layered two-dimensional materials. Nat Mater 16:182–194. https://doi.org/10.1038/nmat4792
20. Du L, Molas MR, Huang Z, Zhang G, Wang F, Sun Z (2023) Moiré photonics and optoelectronics. Science 379. https://doi.org/10.1126/science.adg0014
21. Lin Z, Xiao C, Nguyen D-P, Arwas G, Ciuti C, Yao W (2023) Remote gate control of topological transitions in moiré superlattices via cavity vacuum fields. Proc Natl Acad Sci 120. https://doi.org/10.1073/pnas.2306584120
22. Huang B, McGuire MA, May AF, Xiao D, Jarillo-Herrero P, Xu X (2020) Emergent phenomena and proximity effects in two-dimensional magnets and heterostructures. Nat Mater 19:1276–1289. https://doi.org/10.1038/s41563-020-0791-8
23. Park H, Cai J, Anderson E, Zhang Y, Zhu J, Liu X, Wang C, Holtzmann W, Hu C, Liu Z et al (2023) Observation of fractionally quantized anomalous Hall effect. Nature 622:74–79. https://doi.org/10.1038/s41586-023-06536-0
24. Zibrov AA, Kometter C, Zhou H, Spanton E, Taniguchi T, Watanabe K, Zaletel M, Young A (2017) Tunable interacting composite fermion phases in a half-filled bilayer-graphene Landau level. Nature 549:360–364. https://doi.org/10.1038/nature23893
25. Broholm C, Fisher I, Moore J, Murnane M, Moreo A, Tranquada J, Basov D, Freericks J, Aronson M, MacDonald A, Fradkin E, Yacoby A, Samarth N, Stemmer S, Horton L, Horwitz J, Davenport J, Graf M, Krause J, Pechan M, Perry K, Rhyne J, Schwartz A, Thiyagarajan T, Yarris L, Runkles K (2016) Basic Research needs workshop on quantum materials for energy relevant technology. https://doi.org/10.2172/1616509
26. Song JC, Gabor NM (2018) Electron quantum metamaterials in van der Waals heterostructures. Nat Nanotechnol 13:986–993. https://doi.org/10.1038/s41565-018-0294-9
27. Basov DN, Averitt RD, Hsieh D (2017) Towards properties on demand in quantum materials. Nat Mater 16:1077–1088. https://doi.org/10.1038/nmat5017
28. Zou X, Xu Y, Duan W (2021) 2D materials: rising star for future applications. Innovation 2:100115. https://doi.org/10.1016/j.xinn.2021.100115
29. Basov DN, Fogler MM, de Abajo FJG (1992) Polaritons in van der Waals materials. Science 354. https://doi.org/10.1126/science.aag1992

30. Guo X, Lyu W, Chen T, Luo Y, Wu C, Yang B, Sun Z, García de Abajo FJ, Yang X, Dai Q (2023) Polaritons in Van der Waals Heterostructures. Adv Mater 35. https://doi.org/10.1002/adma.202201856
31. Goyal RK, Maharaj S, Kumar P, Chandrasekhar M (2025) Exploring quantum materials and applications: a review. J Mater Sci: Mater Eng 20:4. https://doi.org/10.1186/s40712-024-00202-7
32. Armitage N (2009) Electrodynamics of correlated electron systems. arXiv preprint arXiv:0908.1126
33. Politano A, Chiarello G (2014) Plasmon modes in graphene: status and prospect. Nanoscale 6:10927–10940. https://doi.org/10.1039/C4NR03143A
34. Calderón MJ, Bascones E (2020) Correlated states in magic angle twisted bilayer graphene under the optical conductivity scrutiny. NPJ Quantum Mater 5:57. https://doi.org/10.1038/s41535-020-00258-6
35. Nikitin AY, Alonso-González P, Vélez S, Mastel S, Centeno A, Pesquera A, Zurutuza A, Casanova F, Hueso LE, Koppens FHL, Hillenbrand R (2016) Real-space mapping of tailored sheet and edge plasmons in graphene nanoresonators. Nature Photonics 10:239–243. https://doi.org/10.1038/nphoton.2016.44
36. Ni GX, Wang H, Wu JS, Fei Z, Goldflam MD, Keilmann F, Özyilmaz B, Castro Neto AH, Xie XM, Fogler MM, Basov DN (2015) Plasmons in graphene moiré superlattices. Nat Mater 14:1217–1222. https://doi.org/10.1038/nmat4425
37. Stinson HT, Wu JS, Jiang BY, Fei Z, Rodin AS, Chapler BC, McLeod AS, Castro Neto A, Lee YS, Fogler MM, Basov DN (2014) Infrared nanospectroscopy and imaging of collective superfluid excitations in anisotropic superconductors. Phys Rev B 90. https://doi.org/10.1103/PhysRevB.90.014502
38. Ju L, Wang L, Cao T, Taniguchi T, Watanabe K, Louie SG, Rana F, Park J, Hone J, Wang F, McEuen PL (2017) Tunable excitons in bilayer graphene. Science 358:907–910. https://doi.org/10.1126/science.aam9175
39. Lujan D, Choe J, Rodriguez-Vega M, Ye Z, Leonardo A, Nunley TN, Chang LJ, Lee S-F, Yan J, Fiete GA, He R (2022) Magnons and magnetic fluctuations in atomically thin MnBi2Te4. Nat Commun 13:2527. https://doi.org/10.1038/s41467-022-29996-w
40. Dai S, Fei Z, Ma Q, Rodin AS, Wagner M, McLeod AS, Liu MK, Gannett W, Regan W, Watanabe K, Taniguchi T, Thiemens M, Dominguez G, Neto AHC, Zettl A, Keilmann F, Jarillo-Herrero P, Fogler MM, Basov DN (2014) Tunable phonon polaritons in atomically thin van der Waals crystals of boron nitride. Science 343:1125–1129. https://doi.org/10.1126/science.1246833
41. Brar VW, Jang MS, Sherrott M, Kim S, Lopez JJ, Kim LB, Choi M, Atwater H (2014) Hybrid surface-phonon-plasmon polariton modes in graphene/monolayer h-BN heterostructures. Nano Lett. 14:3876–3880. https://doi.org/10.1021/nl501096s
42. Prochaska L, Li X, MacFarland DC, Andrews AM, Bonta M, Bianco EF, Yazdi S, Schrenk W, Detz H, Limbeck A, Si Q, Ringe E, Strasser G, Kono J, Paschen S (2020) Singular charge fluctuations at a magnetic quantum critical point. Science 367:285–288. https://doi.org/10.1126/science.aag1595
43. Basov DN, Averitt RD, Van Der Marel D, Dressel M, Haule K (2011) Electrodynamics of correlated electron materials. Rev Mod Phys 83:471. https://doi.org/10.1103/RevModPhys.83.471
44. Sigrist M, Ueda K (1991) Phenomenological theory of unconventional superconductivity. Rev Mod Phys 63:239–311. https://doi.org/10.1103/RevModPhys.63.239
45. Guo X, Bertling K, Donose BC, Brünig M, Cernescu A, Govyadinov AA, Rakić AD (2024) Terahertz nanoscopy: advances, challenges, and the road ahead. Appl Phys Rev 11. https://doi.org/10.1063/5.0189061
46. Xu S, Li Y, Vitalone RA, Jing R, Sternbach AJ, Zhang S, Ingham J, Delor M, McIver J, Yankowitz M, Queiroz R, Millis AJ, Fogler MM, Dean CR, Hone J, Liu M, Basov D (2023) Electronic interactions in Dirac fluids visualized by nano-terahertz spacetime mapping. arXiv preprint arXiv:2311.11502

47. Zhang J, Chen X, Mills S, Ciavatti T, Yao Z, Mescall R, Hu H, Semenenko V, Fei Z, Li H, Dai Q, Liu M (2018) Terahertz nanoimaging of graphene. ACS Photon 5:2645–2651. https://doi.org/10.1021/acsphotonics.8b00190
48. Cocker T, Jelic V, Hillenbrand R, Hegmann F (2021) Nanoscale terahertz scanning probe microscopy. Nat Photon 15:558–569. https://doi.org/10.1038/s41566-021-00835-6
49. Jelic V, Adams S, Hassan M, Cleland-Host K, Ammerman SE, Cocker TL (2024) Atomic-scale terahertz time-domain spectroscopy. Nat Photon 18:898–904. https://doi.org/10.1038/s41566-024-01467-2
50. Sheng S, Abdo M, Rolf-Pissarczyk S, Lichtenberg K, Baumann S, Burgess JA, Malavolti L, Loth S Terahertz spectroscopy of collective charge density wave dynamics at the atomic scale. Nat Phys 1–6. https://doi.org/10.1038/s41567-024-02552-7
51. Roelcke C, Kastner LZ, Graml M, Biereder A, Wilhelm J, Repp J, Huber R, Gerasimenko YA (2024) Ultrafast atomic-scale scanning tunnelling spectroscopy of a single vacancy in a monolayer crystal. Nat Photon 1–8. https://doi.org/10.1038/s41566-024-01390-6
52. Bobzien L, Allerbeck J, Ammerman SE, Torsi R, Robinson JA, Schuler B (2024) Ultrafast state-selective tunneling in two-dimensional semiconductors with a phase-and amplitude-controlled THz-scanning tunneling microscope. APL Mater 12. https://doi.org/10.1063/5.0200845
53. Cocker TL, Jelic V, Gupta M, Molesky SJ, Burgess JA, Reyes GD, Titova LV, Tsui YY, Freeman MR, Hegmann FA (2013) An ultrafast terahertz scanning tunnelling microscope. Nat Photon 7:620–625. https://doi.org/10.1038/NPHOTON.2013.151
54. Müller M, Martín Sabanés N, Kampfrath T, Wolf M (2020) Phase-resolved detection of ultrabroadband THz pulses inside a scanning tunneling microscope junction. ACS Photon 7:2046–2055. https://doi.org/10.1021/acsphotonics.0c00386
55. Gallagher P, Yang C-S, Lyu T, Tian F, Kou R, Zhang H, Watanabe K, Taniguchi T, Wang F (2019) Quantum-critical conductivity of the Dirac fluid in graphene. Science 364:158–162. https://doi.org/10.1126/science.aat8687
56. Parker-Jervis R, Park S, Zonetti S, Sydoruk O, Wu J, Wood C, Li L, Davies A, Linfield E, Cunningham J (2020) On-chip terahertz spectroscopy of magnetoplasmons in a two-dimensional electron gas. In: 2020 45th international conference on infrared, millimeter, and terahertz waves (IRMMW-THz). https://doi.org/10.1109/irmmw-thz46771.2020.9370779
57. Zhao W, Wang S, Chen S, Zhang Z, Watanabe K, Taniguchi T, Zettl A, Wang F (2023) Observation of hydrodynamic plasmons and energy waves in graphene. Nature 614:688–693. https://doi.org/10.1038/s41586-022-05619-8
58. Sterbentz RM, Island JI (2023) On-chip sub-diffraction THz spectroscopy of materials and liquids. In: Meziani YM, Velázquez-Pérez J (eds) IntechOpen. https://doi.org/10.5772/intechopen.110877
59. A. M. Potts, A. K. Nayak, M. Nagel, K. Kaj, B. Stamenic, D. D. John, R. D. Averitt, and A. F. Young () On-chip time-domain terahertz spectroscopy of superconducting films below the diffraction limit. Nano Lett 23:3835–3841. https://doi.org/10.1021/acs.nanolett.3c00412
60. Chen SD, Feng Q, Zhao W, Qi R, Zhang Z, Abeysinghe D, Uzundal C, Xie J, Taniguchi T, Watanabe K, Wang F (2024) Direct measurement of terahertz conductivity in a gated monolayer semiconductor. arXiv preprint arXiv:2409.17633
61. Seo J, Lu Z, Park S, Yang J, Xia F, Ye S, Yao Y, Han T, Shi L, Watanabe K et al (2024) On-chip terahertz spectroscopy for dual-gated van der Waals Heterostructures at cryogenic temperatures. Nano Lett. https://doi.org/10.1021/acs.nanolett.4c04137
62. McIver JW, Schulte B, Stein FU, Matsuyama T, Jotzu G, Meier G, Cavalleri A (2020) Light-induced anomalous Hall effect in graphene. Nat Phys 16:38–41. https://doi.org/10.1038/s41567-019-0698-y
63. Karadi C, Jauhar S, Kouwenhoven LP, Wald K, Orenstein J, McEuen PL, Nagamune Y, Sakaki H (1994) Dynamic response of a quantum point contact. JOSA B 11:2566–2571
64. Zhong Z, Gabor NM, Sharping JE, Gaeta AL, McEuen PL (2008) Terahertz time-domain measurement of ballistic electron resonance in a single-walled carbon nanotube. Nat Nanotechnol 3:201–205. https://doi.org/10.1038/nnano.2008.60

65. Wang E, Adelinia JD, Chavez-Cervantes M, Matsuyama T, Fechner M, Buzzi M, Meier G, Cavalleri A (2023) Superconducting nonlinear transport in optically driven high-temperature K_3C_{60}. Nat Commun 14:7233. https://doi.org/10.1038/s41467-023-42989-7
66. Adelinia JD, Wang E, Chavez-Cervantes M, Matsuyama T, Fechner M, Buzzi M, Meier G, Cavalleri A (2025) Probing optically driven K_3C_{60} thin films with an ultrafast voltmeter. Struct Dyn 12:024503. https://doi.org/10.1063/4.0000295
67. Karnetzky C, Zimmermann P, Trummer C, Duque Sierra C, Wörle M, Kienberger R, Holleitner A (2018) Towards femtosecond on-chip electronics based on plasmonic hot electron nano-emitters. Nat Commun 9:2471. https://doi.org/10.1038/s41467-018-04666-y
68. Yoshioka K, Bernard G, Wakamura T, Hashisaka M, Sasaki KI, Sasaki S, Watanabe K, Taniguchi T, Kumada N (2024) On-chip transfer of ultrashort graphene plasmon wave packets using terahertz electronics. Nat Electron 7:537–544. https://doi.org/10.1038/s41928-024-01197-x
69. Horng J, Chen C-F, Geng B, Girit C, Zhang Y, Hao Z, Bechtel HA, Martin M, Zettl A, Crommie MF et al (2011) Drude conductivity of Dirac fermions in graphene. Phys Review B—Condensed Matter Mater Phys 83:165113. https://doi.org/10.1103/PhysRevB.83.165113
70. Razanoelina M, Bagsican FR, Kawayama I, Zhang X, Ma L, Murakami H, Vajtai R, Ajayan PM, Kono J, Tonouchi M (2016) Probing low-density carriers in a single atomic layer using terahertz parallel-plate waveguides. Opt Express 24:3885–3893. https://doi.org/10.1364/OE.24.003885

Chapter 2
Fundamentals of On-chip THz Spectroscopy

Abstract This chapter presents the fundamental principles of THz spectroscopy and its chip-scale counterpart, on-chip THz spectroscopy. It covers the generation and detection of THz radiation, the confinement of THz fields to the near field through metallic transmission lines, and introduces the new circuit architecture for on-chip THz spectroscopy used in this thesis.

VdW heterostructures host highly tunable quantum phases and collective modes that predominantly emerge in the THz frequency range, making resonant THz optical probes a uniquely powerful tool for investigating the low-energy phenomena of these materials. One such well-established THz optical probe is *far-field* time-domain THz spectroscopy [2].

The principle of far-field time-domain THz spectroscopy is to look at how a THz pulse is modified after interaction with a sample and three steps are required for this: generation of a THz pulse with the desired bandwidth, interaction of the THz pulse with the sample, and finally, detection of modifications to the THz pulse due to the interaction with the sample. The last step is conventionally performed by comparing the modified pulse to a reference THz pulse (see Fig. 2.1a). By taking the Fourier transform of the two transmitted time-domain signals and using Fresnel's equations, linear response functions such as the optical conductivity can be extracted from the sample as a function of frequency.

In the case of far-field time-domain THz spectroscopy applied to vdW heterostructures, a major challenge arises: the crucial second step, allowing the THz pulse to interact effectively with the sample, is hindered by the fact that far-field THz radiation cannot be focused onto vdW heterostructures, which are orders of magnitude smaller than the diffraction limit (see Fig. 2.1b). The diffraction limit at 1 THz is $\sim$150 μm, and at 50 GHz it is $\sim$3 mm, while typical samples are $\sim 10\times 10\,\mu\text{m}^2$ and only a few tens of nanometers thick, making far-field THz

G. Kipp, *Self-Cavity Electrodynamics of Van der Waals Heterostructures*, Springer Theses, https://doi.org/10.1007/978-3-032-31979-1_2

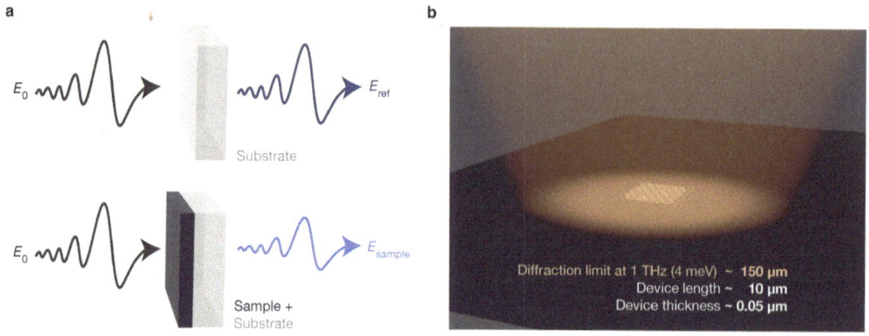

Fig. 2.1 Far-field time-domain THz spectroscopy: (**a**) To extract a sample's optical conductivity using time-domain THz spectroscopy, two measurements are conducted: First, a THz pulse with electric field E_0 is directed onto a bare substrate to record the resulting field E_{ref}, serving as a reference. Then, a second measurement is taken with the same THz pulse incident on the substrate with the sample, yielding a field E_{sample} that reflects the combined response of both substrate and sample. By comparing E_{sample} to E_{ref}, the sample's optical conductivity can be extracted. (**b**) The problem with vdW heterostructures and THz optical probes is that typical device dimensions are far smaller than the THz diffraction limit, precluding far-field measurements as a method to sense the electrodynamics of vdW heterostructures

spectroscopy prohibitively difficult with these materials. Instead, near-field optical probes are needed, where THz radiation is confined to a small region comparable to the sub-wavelength size of the sample. Examples of such methods include THz scanning probes such as THz SNOM [3–5] and THz STM [6–12], and on-chip THz spectroscopy [13–17], the employed method in this thesis.

On-chip THz spectroscopy overcomes the limitations of far-field methods by confining THz light to the near-field with integrated circuits to perform spectroscopy directly on a chip, and on length scales that match the sizes of typical vdW heterostructures. Importantly, on-chip THz spectroscopy, when using a reliable reference, enables the extraction of the real and imaginary parts of the global near-field optical conductivity of an entire vdW heterostructure.

This chapter discusses the fundamentals of on-chip THz spectroscopy and begins by introducing the methods of generating and detecting THz light (see Sect. 2.1). Next, the principles and design of on-chip THz spectroscopy transmission lines are discussed in Sect. 2.2. Section 2.3 introduces an architecture for on-chip THz spectroscopy with in situ referencing and Sect. 2.4 discusses the advances of this circuit architecture over previous approaches.

2.1 Generation and Detection of THz Frequencies

Nowadays, there are many different electronic and optoelectronic techniques to generate and detect THz frequencies [18]. Well-established sources of THz radiation include photoconductive switches [19], optical rectification [20] and quantum cascade lasers [21], as well as free electron lasers [22]. Typical THz detectors are photoconductive switches, too, and electro-optic crystals [23] as well as bolometers [24].

Particularly, optically driven sources have proven attractive, as they are based on few-cycle femtosecond (fs) pulses that enable direct measurement of the THz pulse's time trace. This allows for time-domain THz spectroscopy (TDTS) with full phase and amplitude information, eliminating the need for Kramers-Kronig relations to extract the sample's complex dielectric function [25, 26].

In the framework of this thesis, photoconductive switches were used that serve as both THz sources and detectors when triggered with femtosecond laser pulses. The working principle of a photoconductive switch, often also called an "Auston-switch" (named after its inventor David H. Auston [19]), is illustrated in Fig. 2.2. If a bias voltage is applied across a patch of a semiconducting material, in the "off state", no current flows, because the switch is insulating (Fig. 2.2a). But if the

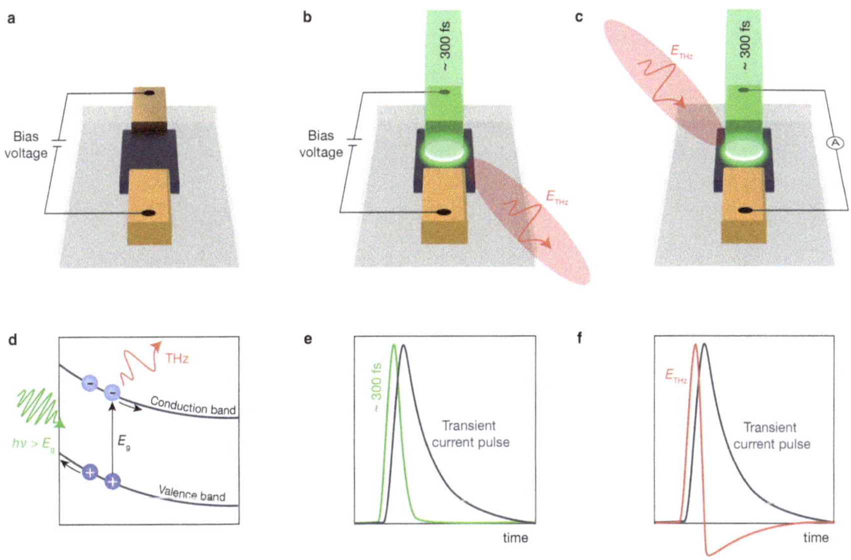

Fig. 2.2 Working principle of a photoconductive switch: (**a**) If a bias voltage is applied via two metal contacts (gold) across a photoconductive switch (dark blue), initially no current flows, as the switch is a semiconductor. (**b**) But if the switch is excited with a ~300 fs short laser pulse, a transient current pulse can flow through the switch allowing the emission of a THz electric field pulse (red). (**c**) A photoconductive switch can also be used as a detector: when a THz pulse is illuminating the switch, the electric field of the THz pulse transiently biases the switch itself. When a short laser pulse is then exciting the switch, a current can flow that is proportional to a slice of the time-domain THz pulse. By varying the delay between THz pulse and green fs pulse, the full electric field profile of the THz pulse can be mapped out. (**d**) If a light pulse with energy $h\nu > E_g$ is illuminating a photoconductive switch (semiconductor), photo-charge carriers are excited from the valence to the conductance band. THz light can be emitted through motion of the charges. (**e**) Temporal profiles of fs beam (green) and transient current pulse $I(t)$ (dark blue). (**f**) The transient current pulse (dark blue) emits an electric field pulse E_{THz} (red). (Adapted from [27, 28])

switch is then excited with a short (∼300 fs) laser pulse, charge carriers are excited from the valence to the conduction band (Fig. 2.2d), such that the switch becomes conductive. Through the applied bias voltage, charges can move and are accelerated, forming a transient current pulse $I(t)$ that flows across the switch (Fig. 2.2e). The duration of the current pulse is set by the duration of the laser pulse (rise time), and the recombination/relaxation time of the excited electrons (falling edge) [29]. As all of this is happening only over the course of a few hundreds of femtoseconds, the transient current pulse can emit a few- or single cycle THz electric field pulse $E_{\mathrm{THz}} \propto \frac{\partial I(t)}{\partial t}$ (Fig. 2.2b), with a full width half maximum of less than a picosecond (Fig. 2.2f) [27, 28].

If a photoconductive switch is used as a detector (see Fig. 2.2c), no bias voltage is applied across the switch. Instead, the THz pulse that shall be detected itself is used as a transient bias for the switch. Thus, if the switch is excited with a ∼300 fs short laser pulse, a transient current pulse can flow and can be detected with an ammeter, the amplitude of which is proportional to a temporal slice of the THz pulse. By then varying the delay between THz pulse and fs beam, the full temporal profile of the THz profile can be mapped out.

Having introduced the working principle of photoconductive switches, the following section shows how a generated THz pulse can be confined to the near-field using metallic transmission lines, enabling the guidance of a THz pulse from a generator to a detection switch and its interaction with a micron-sized vdW heterostructure in between.

2.2 On-chip THz Spectroscopy

This chapter explores the working principle of on-chip THz spectroscopy circuitry and the interplay of the individual components: THz generation via a photoconductive switch, signal propagation and sample interaction through metallic transmission lines, and detection of the sample-modified THz pulse using a second photoconductive switch.

Figure 2.3 illustrates an example of a simple ultrafast circuit for on-chip THz spectroscopy. It consists of a transmission line, that has a signal and a ground line, and two integrated photoconductive switches that serve as THz generator and detector. A vdW heterostructure can be embedded beneath the transmission line in the center between generator and detector switches. When the generator switch on the left is held under a voltage bias and simultaneously illuminated with a femtosecond laser pulse, a charge cloud is injected into the signal line. This charge cloud, in turn, induces an image dipole in the ground line. Because this charge cloud is injected over the course of a few hundred fs, the result is an ultrafast electric field pulse pointing from the signal to the ground line that then propagates down the transmission line. The interaction of the THz pulse and a vdW heterostructure modifies the spectral content of the THz pulse, which leads to a distortion of its electric field. The resulting modified THz pulse can be detected with the second

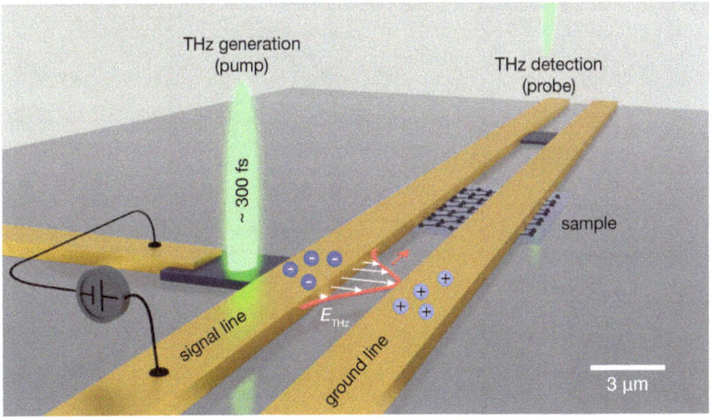

Fig. 2.3 Simple ultrafast circuit for on-chip THz spectroscopy: A laser-triggered photoconductive switch induces a charge cloud that, in turn, induces image dipoles to the ground line, such that electric field lines point from the signal to the ground line. As this happens within a period of ~300 fs, the induced electric field has the shape of a sub-picosecond pulse, capturing THz frequencies. This pulse propagates along the transmission line, where it interacts with the sample. The subsequent distorted pulse can be detected with a second photoconductive switch

photoconductive switch. By changing the delay between pump and probe beam, the full temporal profile of E_{THz} can be mapped out and used to extract the optical conductivity of the sample.

The following sections summarize the principles of transmission lines and designs (Sect. 2.2.1), their representation through Telegrapher's equations (Sect. 2.2.2) and the causes of signal attenuation, along with strategies for mitigating it (Sect. 2.2.3).

2.2.1 *Transmission Lines and Designs*

In electrical and telecommunications engineering, transmission lines guide electromagnetic waves through a closed circuit, such as a cable, over a defined distance. The transmission line type and geometry determines its electrical characteristics, such as impedance, signal propagation speed, attenuation, and dispersion, and thus defines the frequency range for which the best compromise between minimum attenuation and space is achievable [30].

There are various different designs of transmission lines such as coaxial cables, planar transmission lines (coplanar waveguide, coplanar strip), coupled wires (slotline), microstrip, stripline and waveguides (single conductor), some of which are illustrated in Fig. 2.4. Each transmission line has one fundamental polarization of the radio frequency (RF) or THz electric fields, and one fundamental propagation

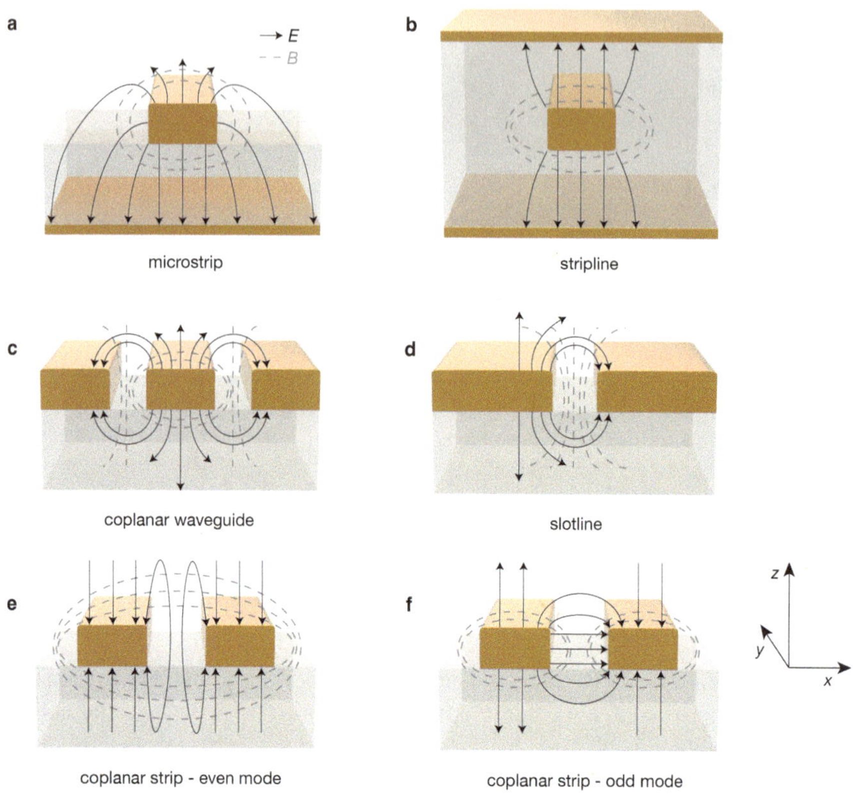

Fig. 2.4 Selection of planar transmission line designs: Cross-section and electric (E) and magnetic (B) field distributions of a (**a**) microstrip, (**b**) stripline, (**c**) coplanar waveguide, (**d**) slotline, (**e**) even mode in a coplanar strip and (**f**) odd mode in a coplanar strip. The transparent grey regions represent the dielectric substrate, while the gold structures indicate the metallic signal and ground lines or planes

mode, which is highly influenced by the geometry and dimension of the particular transmission line design [30, 31].

The three main propagation modes of electromagnetic waves in transmission lines propagating in y-direction are shown in Table 2.1 below [31].

Though there are many transmission line designs, only planar ones are relevant for interfacing two-dimensional vdW heterostructures. Figure 2.4 presents a selection of planar transmission line designs, where the layout of the signal and ground lines significantly alters the electric field distribution. In this thesis, the aim is to probe the in-plane optical conductivity of a vdW heterostructure. Thus, a transmission line design is required where the electric field primarily aligns with the x-direction. This condition is met by a coplanar strip transmission line, which supports the propagation of pure odd modes (see Fig. 2.4e, f).

Table 2.1 Main propagation modes of electromagnetic waves in transmission lines that propagate along the y-direction

Transverse electric mode (**TE**)	$E_y = 0$	$B_y \neq 0$
Transverse magnetic mode (**TM**)	$E_y \neq 0$	$B_y = 0$
Transverse electro-magnetic mode (**TEM**)	$E_y = 0$	$B_y = 0$

Coplanar strips consist of two parallel narrow metal traces (see Fig. 2.4e, f). The fundamental mode of a coplanar strip is considered as a *quasi*-TEM mode, due to the discontinuity of bottom dielectric (substrate) and top dielectric (vacuum). This discontinuity also results in an effective permittivity of the transmission line of $\epsilon_{\text{eff}} = (\epsilon_{\text{substrate}} + 1)/2$ [30].

Coplanar strips allow the propagation of both even modes (out-of-plane electric field between the metal strips) and odd modes (in-plane electric field between the metal strips). However, in order to probe the in-plane optical conductivity of a vdW heterostructure with on-chip THz spectroscopy, the transmission of pure odd modes is desirable, as achieved with the ultrafast circuitry described later in Sect. 2.3. Additionally, the transmission line should be designed to minimize both attenuation and dispersion of the propagating THz signal. Such active design can be guided by a theoretical framework of transmission lines, specifically through the Telegrapher's equations, that are introduced in the following section. These equations are essential not only for optimizing the design of a transmission line for THz pulse propagation, but also for providing the foundation for the Ansatz used to extract the near-field conductivity from a detected THz pulse that has interacted with a 2D sample, as discussed later in Sect. 5.3.

2.2.2 Telegrapher's Equations

The telegrapher's equations describe the change in the voltage and current of a wave propagating down a transmission line circuit with its characteristic impedance Z_0. The two sets of linear differential equations can be applied to every transmission line and are helpful for characterizing and optimizing signal propagation [31, 32].

In the model, the circuit is treated as a lumped element circuit that is comprised of elements of infinitesimal segments dy, described by a resistance (R'), inductance (L'), shunt capacitance (C'), and conductance (G'), all per unit length (see Fig. 2.5) [31]:

- R' captures the series resistance due to the finite conductivity of the metal traces (in Ohms per unit length),
- L' expresses the distributed inductance due to self-inductance of the metal traces which is represented by a series inductor (in henries per unit length),

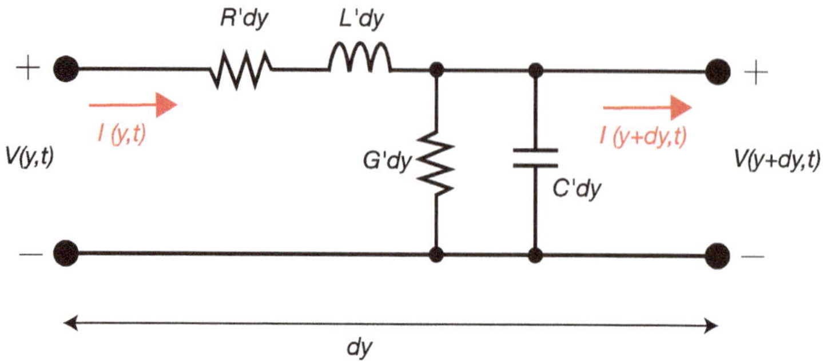

Fig. 2.5 Lumped element transmission line model: To model the time-dependent current and voltage along transmission lines, two conductors can be represented as an infinite sum of short segments, where each infinitesimal segment is comprised of an equivalent circuit with resistance R', inductance L', shunt capacitance C', and shunt conductance G', per unit length

- C' is the capacitance between the two metal traces representing a shunt capacitor, as some electrical energy is stored (in Farads per unit length),
- G' comprises the conductance which is responsible for the loss in the dielectric material (*lossy by nature*) between the two metal traces and is represented by a shunt resistor between the signal and the ground line (in Siemens per unit length).

Next, Kirchhoff's circuit laws ((1) Overall sum of currents meeting at one node amounts to zero. (2) Sum of voltages around a closed loop is zero) can be used to derive the following equations:

$$\tilde{V}(y,t) = R'dy\tilde{I}(y,t) + L'dy\frac{\partial}{\partial t}\tilde{I}(y,t) + \tilde{V}(y+dy,t), \tag{2.1}$$

$$\tilde{I}(y,t) = G'dy\tilde{V}(y,t) + C'dy\frac{\partial}{\partial t}\tilde{V}(y,t) + \tilde{I}(y+dy,t). \tag{2.2}$$

By applying the mathematical definition of the derivative (as shown in Ref. [33]), the two Telegrapher's equations can be obtained [31]:

$$\frac{d}{dy}\tilde{V}(y,t) = -(i\omega L' + R')\tilde{I}(y), \tag{2.3}$$

$$\frac{d}{dy}\tilde{I}(y,t) = -(i\omega C' + G')\tilde{V}(y). \tag{2.4}$$

Differentiating again gives then the two following differential equations:

$$\frac{d^2\tilde{V}(y)}{dy^2} = (R' + i\omega L')(G' + i\omega C')\tilde{V}(y) = \gamma^2\tilde{V}(y), \tag{2.5}$$

$$\frac{d^2\tilde{I}(y)}{dy^2} = (R' + i\omega L')(G' + i\omega C')\tilde{I}(y) = \gamma^2\tilde{I}(y). \tag{2.6}$$

The solution to these equations can be found as a superposition of traveling, attenuated waves propagating to the left and right along y, of the form,

$$\tilde{V}(y) = \tilde{V}^{+}(y)e^{-\gamma y} + \tilde{V}^{-}(y)e^{\gamma y} \tag{2.7}$$

$$\tilde{I}(y) = \tilde{I}^{+}(y)e^{-\gamma y} + \tilde{I}^{-}(y)e^{\gamma y}, \tag{2.8}$$

where

$$\gamma = \alpha + i\beta = \sqrt{(R' + i\omega L')(G' + i\omega C')} \tag{2.9}$$

is defined as the propagation constant of the transmission lines, and ω corresponds to the angular frequency $\omega = 2\pi\nu$. The real part of γ is the attenuation constant α (reduction of signal amplitude) and the imaginary part of γ corresponds to the phase constant β.

Another important parameter that characterizes a transmission line is the characteristic impedance Z_0. Z_0 is set by the geometry and material of a transmission line and is independent of the transmission line length. One can think about Z_0 as being the resistance of the transmission line, if it were infinitely long (thus neglecting reflections) [31].

From telegrapher's equations, the characteristic impedance can be written as:

$$Z_0 = \sqrt{\frac{R' + i\omega L'}{G' + i\omega C'}}. \tag{2.10}$$

Equation 2.10 shows, that in the limit of very small frequencies ($\omega \rightarrow 0$), the characteristic impedance reaches the value $Z_0 = \sqrt{R'/G'}$. In the opposite limit of very large frequencies (THz frequencies), where $\omega \gg R'/L'$ and $\omega \gg G'/C'$, the so called weak damping limit is reached and the characteristic impedance approaches the value

$$Z_0 = \sqrt{\frac{L'}{C'}}. \tag{2.11}$$

The characteristic impedance of a coplanar strip is given by

$$Z_0 = \frac{120\pi K(k)\sqrt{\mu_{\mathrm{re}}}}{\sqrt{\epsilon_{\mathrm{re}}}K(k')}, \tag{2.12}$$

where $K(k)$ and $K(k')$ are the elliptical integrals of $k = W_2/(2W_1 + W_2)$ (W_1 is the width of each metal trace, W_2 is the width of the coplanar strip gap) and $k' = \sqrt{1-k^2}$, $\mu_{\mathrm{re}} = (\mu_r + 1)/2$ and $\epsilon_{\mathrm{re}} = (\epsilon_r + 1)/2$ [30]. Calculating the characteristic impedance with Eq. 2.12 of a coplanar strip with varying k-values reveals that Z_0 is constant if both W_1 and W_2 are varied by the same amount, Z_0 increases if W_1 is decreased, or if W_2 is increased, or if ϵ_r is decreased.

More precise values of the characteristic impedance of a specific coplanar strip geometry can be obtained using RF and Microwave design tools such as Ref. [34].[1]

The actual value of the characteristic impedance does not matter. However, the impedance of the sample has to be comparable to the impedance of the coplanar strip. If the sample is highly metallic or superconducting, its impedance will be much lower than that of the transmission line, resulting in significant signal reflection and minimal transmitted signal, making it challenging to detect useful information. In contrast, if the sample has a low conductivity, its impedance will be higher, and the transmitted THz pulse will remain largely undistorted, though a high signal-to-noise ratio will be required to measure the THz electrodynamics accurately. Optimal signal resolution is achieved when the impedance of the sample is half the characteristic impedance of the coplanar strip ($Z_0 = 2Z_{\text{sample}}$), defining the point at which 50 % of the signal is reflected and 50 % is transmitted at the sample (for more details see Appendix A.8) [35].

A further thing to consider when designing a transmission line is to minimize losses that could lead to signal attenuation, as discussed in the following section.

2.2.3 *Signal Attenuation in Transmission Lines*

In transmission lines, particularly high-frequency signals such as THz pulses suffer from attenuation due to different phenomena such as radiation losses, conductor losses or dielectric losses. Particularly for THz frequencies and transmission distances longer than a few millimeters, which is desirable for on-chip THz spectroscopy to detect lower frequencies,[2] attenuation becomes especially important when designing a circuit.

Let's therefore briefly discuss Cherenkov radiation and the skin effect, which are the dominant sources for signal attenuation in transmission lines.

2.2.3.1 Cherenkov Radiation

For coplanar strips patterned on a substrate with dielectric constant $\epsilon_r > 1$, the propagating signal has a phase velocity of $v_{\text{signal}} = \sqrt{2}c/(\sqrt{\epsilon_r + 1})$, while light propagating in the substrate has a phase velocity of $v_{\text{light}} = c/\sqrt{\epsilon_r}$, where c is the speed of light in vacuum [30]. Thus, the signal in coplanar strip transmission

---

[1] A coplanar strip with $W_1 = W_2 = 3\,\mu\text{m}$ (W_1 is the width of each metal trace, W_2 is the width of the coplanar strip gap) placed on a 2 mm thick sapphire substrate as used in this thesis has a characteristic impedance of $Z_0 = 103\,\Omega$ at a frequency of 0.5 THz.

[2] The THz pulses in the circuits built in this thesis propagate with a velocity of $\sim 0.4c$, where c is the speed of light, such that in a 1 mm long transmission line, a reflection-free pulse detection is possible in the period of 16 ps, allowing a minimum detectable frequency of 60 GHz. In order to allow spectral resolution down to 10 GHz, a 6 mm long transmission line would be required.

lines propagates faster than light does in the substrate, producing an electromagnetic shock wave resembling Cherenkov radiation [35, 36].

Indeed, Cherenkov radiation is the dominant source of radiation losses in coplanar strips [37]. The radiation loss at frequency ν on a substrate with dielectric constant ϵ_r is given by [36]:

$$\alpha_{\text{radiation}} = \left(\frac{\pi}{2}\right)^5 \nu^3 \frac{\sqrt{2}(3-\sqrt{8})(2W_1+W_2)^2)}{c^3 K(k)K'(k)} \sqrt{\epsilon_r}^3 (1-1/\epsilon_r)^2(1+1/\epsilon_r)^{1/2}, \tag{2.13}$$

where W_1 is the width of each metal trace, W_2 is the width of the coplanar strip gap, $k = W_2/(2W_1 + W_2)$, and $K(k)$ and $K'(k)$ are the elliptical integrals of k. Equation 2.13 shows, that the attenuation due to Cherenkov radiation increases rapidly with increasing frequency $\alpha_{\text{radiation}} \propto \nu^3$ and also significantly with increasing coplanar strip width. Further, the metal trace width W_1 has a stronger effect on the attenuation than the coplanar strip gap width W_2.

Thus, by making the coplanar strips and the sample small (~few microns), Cherenkov radiation can be suppressed quite successfully. However, if the coplanar strip becomes too small, resistive losses would start to significantly reduce the bandwidth [37].

Another source of signal attenuation to consider is the skin effect, which is briefly introduced in the next section.

2.2.3.2 Skin Effect

In the transmission line model, the parameters R', L', G' and C' were assumed to be constant as a function of frequency. However, in reality, the parameters are frequency-dependent, as the so called skin effect changes the effective metal trace resistivity dependent on the frequency of the propagating signal. This is due to the penetration of high-frequency alternating currents into the surface of a conductor only up to a certain depth, the skin depth. This effect causes the current density j to be mostly confined near the surface of the conductor, and j decreases when approaching larger depths [31].

The skin effect arises due to circulating eddy currents: electric current loops induced by a varying magnetic field as provided by an alternating current in a conductor (Faraday's induction law) cancel the current flow inside the conductor and enhance it at the conductor's surface.

The skin depth δ, defined as the depth at which the current density decayed to $1/e$ relative to the surface current density, is defined in the good conductor limit $(\sigma \gg \epsilon_0\epsilon_r\omega)$ by

$$\delta = \sqrt{\frac{2\rho}{\omega\mu_0\mu_r}} = \sqrt{\frac{2}{\omega\mu_0\mu_r\sigma}}, \tag{2.14}$$

Table 2.2 Skin depths in gold ($\sigma = 4.6 \cdot 10^7$ S/m, $\mu_r = 1$) and graphite ($\sigma = 46{,}000$ S/m, $\mu_r = 1$) for three frequencies ω

ω	Gold	Graphite
10 GHz	0.74 μm	23.47 μm
100 GHz	0.23 μm	7.42 μm
1 THz	0.07 μm	2.35 μm

where σ is the specific conductivity and ρ is the specific resistivity of the conductor, μ_0 is the permeability of free space ($\mu_0 = 4\pi \cdot 10^{-7}$ Vs/Am) and μ_r is the permeability of the conductor [38]. As Eq. 2.14 shows, δ becomes smaller for increasing frequencies ω, making the skin effect more dominant, as the effective cross-section of the conductor is decreased. The resulting attenuation of the propagating signal due to the skin effect is (in Nepers per meter; 1 Neper = 8.69 dB):

$$\alpha_{\text{skin effect}} = \frac{1}{\delta}. \tag{2.15}$$

Example skin depths in gold and graphite are given in Table 2.2.

In on-chip THz spectroscopy, THz pulses consist of a frequency spectrum (~0.05–1 THz). This leads to additional dispersion due to the skin effect. Furthermore, if the skin effect is strong ($\delta \ll W_1$), and the effective resistance of the conductor R is increased as the current only propagates in a fraction of the transmission line's cross-section, conductor losses ($\propto I^2R$) become dominant [39]. Conductor losses appear as heat and reduce the efficiency of a transmission line. As high frequencies propagate at a higher resistance due to the more prominent skin effect, they suffer more from power losses than lower frequency components of a signal, leading to a reduced spectral sensitivity range.

For very narrow transmission lines ($\delta \approx W_1$), conductor losses due to the skin effect become severe. In this case, the cross-sectional area available for currents to flow is reduced not only due to the skin effect, but also due to the reduced geometric cross-section, such that the resistance R rises dramatically, increasing the conductor losses.

In summary, the skin effect can be kept small by using wider transmission lines ($W_1 \gg \delta$), using highly conductive metals,[3] and operating at lower frequencies (not an option for on-chip THz spectroscopy). Thus, wide transmission lines are favorable to reduce conductor losses, but narrow transmission lines are favorable to reduce radiation losses (see Sect. 2.2.3.1), which raises the question: Where lies the optimum?

The optimum coplanar strip geometry was found through various full 3D electromagnetic simulations by Dr. Toru Matsuyama and is somewhere around a $3 - 3 - 3$ geometry, meaning that both metal trace width W_1 and coplanar strip gap width W_2 are 3 μm, which is an ideally suited geometry to interface vdW

[3] Although a higher conductivity reduces the skin depth and enhances the skin effect, the overall conductor losses are still reduced to the reduced bulk resistivity of good conductors.

heterostructures (typical dimension: ∼10 × 10 μm^2). This geometry enables the propagation of a *quasi*-TEM mode with minimal dispersion and attenuation over a propagation distance of a few millimeters.

2.3 On-chip THz Spectroscopy Circuitry with In situ Referencing

Based on the optimum coplanar strip geometry ($W_1 = W_2 = 3\,\mu\text{m}$), an on-chip THz spectroscopy circuitry was designed and optimized [29, 40, 41], as depicted in Fig. 2.6. The key features are:

- A symmetric THz antenna (photoconductive switch) at the center of the device that emits a single-cycle THz field: Through the capacitive nature of the THz antenna, there is no spectral sensitivity at zero frequency, in contrast to DC

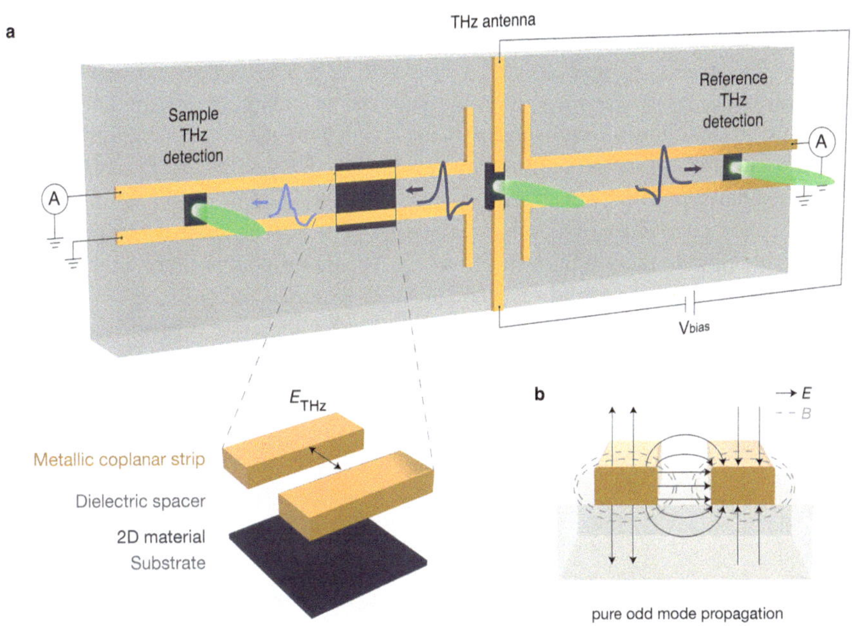

Fig. 2.6 On-chip THz spectroscopy circuitry with in situ referencing: (**a**) Circuit design for on-chip THz spectroscopy of 2D materials: Two identical THz pulses (dark blue) are emitted from a voltage-biased THz antenna and coupled into two symmetric coplanar strip transmission line arms. The THz pulse interacts with the sample (dark → light blue) and is measured using a photoconductive switch triggered by a delayed probe beam on the left. The signal can be compared to the reference signal measured on the detector on the right. (**b**) The circuitry excites and propagates pure odd modes, where the electric field is confined between the two metal strips [40]. (Figure adapted from [1] under a CC BY 4.0 license)

coupled THz generator switches, for which the spectral sensitivity peaks at zero frequency. However, the spectral sensitivity is maximum at a finite frequency, increasing the overall spectral sensitivity range to higher frequencies. This is particularly useful in on-chip THz spectroscopy, where the low frequency resolution is already limited due to the limited transmission line length.[4]

- In situ referencing: Due to the symmetry of the THz antenna, two identical pulses can be emitted to two opposite transmission line arms. One arm can remain empty to serve as a reference, while the other measurement arm contains the 2D material. Thereby, the Fourier transforms of the recorded time-domain signals on sample and reference side can be used to calculate the transmission coefficient, which in turn can be used for computing the real and imaginary parts of the optical near-field conductivity of the sample.
- Pure odd mode propagation (see Sect. 2.2.1): Previous on-chip THz experiments used a side-coupled photoconductive switch for THz generation, inherently exciting a mixture of odd and even modes in the coplanar strip. Through the symmetric launching of the THz pulse with the THz antenna and the narrow coplanar strip geometry ($W_1 = W_2 = 3\ \mu$m) of the circuit architecture, the even mode is suppressed, enabling the propagation of a pure odd, *quasi*-TEM mode (see Fig. 2.6b) [40]. This pure odd mode is ideal for probing the pure in-plane conductivity of a sample, as its electric field is primarily directed between the metal strips (x-direction), with evanescent out-of-plane components above and below (z-direction). Moreover, the *quasi*-TEM nature ensures a negligible field component along the propagation direction of the THz pulse (y), a crucial aspect in the analytical theory discussed in Sect. 6.4.
- Easy photoconductive switch fabrication protocol: α-silicon switches are simply evaporated onto the sapphire substrate via e-beam evaporation (see Sect. 3.6).
- Simultaneous DC transport measurements: The sample embedded in the ultrafast circuitry can extend the coplanar strip width, allowing placements of DC contacts. This enables gating of 2D materials during on-chip THz spectroscopy, while simultaneously performing DC transport measurements.
- 2 mm thick sapphire substrates with a particularly smooth surface: The thickness of the substrate reduces the impact of back reflections from the laser beams that excite the photoconductive switches.

---

[4] Typical substrates are 10 mm × 10 mm, giving a diagonal axis of 14 mm. In presence of a reference measurement arm, the length per measurement arm is then already only 7 mm. For avoiding back reflections from the transmission line end to distort the signal at the detector switch, the length between THz antenna and detection switch and detection switch and transmission line end should be similar, such that we remain with a maximum propagation distance of 3.5 mm, minus unusable substrate edges.

2.4 Discussion

This chapter introduced on-chip THz spectroscopy, the experimental platform of this thesis that enables the direct measurement of the complex optical near-field[5] conductivity of micron-sized vdW heterostructures.

Previous and recent on-chip THz spectroscopy experiments, such as Refs. [13, 14, 16, 17], measured relative changes in the conductivity in the absence of an absolute reference. This approach was partially driven by the challenge of producing identical photoconductive switches from typically used materials such as GaAs. GaAs switches must always be transferred from the growth substrate to the substrate where the ultrafast circuitry is processed, which can introduce significant strain and result in switches of varying quality. Thus, THz signals detected on two different GaAs switches, such as on separate reference and sample measurement arms, can't be compared, preventing a reliable reference measurement.

However, the circuitry used in this thesis is based on highly reproducible silicon switches, that can simply be evaporated on the sapphire substrate, where the circuitry is processed (see Sect. 3.6). This enables the introduction of a separate reference transmission line arm to measure a reference trace, a requirement of time-domain spectroscopy (see Fig. 2.1a), to actually extract the accurate THz complex conductivity without relying on the aforementioned assumption. This feature significantly expands the application possibilities of this technique as it enables the characterization and use of reliable metallic graphite gates, in addition to probing single-layer heterostructures and samples without gate-tunable insulating states. Importantly, both real and imaginary parts of the optical conductivity of the entire heterostructure can be probed. This is vital for accurate measurements of parameters such as superfluid density, particularly in systems where electronic correlations enhance the interactions between electrons and thereby shift spectral weight to high frequencies. In such cases, the usefulness of the Kramers-Kronig transformations to extract complex quantities would be limited, as these relations require knowledge of the entire frequency spectrum [42, 43].

While the dimensions of the transmission lines can generally be varied between sub-micrometer up to hundreds of micrometer widths, it was found through full 3D electromagnetic simulations that the optimum circuitry with minimum dispersion and attenuation consists of a coplanar strip geometry where both the metal traces and the coplanar strip gap are $\sim$3 μm wide, resulting in an overall transmission line width of $\sim$9 μm, which matches typical length scales of vdW heterostructures. If much smaller or larger heterostructures shall be interfaced or if the ratio of metal width and gap width shall be changed, it is desirable to only change the transmission line geometry locally, around the sample area, both on the sample and reference arm,

[5] Since the electric field is predominantly confined to the x-direction, on-chip THz spectroscopy operates in the near-field regime, where a portion of the THz pulse is evanescent, causing the electric field to decay rapidly along the z-direction.

to minimise attenuation and dispersion of the THz pulse propagating in the several millimeter long coplanar strips (see Sect. 2.2.3).

Finally, another advance of the circuitry architecture is the symmetric THz antenna placed in the center of the coplanar strip transmission lines. The antenna enables the excitation of pure odd modes (where the electric field is mostly confined in-plane between the metal traces), a prerequisite of capturing the pure in-plane conductivity of vdW heterostructures [32]. In contrast, previous on-chip THz experiments always used photoconductive switches placed outside the coplanar strips to generate THz pulses, which weakly excite even modes (where the electric field is pointing out-of-plane between the coplanar strips). This presence of an admixture of even modes in the THz pulse complicates the excitation mechanism of an integrated 2D material, potentially making it more challenging to reliably extract the intrinsic conductivity of a vdW heterostructure.

In summary, the on-chip THz spectroscopy platform used in this thesis incorporates an updated circuitry architecture designed to minimize dispersion and attenuation, guiding only odd modes, and allowing spectral resolution in the frequency range of ~0.05–1 THz. It further features a separate transmission line reference arm, allowing precise measurements of the absolute optical near-field conductivity of a vdW heterostructure, while simultaneously enabling DC transport measurements. Such a strategy opens up a path to probing various different (gate-tunable) vdW heterostructures in the future to investigate the dynamics of charge carriers, interband transitions, superfluid densities or resonant excitations of collective modes as a function of temperature, doping and magnetic field.

Having discussed the fundamentals of on-chip THz spectroscopy, the following section presents how a vdW heterostructure and the surrounding optoelectronic circuitry is fabricated.

References

1. Kipp G, Bretscher HM, Schulte B, Herrmann D, Kusyak K, Day MW, Kesavan S, Matsuyama T, Li X, Langner SM, Hagelstein J, Sturm F, Potts AM, Eckhardt CJ, Huang Y, Watanabe K, Taniguchi T, Rubio A, Kennes DM, Sentef MA, Baudin E, Meier G, Michael MH, McIver JW (2025) Cavity electrodynamics of van der Waals heterostructures. Nat Phys 21:1926–1933. https://doi.org/10.1038/s41567-025-03064-8
2. Grischkowsky D, Keiding S, van Exter M, Fattinger C (1990) Far-infrared time-domain spectroscopy with terahertz beams of dielectrics and semiconductors. J Opt Soc Am B 7:2006–2015. https://doi.org/10.1364/JOSAB.7.002006
3. Guo X, Bertling K, Donose BC, Bruünig M, Cernescu A, Govyadinov AA, Rakić AD (2024) Terahertz nanoscopy: advances, challenges, and the road ahead. Appl Phys Rev 11. https://doi.org/10.1063/5.0189061
4. Xu S, Li Y, Vitalone RA, Jing R, Sternbach AJ, Zhang S, Ingham J, Delor M, McIver J, Yankowitz M, Queiroz R, Millis AJ, Fogler MM, Dean CR, Hone J, Liu M, Basov D (2023) Electronic interactions in Dirac fluids visualized by nano-terahertz spacetime mapping. arXiv preprint arXiv:2311.11502
5. Zhang J, Chen X, Mills S, Ciavatti T, Yao Z, Mescall R, Hu H, Semenenko V, Fei Z, Li H, Dai Q, Liu M (2018) Terahertz nanoimaging of graphene. ACS Photonics 5:2645–2651. https://doi.org/10.1021/acsphotonics.8b00190

6. Cocker T, Jelic V, Hillenbrand R, Hegmann F (2021) Nanoscale terahertz scanning probe microscopy. Nat Photonics 15:558–569. https://doi.org/10.1038/s41566-021-00835-6
7. Jelic V, Adams S, Hassan M, Cleland-Host K, Ammerman S, Cocker T (2024) Atomic-scale terahertz time-domain spectroscopy. Nat Photonics 18:898–904. https://doi.org/10.1038/s41566-024-01467-2
8. Sheng S, Abdo M, Rolf-Pissarczyk S, Lichtenberg K, Baumann S, Burgess JA, Malavolti L, Loth S (2024) Terahertz spectroscopy of collective charge density wave dynamics at the atomic scale. Nat Phys:1–6. https://doi.org/10.1038/s41567-024-02552-7
9. Roelcke C, Kastner LZ, Graml M, Biereder A, Wilhelm J, Repp J, Huber R, Gerasimenko YA (2024) Ultrafast atomic-scale scanning tunnelling spectroscopy of a single vacancy in a monolayer crystal. Nat Photonics:1–8. https://doi.org/10.1038/s41566-024-01390-6
10. Bobzien L, Allerbeck J, Ammerman S, Torsi R, Robinson J, Schuler B (2024) Ultrafast state-selective tunneling in two-dimensional semiconductors with a phase-and amplitude-controlled THz-scanning tunneling microscope. APL Mater 12. https://doi.org/10.1063/5.0200845
11. Cocker TL, Jelic V, Gupta M, Molesky SJ, Burgess JA, Reyes GDL, Titova LV, Tsui YY, Freeman MR, Hegmann FA (2013) An ultrafast terahertz scanning tunnelling microscope. Nat Photonics 7:620–625. https://doi.org/10.1038/NPHOTON.2013.151
12. Müller M, Martín Sabanês N, Kampfrath T, Wolf M (2020) Phase-resolved detection of ultrabroadband THz pulses inside a scanning tunneling microscope junction. ACS Photonics 7:2046–2055. https://doi.org/10.1021/acsphotonics.0c00386
13. Gallagher P, Yang C-S, Lyu T, Tian F, Kou R, Zhang H, Watanabe K, Taniguchi T, Wang F (2019) Quantum-critical conductivity of the Dirac fluid in graphene. Science 364:158–162. https://doi.org/10.1126/science.aat8687
14. Zhao W, Wang S, Chen S, Zhang Z, Watanabe K, Taniguchi T, Zettl A, Wang F (2023) Observation of hydrodynamic plasmons and energy waves in graphene. Nature 614:688–693. https://doi.org/10.1038/s41586-022-05619-8
15. Potts AM, Nayak AK, Nagel M, Kaj K, Stamenic B, John DD, Averitt RD, Young AF (2023) On-chip time-domain terahertz spectroscopy of superconducting films below the diffraction limit. Nano Letters 23:3835–3841. https://doi.org/10.1021/acs.nanolett.3c00412
16. Chen S-D, Feng Q, Zhao W, Qi R, Zhang Z, Abeysinghe D, Uzundal C, Xie J, Taniguchi T, Watanabe K, Wang F (2024) Direct measurement of terahertz conductivity in a gated monolayer semiconductor. arXiv preprint arXiv:2409.17633
17. Seo J, Lu Z, Park S, Yang J, Xia F, Ye S, Yao Y, Han T, Shi L, Watanabe K et al (2024) On-chip terahertz spectroscopy for dual-gated van der Waals heterostructures at cryogenic temperatures. Nano Letters. https://doi.org/10.1021/acs.nanolett.4c04137
18. Zouaghi W, Thomson MD, Rabia K, Hahn R, Blank V, Roskos HG (2013) Broad-band terahertz spectroscopy: principles, fundamental research and potential for industrial applications. Eur J Phys 34:S179. https://doi.org/10.1088/0143-0807/34/6/S179
19. Auston DH (1975) Picosecond optoelectronic switching and gating in silicon. Appl Phys Lett 26:101–103. https://doi.org/10.1063/1.88079
20. Bass M, Franken PA, Ward JF, Weinreich G (1962) Optical rectification. Phys Rev Lett 9:446–448. https://doi.org/10.1103/PhysRevLett.9.446
21. Faist J, Capasso F, Sivco DL, Sirtori C, Hutchinson AL, Cho AY (1994) Quantum cascade laser. Science 264:553–556. https://doi.org/10.1126/science.264.5158.553
22. Schöllkopf W, Gewinner S, Junkes H, Paarmann A, von Helden G, Bluem HP, Todd AMM (2015) The new IR and THz FEL facility at the Fritz Haber Institute in Berlin. In: Biedron SG (ed) Advances in X-ray free-electron lasers instrumentation III, vol 9512. International Society for Optics and Photonics. SPIE, p. 95121L. https://doi.org/10.1117/12.2182284
23. Wu Q, Zhang X-C (1995) Free-space electro-optic sampling of terahertz beams. Appl Phys Lett 67:3523–3525. https://doi.org/10.1063/1.114909
24. Simoens F (2014) THz bolometer detectors. In: Perenzoni M, Paul DJ (eds) Physics and applications of terahertz radiation. Springer Netherlands, Dordrecht, pp 35–75. https://doi.org/10.1007/978-94-007-3837-9_2

25. Hoffmann MC, Fülöp JA (2011) Intense ultrashort terahertz pulses: generation and applications. J Phys D: Appl Phys 44:083001. https://doi.org/10.1088/0022-3727/44/8/083001
26. Han PY, Zhang X-C (2001) Free-space coherent broadband terahertz time-domain spectroscopy. Meas Sci Technol 12:1747. https://doi.org/10.1088/0957-0233/12/11/301
27. Sakai Y, Kawayama I, Nakanishi H, Tonouchi M (2015) Visualization of GaN surface potential using terahertz emission enhanced by local defects. Sci Rep 5:13860. https://doi.org/10.1038/srep13860
28. Kumar M (2016) The development of on-chip THz time-domain spectroscopy. PhD Thesis. University of Leeds
29. Schulte BF (2022) Ultrafast transport in tunable graphene heterostructures. PhD Thesis. University of Hamburg
30. DiPaolo F (2000) Networks and devices using planar transmission lines. CRC Press LLC
31. Kipp G (2019) Towards on-chip THz spectroscopy of quantum materials and heterostructures. Master's thesis. University of Hamburg
32. Di Paolo F (2018) Networks and devices using planar transmissions lines. CRC Press
33. Talluri SR (2016) Analysis of transmission line models with non linear elements for opposite phase and group velocities. PhD Thesis. Jaypee University of Information Technology
34. Coplanar strips calculator (2025). https://zalophusdokdo.github.io
35. Potts AM (2023) On-chip terahertz time domain spectroscopy for sub-diffraction van der Waals heterostructures. PhD Thesis. University of California, Santa Barbara
36. Grischkowsky D, Duling III IN, Chen JC, Chi C (1987) Electromagnetic shock waves from transmission lines. Phys Rev Lett 59:1663–1666. https://doi.org/10.1103/PhysRevLett.59.1663
37. Grischkowsky D (2000) Optoelectronic characterization of transmission lines and waveguides by terahertz time-domain spectroscopy. IEEE J Sel Top Quantum Electron 6:1122–1135. https://doi.org/10.1109/2944.902161
38. Wheeler H (1942) Formulas for the Skin Effect. Proc IRE 30:412–424. https://doi.org/10.1109/JRPROC.1942.232015
39. Pryer D, Gale P (2003) 31 - Cables. In: Laughton M, Warne D (eds) Electrical engineer's reference book, 16th edn. Newnes, Oxford, pp. 31-1–31-44. https://doi.org/10.1016/B978-075064637-6/50031-9
40. Kusyak K, Schulte B, Matsuyama T, Kipp G, Bretscher HM, Day MW, Meier G, Potts AM, McIver JW (2025) Monolithic optoelectronic circuit design for on-chip terahertz applications. APL Photonics 10:076117. https://doi.org/10.1063/5.0278183
41. Kusyak K (2022) Increasing the Bandwidth of on-Chip THz Spectroscopy. Master's thesis. University of Hamburg
42. Armitage N (2009) Electrodynamics of correlated electron systems. arXiv preprint arXiv:0908.1126
43. Basov DN, Averitt RD, Van Der Marel D, Dressel M, Haule K (2011) Electrodynamics of correlated electron materials. Rev Mod Phys 83:471. https://doi.org/10.1103/RevModPhys.83.471

Chapter 3
Fabrication Protocol for On-chip THz Devices

Abstract This chapter outlines the fabrication protocol developed for on-chip THz devices. It describes the process from the exfoliation of van der Waals materials to the precise patterning of their geometry using atomic force microscopy lithography, detailing all steps that culminate in a fully functional on-chip THz device with an embedded, gate-tunable van der Waals heterostructure. The chapter concludes with a summary of the complete fabrication workflow, including practical insights and tips that contribute to the realization of high-quality devices.

The fabrication protocol for building high-quality gate-tunable vdW heterostructures and surrounding circuitry was initiated with the support of Dr. Benedikt Schulte [1] and continuously optimized throughout this work. In contrast to typical DC transport devices, vdW heterostructures for on-chip THz spectroscopy measurements have to be carefully shaped and electrically isolated from the circuitry, increasing the level of complexity in the fabrication process.

Within this thesis, two different types of vdW heterostructures are investigated with the on-chip THz spectroscopy circuitry introduced in Sect. 2.3. Firstly, graphite heterostructures consisting of a few-nm thick graphite layer and an insulating hexagonal boron nitride (hBN) layer are probed (see Chap. 6). Secondly, gate-tunable graphene heterostructures are measured (see Chap. 8) that consist of a graphene monolayer encapsulated in insulating hexagonal boron nitride (hBN) on a graphite gate.

The hBN encapsulation of the graphene and graphite flakes offers multiple advantages: First, it enables the strain-free and clean pick-up and transfer of graphene and graphite flakes, as it has a similar (honeycomb) lattice structure as graphene. Secondly, it protects the intrinsic properties, like high mobility, of the graphene from the deleterious effects of the environment [2]. Thirdly, the hBN flakes electrically isolate graphene and graphite flakes from one another (in case of a gate-tunable heterostructure) and from the ultrafast circuitry that is patterned on top of the heterostructure, as shown in the following sections.

G. Kipp, *Self-Cavity Electrodynamics of Van der Waals Heterostructures*,
Springer Theses, https://doi.org/10.1007/978-3-032-31979-1_3

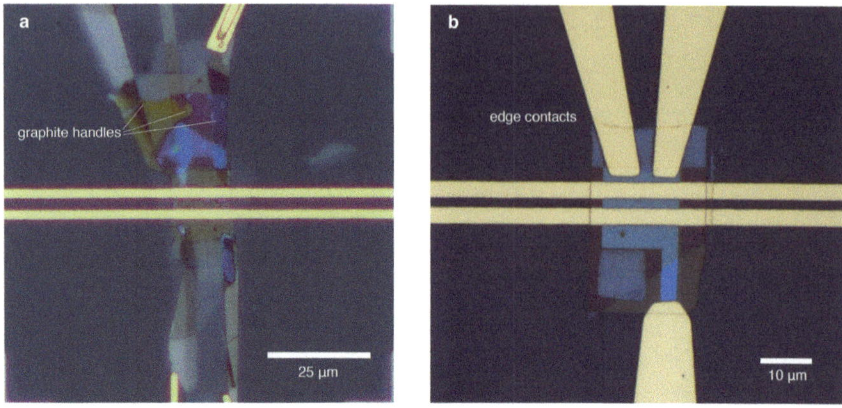

Fig. 3.1 Comparison of devices built in 2020 and 2023: (**a**) In 2020, nine different, randomly shaped flakes were assembled, because narrow graphite handles were required to contact graphene and graphite back gate flakes underneath the protecting hBN layers. Due to the inhomogenous flakes and high amount of deposited glue, the stack hosted a large number of bubbles. (**b**) With an improved fabrication protocol, a high quality vdW heterostructure with edge contacts consisting of only four flakes was fabricated in 2023, employing precisely cut flakes

At the early stage of the device fabrication (2020), nine different, randomly shaped flakes were assembled to fabricate a gate-tunable graphene heterostructure. Problems with the first fabrication procedure included a high failure rate due to the high number of assembled flakes and a significant amount of encapsulated glue, leading to bubbles, reducing the device quality (see Fig. 3.1a). In this thesis, the number of stacked flakes was reduced to four by employing reactive ion etching and AFM cutting methods [3], eliminating the need for graphite handles to contact graphene and graphite beneath the protective hBN layer. These pre-cut flakes, along with the use of clean 1D edge contacts and other methodological improvements, resulted in the fabrication of well-defined, bubble-free, high-quality vdW heterostructures for on-chip THz spectroscopy (see Fig. 3.1b).

Given the similarity in fabrication processes of pure graphite heterostructures and gate-tunable graphene-graphite heterostructures, with the first being simpler to implement, the fabrication protocol for complete gate-tunable graphene-graphite heterostructures is discussed in the following.

This chapter is structured as follows: the exfoliation of vdW materials is described in Sect. 3.1 and the precision cutting of vdW materials through atomic force microscopy is described in Sect. 3.2. The heterostructure assembly is explained in Sect. 3.3 and the transfer of the heterostructure to a substrate is discussed in Sect. 3.4. The method of reactive ion etching (RIE) is introduced in Sect. 3.5 and the fabrication of the ultrafast circuitry and contacting is shown in Sect. 3.6.

3.1 Exfoliation of van der Waals Materials

The first step in integrating vdW heterostructures into on-chip THz spectroscopy circuitry is obtaining the individual material layers that will form the target heterostructure. This requires clean, step- and fold-free graphene monolayers, graphite flakes 7–25 nm thick, and hexagonal boron nitride (hBN) crystals 15–100 nm thick, with typical lateral dimensions of $\sim 15 \times 20\ \mu m^2$ for graphene and $\sim 40 \times 40\ \mu m^2$ for graphite and hBN. These flakes are obtained via mechanical exfoliation, a technique introduced by K. S. Novoselov and A. K. Geim in 2004 [4]. Owing to the weak vdW forces between 2D material layers, individual sheets can be easily separated using adhesive tape.

Bulk "Flake Graphite" was obtained from *NGS Trading & Consulting GmbH*, from which graphene and graphite flakes were exfoliated using an adhesive strip of *Scotch Magic tape* (see Fig. 3.2a). A small, thin, and visually uniform graphite crystal is placed near one end of the tape, while the opposite end is folded over it in a crisscross manner to split the crystal and reduce its thickness. This process is repeated multiple times, each time placing the resulting graphite pieces side-by-side

Fig. 3.2 Exfoliation of graphene, graphite and hBN crystals: (**a**) Bulk graphite flakes are used to (**b**) exfoliate graphene and graphite crystals with an adhesive tape to (**c**) obtain high-quality graphene and graphite flakes on a Si/SiO_2 substrate. (**d**) With a similar procedure, bulk hBN crystals are used to (**e**) exfoliate thin hBN crystals with an adhesive tape to (**f**) obtain high-quality hBN flakes on a Si/SiO_2 substrate

on opposing ends of the tape. Through successive exfoliations, a uniform area of few-layer or even monolayer graphite can be created, typically covering a region of approximately $1 \times 1\ \text{cm}^2$ (see Fig. 3.2b).

The next step is to transfer the crystals from the tape to a substrate to search for ideal flakes. Silicon substrates with a 285 nm SiO_2 layer produced by Silicon Materials are manually cut into $\sim 1 \times 1\ \text{cm}^2$ pieces from a larger wafer. To transfer the crystals from the tape to a substrate, the tape sections bearing the thin graphite flakes are pressed onto the pre-heated (104° C) silicon substrates. To ensure a clean transfer with minimal adhesive residue, the tape is gently pressed onto the preheated substrate, which is then maintained at temperature for 3–60 s. After cooling to room temperature, the tape is slowly peeled away. Van der Waals and capillary forces ensure that few-layer graphite and graphene monolayers remain adhered to the substrate surface (see Fig. 3.2c) [4].

Insulating hBN flakes are obtained through a slightly modified exfoliation procedure. Since 2021, high-quality hBN crystals produced by Kenji Watanabe and Takashi Taniguchi have been used in place of those from *2D Semiconductors* (see Fig. 3.2d). Unlike the process for graphene and graphite, which uses a single strip of *Scotch Magic Tape*, the hBN exfoliation begins with a stickier *Scotch Magic Greener Tape* (the "mother tape"), from which six "daughter tapes" are created using regular *Scotch Magic Tape*. The Greener tape's stronger adhesion helps fragment and redistribute two to three small hBN bulk crystals into a more uniform area of thin, few-layer flakes, similar to the earlier method for graphene and graphite (see Fig. 3.2e). The hBN crystals are transferred onto silicon substrates at room temperature and the tape removal should be carried out as slowly as, or even more gradually than, in the graphene exfoliation process to ensure successful transfer without damaging the flakes.

The SiO_2 substrates with the exfoliated graphene, graphite or hBN flakes are then examined under an optical microscope to estimate flake thickness and assess surface cleanliness. The apparent color of a flake varies with its thickness due to optical interference effects arising from reflections at multiple interfaces: SiO_2–graphite–air and SiO_2–Si. Even bare substrates exhibit characteristic colors depending on the oxide thickness. For example, a 285 nm SiO_2 layer appears purple, caused by the phase difference between reflected light waves. When a vdW material flake is present, it modifies the interference pattern, making even monolayers visible through subtle color changes. A graphene monolayer on a 285 nm SiO_2 substrate appears bright purple (see Fig. 3.2c), as it absorbs $\sim$6–8% of green light, allowing for reliable identification of mono-, bi-, and few-layer graphene. With increasing thickness, the flake's color gradually shifts from purple to dark blue, bright blue, turquoise, green, and finally yellow [5, 6].

Suitably dimensioned and clean looking graphite and hBN flakes are further checked for accurate thicknesses, morphology, and finer cleanliness tests with an atomic force microscope (AFM). Flakes that succeeded in all of these tests are considered for the assembly of a vdW heterostructure.

3.2 Atomic Force Microscopy Lithography

As will be discussed in Chap. 6, the geometry of graphene and graphite flakes significantly influences the observed THz conductivity spectra. Consequently, defining flake shapes prior to heterostructure assembly becomes highly desirable. Atomic force microscopy (AFM) anodic oxidation lithography (AOL) has emerged as an effective method for precisely and cleanly shaping graphene and graphite flakes. This technique was developed by Dorothee Herrmann and Sivasruthi Kesavan as part of their Master's theses [7, 8].

Conventional anodic oxidation lithography (AOL) operates by applying a voltage between an AFM tip in contact mode and a conductive substrate, generating intense electric fields exceeding 10^7 V/m in the tip–substrate gap [9]. When the ambient relative humidity is at least 50 %, capillary forces induce the formation of a water meniscus between the tip and the surface (Fig. 3.3a) [10]. The electric field further attracts water molecules, enlarging the meniscus (Fig. 3.3c) [11]. Its size and shape can thus be tuned via the applied voltage and humidity. To maintain stable humidity during lithography, a humidity control system, developed as part of my Bachelor's thesis, was integrated into the AFM setup [12].

The intense electric field across the water meniscus leads to the direct decomposition of water molecules into ions (H^+, H_3O^+, OH^-, O^{2-}) [13]. When OH^- and O^{2-} ions are driven toward the sample (such as a graphene or graphite flake) by the strong positive electric field at the substrate, localized oxidation can occur:

$$C^{4+} + 2O^{2-} \rightarrow CO_2. \tag{3.1}$$

Since oxidation occurs only at the AFM tip position, moving the tip enables precise cutting of flakes [14]. Conventional AOL, however, requires a DC voltage between the tip and a conductive substrate or pre-contacted sample. As these conditions are not met by unprocessed graphene or graphite flakes on Si/SiO_2, a more suitable approach in such cases is electrode-free AOL.

In electrode-free AOL, a high-frequency (150 kHz) AC voltage (10 V) is applied between a Platinum-Iridium–coated AFM tip (SCM-PIT-V2 from *Bruker*) and the conductive silicon substrate (see Fig. 3.3). During the positive half-cycle, the graphene or graphite flake becomes positively charged, attracting negatively charged ions from the water meniscus and enabling oxidation, as in conventional AOL. During the negative half-cycle, positively charged H^+ and H_3O^+ ions are drawn to the flake, potentially inducing weak hydrogenation. However, oxidation remains the dominant process, making precise cutting of graphene and graphite flakes possible without pre-contacting the sample [3, 7, 8].

By scanning the AFM tip along a defined path, precise oxidation cuts can be made (see Fig. 3.3d). As discussed in Chap. 6, shaping flakes into rectangles is advantageous. To enable electrical contact to the underlying graphite gate beneath the graphene layer (see Fig. 1.2), the graphite flake is extended with a narrow handle structure (Fig. 3.3f, g).

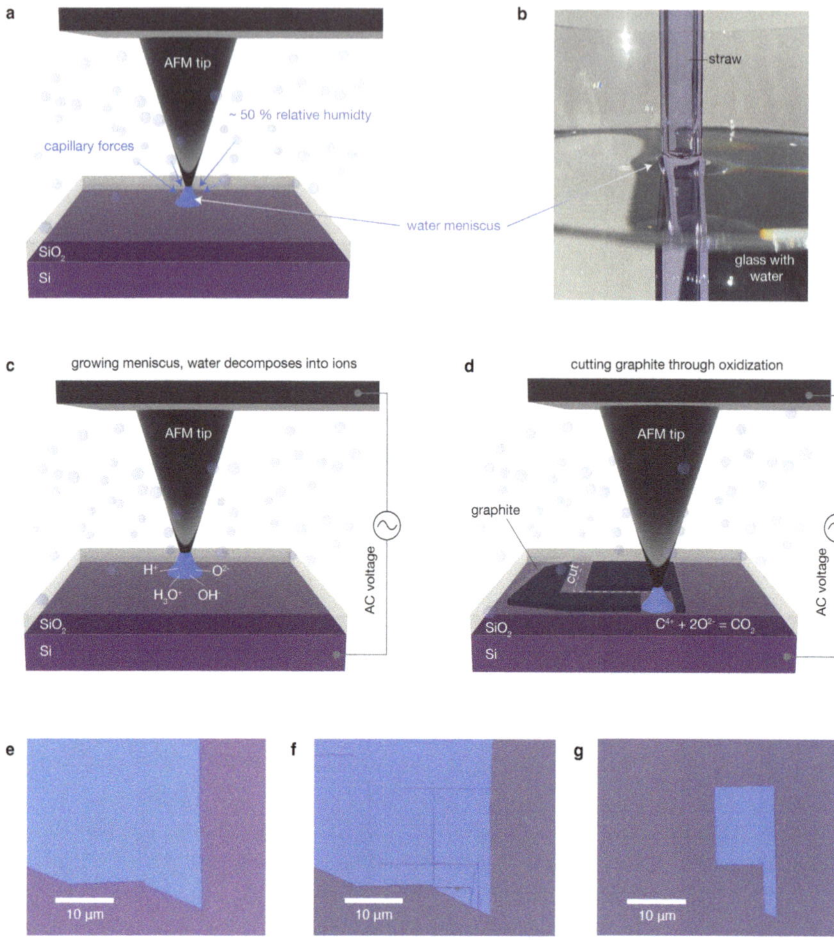

Fig. 3.3 AFM lithography through anodic oxidation: (**a**) An AFM tip in close proximity to a Si/SiO_2 substrate placed in a humid environment enables the formation of a water meniscus due to capillary forces, "soaking" in water molecules from the humid air into the narrow space. (**b**) This phenomenon is comparable to the increased and bent water level that forms in a straw placed in a water glass. (**c**) When applying an AC voltage between the AFM tip and Si substrate, the meniscus further grows, but also, due to the strong electric field that builds up between tip and sample, some of the water molecules are decomposed into ions (H^+, H_3O^+, OH^-, O^{2-}). (**d**) These ions can then oxidize the carbon, such that a cut appears when the tip is moved along a defined path over the graphite flake. The same approach works for graphene. (**e–f**) Optical micrographs of (**e**) exfoliated and clean graphite flake with a thickness of 17 nm, (**f**) graphite flake with several cuts and (**g**) cut graphite flake after pick-up of undesired graphite flake residues

After completing all desired cuts, a stamp carrying an hBN flake with a straight edge can be used to remove the unwanted graphene or graphite fragments. This leaves behind a cleanly defined graphite flake, as shown in Fig. 3.3g, using the method detailed in the following section.

3.3 VdW Heterostructure Assembly

The dry-transfer method, employed under ambient conditions, is used to pick up unwanted flake parts created through AFM lithography (see Fig. 3.3c, d) and, more importantly, for assembling high-quality graphene and/or graphite heterostructures for the on-chip THz measurements discussed in Chaps. 6 and 8.

The dry-transfer technique for vdW materials relies on a polymer stamp coated with a thin thermoplastic film.[1] The polymer stamp is typically fabricated by curing a droplet of polydimethylsiloxane (PDMS)[2] on a microscope glass slide and topping it with a thin film of either polypropylene carbonate (PPC) or polycarbonate (PC).[3]

The preparation of the thin film depends on the material: PPC films are spin-coated, while PC films are formed by spreading a droplet between two glass slides and gliding them relative to each other. The resulting films are then cut into $\sim 1 \times 1$ cm^2 squares with a razor blade. A piece of Kapton tape with a circular hole is used to pick up the film, which is then aligned and pressed onto the PDMS droplet to form the complete stamp. After heating the assembled stamp on a hotplate (5 min at 120 °C for PC), it is ready for use in vdW material transfer.

At the transfer station (Fig. 3.4), equipped with a hot plate and a vacuum chuck mounted on a micromanipulator, the stamp is aligned with the flake in the x-y plane and released in the z-direction. The stamp is then pressed onto the heated substrate and the flake. Heating the substrate to 100–120 °C enhances vdW adhesion between the stamp and the flake, allowing successful stacking and transfer when the stamp is lifted [15].

Initially, PPC-based stamps were used, but PC stamps proved superior, achieving a higher pickup success rate ($\sim$90%), faster flake transfer (1–4 minutes instead of up to an hour), and much cleaner stacks with fewer bubbles or folds. As a result, nearly all vdW heterostructures presented in the following chapters were assembled using PC-based stamps.

The assembly of a complete graphene/graphite vdW heterostructure is illustrated in Fig. 3.5. A PDMS/PC stamp is brought into contact with a carefully AFM-inspected hBN flake on a heated Si/SiO_2 substrate (100–120 °C), ideally in ambient conditions with moderate humidity ($\sim$35–40%).[4] The stamp touches down $\sim 300\,\mu$ m away from the flake center to minimize strain. Upon slow retraction, the hBN flake adheres to the stamp and can be re-aligned over a rectangular graphene flake ($\sim 10\,\mu$ m $\times$ $15\,\mu$ m), ensuring full graphene coverage. The same procedure is followed to sequentially pick up a second hBN flake (hBN2) and the AFM-cut graphite flake. The graphite is aligned such that its rectangular section overlaps the

[1] A thermoplastic material softens or melts at elevated temperatures but solidifies upon cooling.

[2] Fabricated from SYLGARD 184 Silicone Elastomer by *Dow Corning*.

[3] Both PPC and PC are supplied by *Sigma-Aldrich*.

[4] Low humidity can cause electrostatic pick-up before contact, while high humidity promotes water bubble formation between layers, degrading stack quality.

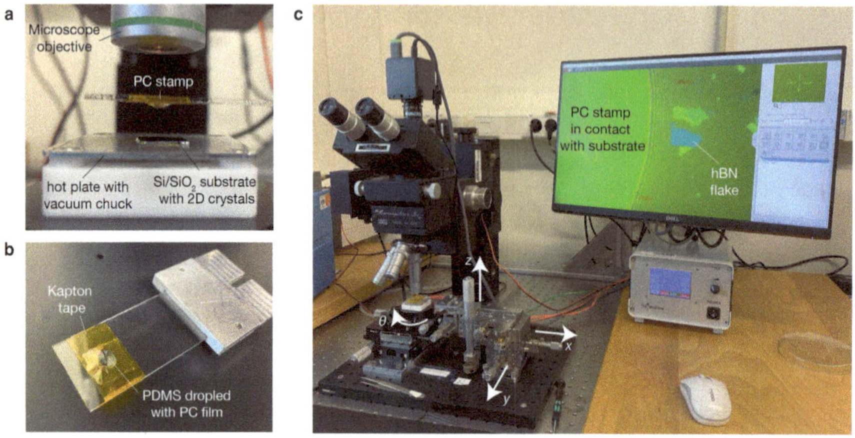

Fig. 3.4 Transfer station: (**a**) A polycarbonate (PC) stamp is positioned under a microscope objective, above a Si/SiO_2 substrate that holds exfoliated 2D crystals such as hBN, graphene, or graphite flakes intended for pickup. The substrate is mounted on a micromanipulator equipped with a heated stage and vacuum chuck. (**b**) Top view of the PC stamp: A thin PC film is secured on top of a PDMS droplet using yellow *Kapton* tape. (**c**) The PC stamp above the substrate, which is mounted on the micromanipulator. The manipulator stages allow precise alignment of the stamp above the 2D crystals in the x-, y-, and z-directions and enable rotation of the substrate by an angle θ. The monitor on the right is connected to the microscope camera and shows that the stamp is already in contact with the substrate on the left half of the imaged frame. By further pressing the stamp down, the wavefront of the stamp (circular diffraction pattern) moves further to the right enabling the pick-up of the labeled hBN flake

graphene flake and the graphite handle extends the graphene flake to allow later contacting of the graphite flake without shortening the graphite with the graphene.

To fabricate a pure graphite heterostructure without a graphene layer, the steps marked in grey in Fig. 3.5 are omitted.

To remove unwanted flake fragments after AFM lithography (Fig. 3.3c, d), a single hBN flake with a straight edge is used. The cut flake is carefully aligned on the transfer stage so that the hBN contacts only the undesired sections, avoiding any overlap with the pristine graphite region intended for the final stack. This prevents contamination and ensures clean stacking.[5]

To avoid premature pickup, the hBN must never touch the target flake. Thus, if residual fragments are scattered around it, the substrate must be rotated and multiple pickup steps performed. This process yields a cleanly defined graphite flake, as shown in Fig. 3.3d.

[5] Contact between the graphite and PC-coated stamp prior to stacking was found to introduce bubbles between vdW layers, likely due to PC residue.

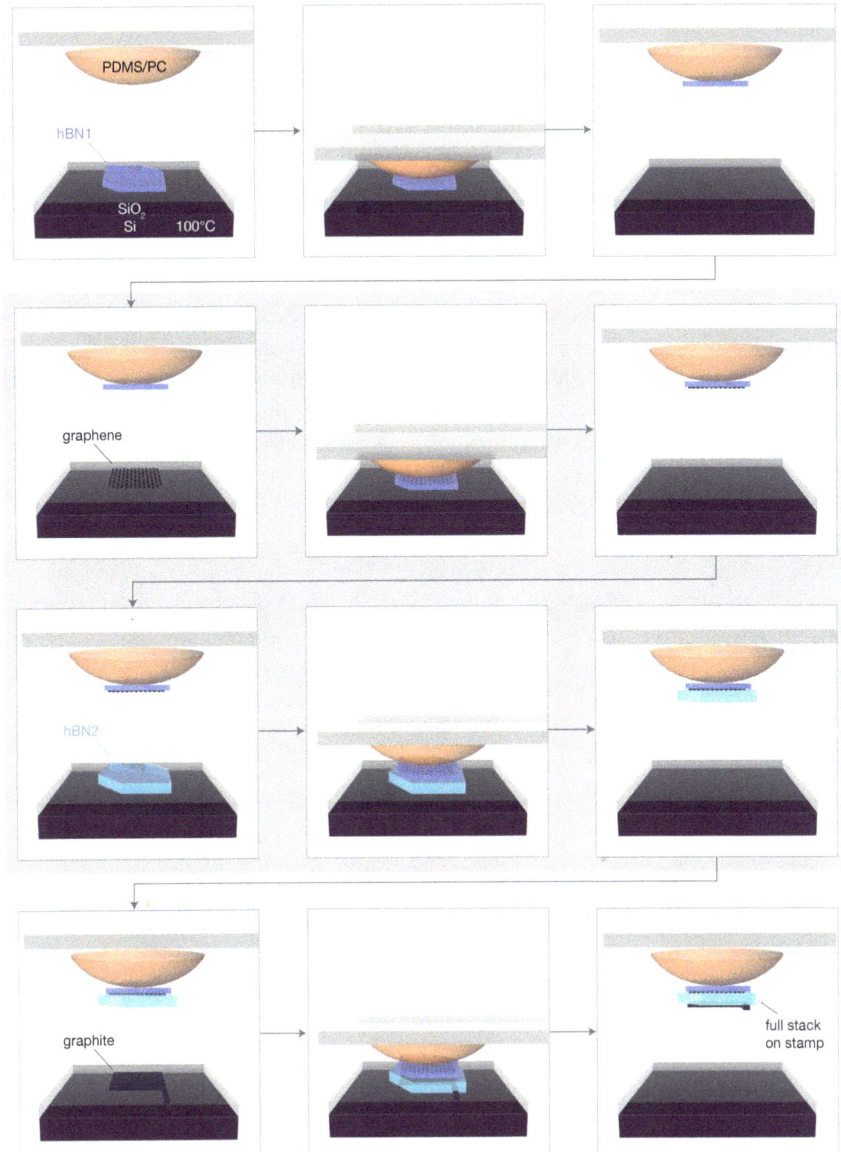

Fig. 3.5 Stacking of a vdW heterostructure: A PDMS/PC stamp is used to pick up a first hBN flake (hBN1), a rectangular cut graphene flake, a second hBN flake (hBN2) and a rectangular graphite flake with handle from a heated Si/SiO_2 substrate. In case that a bare graphite heterostructure shall be built, the steps underlaid in the grey shade have to be left out

3.4 Heterostructure Transfer to Substrate

Before the ultrafast circuitry can be patterned on top of the stacked vdW heterostructure, the stack first has to be transferred from the stamp to a substrate. In order to perform later on-chip THz spectroscopy measurements on the built vdW heterostructure, a non-conductive substrate is required which has a nearly constant dielectric function over the relevant GHz–THz frequency spectral sensitivity range. A sapphire substrate[6] with $\epsilon \sim 10$ for the relevant frequency range turned out to be a good candidate.

The subsequent optical lithography process for fabricating a 9 μ m wide, 10 mm long gold coplanar strip transmission line on the substrate with an integrated vdW heterostructure demands a high degree of surface flatness and sharp, well-defined substrate edges. Any unevenness or rounded edges can cause the laser writer's beam to shift focus across the substrate, resulting in inconsistent feature dimensions and complicating the development or lift-off steps. To ensure optimal conditions, the costly "sharp edges" substrate option is purchased, and each substrate is carefully inspected for flatness.

Once a flat sapphire substrate is selected, it is ultrasonicated in acetone followed by isopropanol, then rinsed with purified water. Residual water is removed by gently swiping the substrate over a cleanroom tissue. To eliminate remaining moisture and enhance adhesion to the stack, the substrate is treated with oxygen plasma for 30 s at 70 W at a pressure of 250 mTorr (Fig. 3.6) [1].

The cleaned, even substrate is then positioned on the transfer station and aligned so that the stack sits approximately 1 mm off-center along the substrate's diagonal axis. This diagonal placement maximizes the length of the coplanar transmission lines in subsequent fabrication steps, effectively extending the time window before reflections in on-chip THz measurements. Once properly aligned, the stack is transferred onto the sapphire substrate at 120 °C.

During stacking and substrate placement, bubbles may form between heterostructure layers or between the stack and substrate (Fig. 3.1), even when flakes and substrate appear clean [16]. To minimize bubbles, the stamp is repeatedly pressed and lifted ("bubble pressing"), sweeping its wavefront across the stack to push out trapped air. The stamp is then fully pressed onto the substrate, which is heated to 200 °C to melt the PC layer. After 5 to 10 min, the stamp is lifted, leaving the PC-covered stack on the substrate. The sample is then immersed, face-up, in PG remover at 65 °C for 24 h to dissolve the PC. This orientation prevents the stack from detaching before metal contacts secure it. During subsequent lift-off steps, the sample is placed face-down on a holder in PG remover to allow unwanted metal and dirt to fall away while the stack remains anchored by the edge contacts (Sect. 3.6). Finally, the sample is rinsed with acetone, isopropanol, and water, and gently dried with a nitrogen stream.

[6] 1×1 cm^2, 2-mm thick, c-cut, two-side epi-polished, sharp edges and two notches from *SurfaceNet GmbH*.

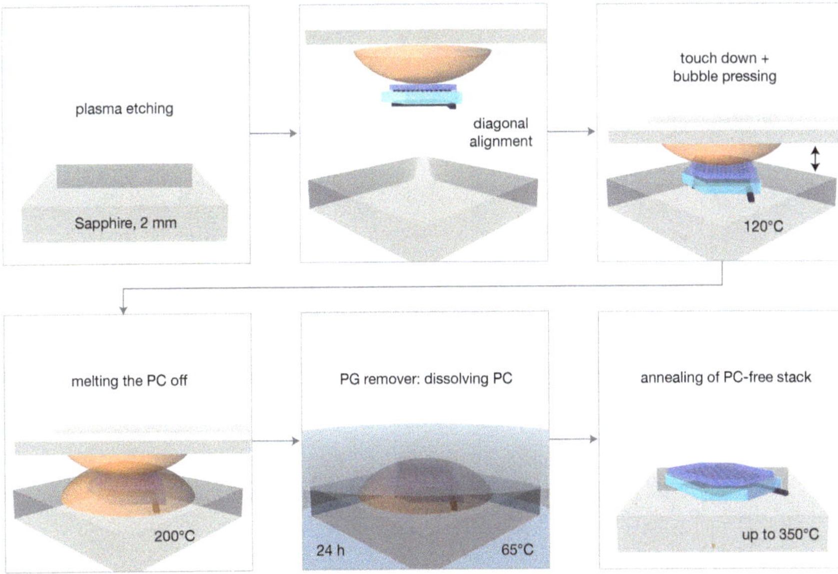

Fig. 3.6 Transfer of a stacked heterostructure to a substrate. A 2 mm thick sapphire substrate is cleaned with plasma etching. The heterostructure on the stamp is then aligned diagonally, 1 mm away from the substrate center and placed onto the substrate. Bubbles that have formed between individual flakes or the stack and the substrates can be removed with a procedure called bubble pressing. By heating the substrate up to 200 °C, the PC of the stamp can be melted off, leaving the stack covered with PC on the sapphire substrate. In a PG remover bath, the PC is washed off. The clean and dry stack is then further placed into an annealing chamber, where the sample is heated up in vacuum to further improve the adhesion between stack and substrate

In a last step, the dry stack on the sapphire substrate is placed in an annealing chamber where the stack is heated up to 150 °C for 30 min while being in vacuum, then cooled down back to room temperature and then heated up to 350 °C for 60 min before being cooled down again. This procedure further improves the adhesion between stack and substrate and can also further reduce the amount of enclosed bubbles.

3.5 Reactive Ion Etching

In the next step, reactive ion etching (RIE) is used to remove unwanted flake fragments surrounding the heterostructure that could interfere with the later fabrication of on-chip THz spectroscopy circuitry. RIE is also employed to selectively etch regions of the top hBN layer, exposing the graphene edges for the formation of one-dimensional electrical contacts.

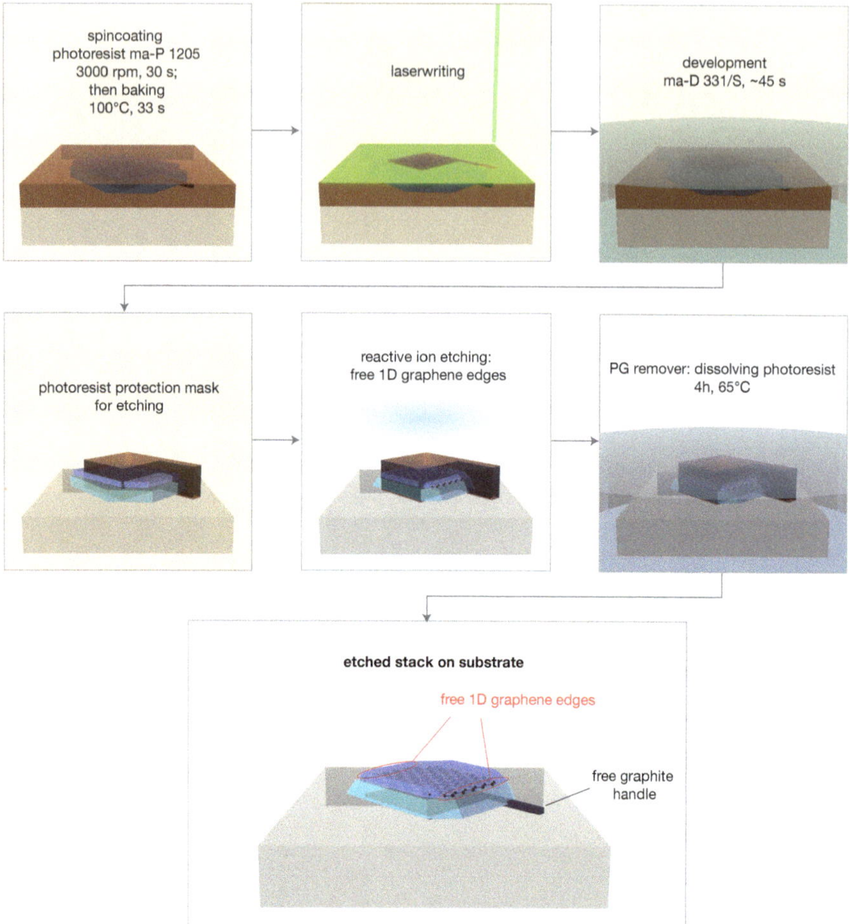

Fig. 3.7 Optical lithography and reactive ion etching: The photoresist ma-P 1205 is spincoated onto the sapphire substrate with the vdW heterostructure. The laserwriter illuminates all areas that are not to be protected from the etching process. The sample is then developed in ma-D 331/S to wash away the illuminated parts of the photoresist. A well defined photoresist protection mask can be obtained. The sample is then placed in the RIE chamber to etch the stack in a perfect rectangular shape and to create free 1D graphene edges for later graphene edge contacting. By placing the sample into PG remover after the etching process, the remaining photoresist can be dissolved and the clean and etched stack is obtained. Process steps shaded in yellow are light sensitive and performed in a yellow-light clean room

Prior to etching, all regions of the heterostructure intended for on-chip THz spectroscopy must be protected using a photoresist mask patterned via optical laser lithography in a cleanroom environment. The photoresist ma-P 1205 (provided by *micro resist technology*) is spin-coated onto the sapphire substrate at 3000 rpm for 30 s, followed by a soft bake at 100 °C for 33 s (see Fig. 3.7).

A corresponding mask, designed in *KLayout*, is transferred to the laser writer.[7] All regions to be etched are selectively exposed to a focused green laser beam (see Fig. 3.7). After exposure, the sample is developed in ma-D 331/S for approximately 45 s, removing the illuminated regions while leaving the target areas protected beneath the photoresist.

Next, all unprotected flakes on the sapphire substrate, including unwanted parts of the stack, are removed via reactive ion etching (RIE). In this process, the sample is exposed to a plasma generated from a mixture of oxygen (O_2) and sulfur hexafluoride (SF_6). The reactive plasma chemically etches graphene, graphite, and hBN.

By adjusting the SF_6/O_2 ratio, gas pressure and flow, and the frequencies of the capacitively and inductively coupled RF sources (RF and ICP), both the etch rate and selectivity between materials can be finely controlled. Pure SF_6 aggressively etches hBN, while O_2 targets graphene and graphite. A balanced mixture of 90 sccm SF_6 and 10 sccm O_2 has proven effective for etching both materials smoothly, with a slightly faster rate for hBN that was found ideal for forming clean one-dimensional edge contacts [1].

After etching, the sample is placed in PG remover to dissolve the photoresist and to obtain the cleanly etched vdW heterostructure with free 1D graphene edges (see Fig. 3.7) allowing to contact the graphene by patterning Cr/Au contacts over the edges, as explained in the next section.

3.6 Edge Contacts and Ultrafast Circuitry Processing

One-dimensional edge contacts to two-dimensional materials were invented by L. Wang in 2013 [17]. The principle is that a 2D material such as graphene, which is usually protected by a thin hBN layer, is etched with RIE to expose free 1D graphene edges, which in turn can be covered with a metal layer to contact the 2D material. Such contacts are essential for enabling electrostatic gating and conducting DC transport measurements.

To enable electrostatic gating of graphene, at least one functional edge contact to both the graphene and the graphite layers is required. For DC transport measurements, however, a minimum of two graphene edge contacts, preferably positioned on opposite sides of the sample, is necessary.

To define the metallic contacts covering the exposed 1D edges, optical lithography is employed once again. However, a single-layer photoresist is insufficient

[7] Laser writer: μPG 101 by *Heidelberg Instruments*. Exposure: 43 % of 10 mW.

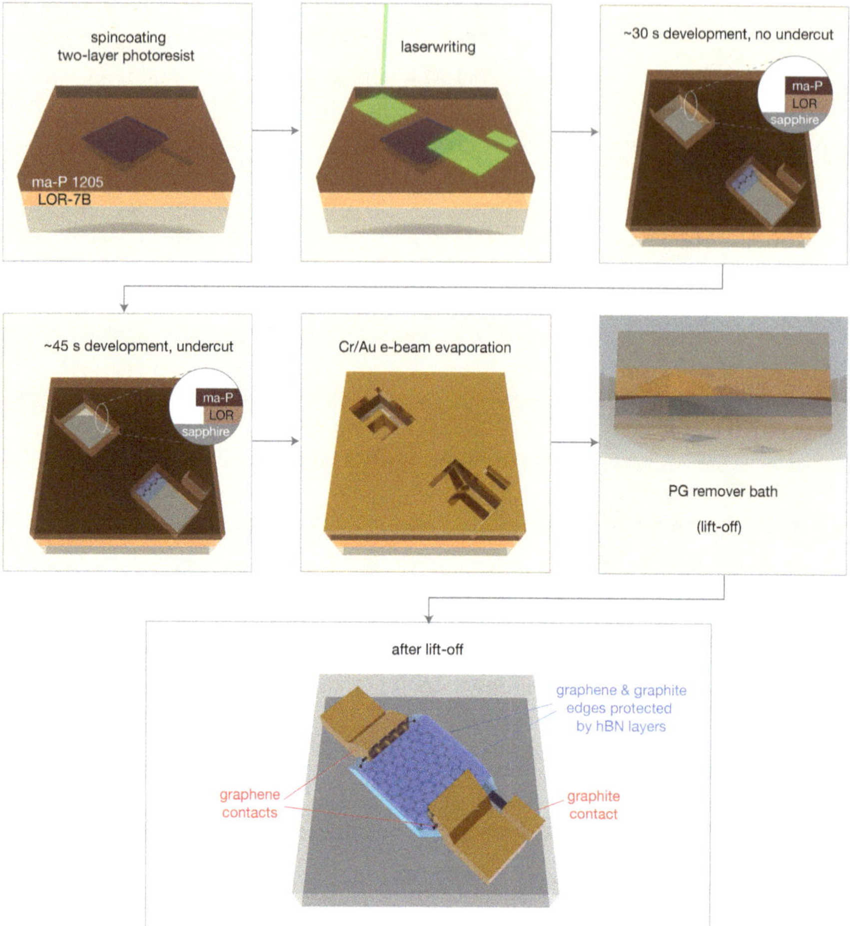

Fig. 3.8 Edge contacts: Edge contacts to the graphene and graphite flakes are fabricated using a bilayer photoresist mask. First, LOR-7B is spin-coated and baked onto the substrate, followed by a top layer of ma-P 1205, also spin-coated and baked. The contact areas are then defined by laser exposure and developed for ∼30 s in ma-D 331/S, removing the exposed regions. At this stage, both resists are etched equally, resulting in vertical sidewalls. To ensure a clean lift-off and prevent tear-off edges, development is extended to ∼45 s, creating an undercut where the LOR layer recedes more than the ma-P layer. A Cr/Au layer is then deposited via e-beam evaporation. Finally, the lift-off process is performed by immersing the sample upside down in PG remover, allowing excess metal to detach cleanly. The result is a graphene and graphite heterostructure with well-defined metallic edge contacts, as shown in the bottom panel

in this case (see Fig. 3.8). Metal deposited onto such a mask may form continuous films across the resist and substrate, causing so-called tear-off edges during the lift-off process. To prevent this, a bilayer resist system is used, consisting of LOR-7B

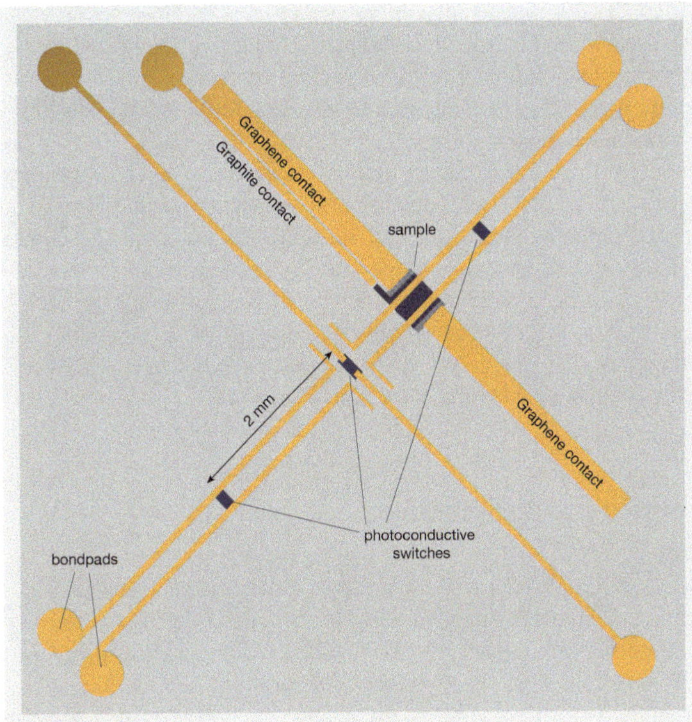

Fig. 3.9 On-chip THz spectroscopy circuit placement on a 1×1 cm^2 sapphire substrate: First, three 5×20 µm^2 α-silicon patches, spaced 2 mm apart, are evaporated onto the substrate to serve as photoconductive switches, with the THz antenna positioned at the center of the substrate. Next, metallic gold coplanar strip transmission lines, THz antenna arms and contacts as well as graphene and graphite contacts extensions with bondpads (large, circular metal areas allowing an easy contacting with wires) are evaporated onto the sample

beneath ma-P 1205. Upon extended development, the lower LOR-7B dissolves more than the top layer, creating an undercut that ensures clean separation between metal on the resist and metal on the substrate. For contacting graphene and graphite flakes, a Cr/Au stack is deposited by electron-beam evaporation, typically 7 nm of chromium followed by 30–100 nm of gold.

Next, the ultrafast circuitry is fabricated on the sample. First, three amorphous silicon (α-Si) patches are evaporated to form the THz antenna and photoconductive detector switches (see Fig. 3.9). α-Si was selected for its ease of deposition in the on-site cleanroom at MPSD Hamburg, in contrast to commonly used materials like GaAs. With a bandgap of $E_g = 1.17$ eV [18], α-Si is well-suited for excitation by the green laser beams (520 nm, $E = 2.38$ eV) used in the experimental setup (see Sects. 4.1 and 2.1).

The silicon is evaporated at a high rate ($10\,\text{Å}\,\text{s}^{-1}$) [19], which results in a reduction of the switch carrier recombination time due to the induced disorder. Since the recombination time governs the falling edge of the transient current pulse generated in the switch (see Sect. 2.1), shortening the recombination time extends the measurement bandwidth.

Subsequently, metallic leads to the THz antenna, contact extensions to the Cr/Au edge contacts, and two coplanar strip transmission lines (for sample and reference) are patterned. These transmission lines follow a 3–3–3 μm conductor-gap-conductor geometry and are composed of 10 nm Ti and 275 nm Au.

Finally, the sample is mounted onto a holder using silver epoxy. Each bondpad and edge contact is wire-bonded with fine gold wires while grounding the sample holder with conductive tape to prevent static discharge damage. After bonding, the device is installed in a continuous-flow optical cryostat.

3.7 Summary and Discussion

In summary, the on-chip THz spectroscopy devices of this thesis consist of vdW heterostructures of typical dimensions ($\sim 10 \times 10\,\mu\text{m}^2$) interfaced with the circuitry architecture introduced in Sect. 2.3. While the vdW heterostructure fabrication procedure follows established dry-transfer techniques from the literature [17, 20, 21], special attention is given to shaping the conductive layers (achieved via precision AFM lithography, Sect. 3.2) and ensuring electrical isolation through hBN encapsulation. Further, fabricating the ultrafast circuitry on top of the precisely shaped and stacked heterostructure requires a clean substrate at the sub-micrometer scale. This ensures that no contaminants, such as dirt particles, unintended vdW materials accidentally picked up with the target flakes, or silicon and gold debris from the lift-off steps during the fabrication of photoconductive switches and edge contacts, interfere with the transmission lines (>10 mm long). Such contaminants could otherwise lead to breakage or electrical shorting of the coplanar strip transmission lines or distortion of the THz pulse apart from the sample. The individual fabrication steps and challenges are summarized in Fig. 3.10.

Once a high-quality on-chip THz spectroscopy device with an embedded vdW heterostructure is built, the next step is to measure the device.

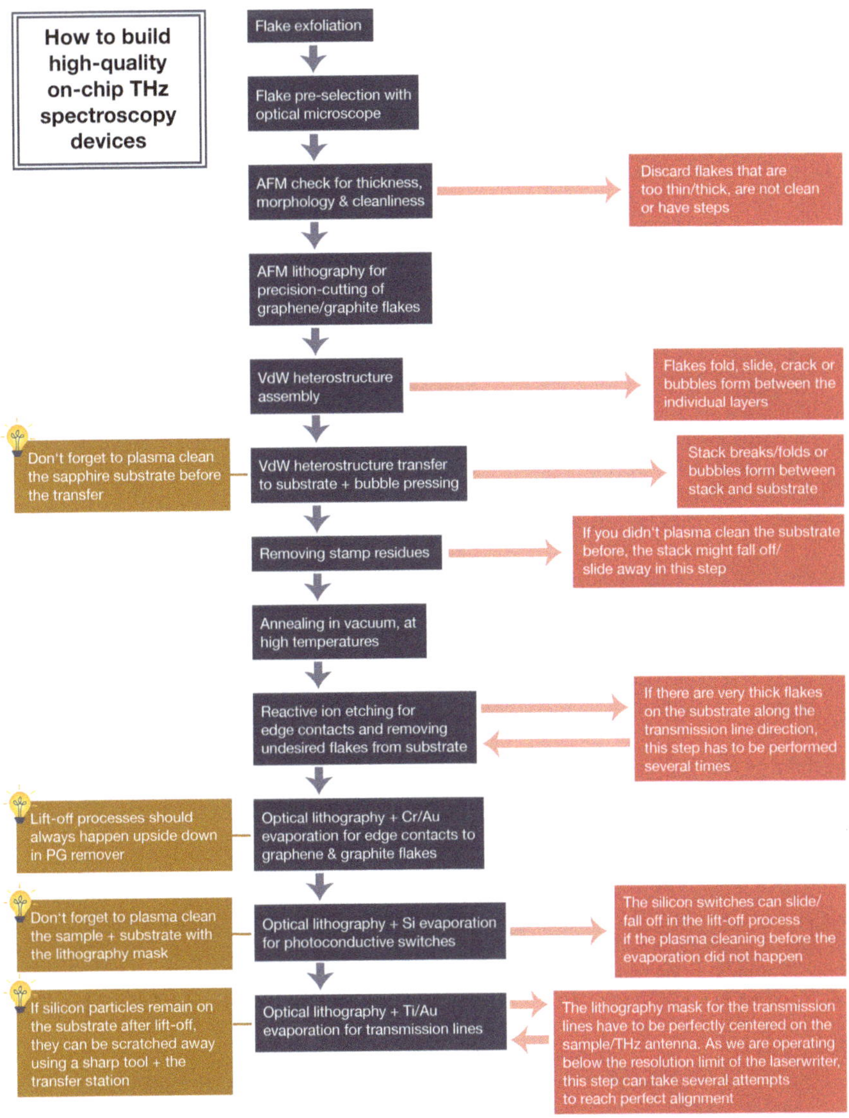

Fig. 3.10 Fabrication protocol for on-chip THz spectroscopy devices interfacing vdW heterostructures: Individual fabrication steps are outlined with the blue boxes, important notes are highlighted in yellow and the red boxes describe challenges or ways that lead to a failure

References

1. Schulte BF (2022) Ultrafast transport in tunable graphene heterostructures. PhD Thesis. University of Hamburg

2. Dean CR, Young AF, Meric I, Lee C, Wang L, Sorgenfrei S, Watanabe K, Taniguchi T, Kim P, Shepard KL, et al (2010) Boron nitride substrates for high-quality graphene electronics. Nat Nanotechnol 5:722–726. https://doi.org/10.1038/nnano.2010.172
3. Li H, Ying Z, Lyu B, Deng A, Wang L, Taniguchi T, Watanabe K, Shi Z (2018) Electrode-free anodic oxidation nanolithography of low-dimensional materials. Nano Letters 18:8011–8015. https://doi.org/10.1021/acs.nanolett.8b04166
4. Novoselov KS, Geim AK, Morozov SV, Jiang D, Zhang Y, Dubonos SV, Grigorieva IV, Firsov AA (2004) Electric field effect in atomically thin carbon films. Science 306:666–669. https://doi.org/10.1126/science.1102896
5. Roddaro S, Pingue P, Piazza V, Pellegrini V, Beltram F (2007) The optical visibility of graphene: interference colors of ultrathin graphite on SiO2. Nano Letters 7:2707–2710. https://doi.org/10.1021/nl0711581
6. Blake P, Hill EW, Castro Neto AH, Novoselov KS, Jiang D, Yang R, Booth TJ, Geim AK (2007) Making graphene visible. Appl Phys Lett 91:063124. https://doi.org/10.1063/1.2768624
7. Herrmann D (2023) Plasmon resonances of graphite heterostructures probed by on-chip THz spectroscopy. Master's thesis. University of Hamburg
8. Kesavan S (2024) Engineering THz plasmonic cavity modes in microstructured van der Waals heterostructures. Master's thesis. University of Hamburg
9. Garcia R, Knoll AW, Riedo E (2014) Advanced scanning probe lithography. Nat Nanotechnol 9:577–587. https://doi.org/10.1038/nnano.2014.157
10. Ondarçuhu T, Fabié L (2013) Capillary forces in atomic force microscopy and liquid nanodispensing. In: Lambert P (ed) Surface tension in microsystems: engineering below the capillary length. Springer Berlin, Heidelberg, pp. 279–305. https://doi.org/10.1007/978-3-642-37552-1_14
11. Cramer T, Zerbetto F, García R (2008) Molecular mechanism of water bridge buildup: field-induced formation of nanoscale menisci. Langmuir 24:6116–6120. https://doi.org/10.1021/la800220r
12. Kipp G (2017) Structural changes in protein crystals induced by systematic variation of the relative humidity. Bachelor's Thesis. University of Hamburg
13. Červenka J, Kalousek R, Bartošík M, Škoda D, Tomanec O, Šikola T (2006) Fabrication of nanostructures on Si(100) and GaAs(100) by local anodic oxidation. Appl Surface Sci 253:2373–2378. https://doi.org/10.1016/j.apsusc.2006.03.095
14. Byun I-S, Yoon D, Choi JS, Hwang I, Lee DH, Lee MJ, Kawai T, Son Y-W, Jia Q, Cheong H, Park BH (2011) Nanoscale lithography on monolayer graphene using hydrogenation and oxidation. ACS Nano 5:6417–6424. https://doi.org/10.1021/nn201601m
15. Pizzocchero F, Gammelgaard L, Jessen BS, Caridad J, Wang L, Hone J, Bøggild P, Booth TJ (2016) The hot pick-up technique for batch assembly of van der Waals heterostructures. Nat Commun 7:11894. https://doi.org/10.1038/ncomms11894.
16. Khestanova E, Guinea F, Fumagalli L, Geim AK, Grigorieva IV (2016) Universal shape and pressure inside bubbles appearing in van der Waals heterostructures. Nat Commun 7:12587. https://doi.org/10.1038/ncomms12587
17. Wang L, Meric I, Huang PY, Gao Q, Gao Y, Tran H, Taniguchi T, Watanabe K, Campos LM, Muller DA, Guo J, Kim P, Hone J, Shepard KL, Dean CR (2013) One-dimensional electrical contact to a two-dimensional material. Science 342:614–617. https://doi.org/10.1126/science.1244358
18. Gross R, Marx A (2014) Festkörperphysik. De Gruyter Oldenbourg, München. https://doi.org/10.1524/9783110358704
19. Kusyak K (2022) Increasing the bandwidth of on-chip THz spectroscopy. Master's thesis. University of Hamburg
20. Geim AK, Grigorieva IV (2013) Van der Waals heterostructures. Nature 499:419–425. https://doi.org/10.1038/nature12385
21. Guo H-W, Hu Z, Liu Z-B, Tian J-G (2021) Stacking of 2D materials. Adv Funct Mater 31:2007810. https://doi.org/10.1002/adfm.202007810

Chapter 4
Measuring On-chip THz Devices

Abstract In this chapter I describe the experimental setup and measurement protocols, including sample preparation, calibration procedures, DC transport measurements, and the on-chip THz spectroscopy workflow. I also outline the methods used for data processing and analysis.

This chapter introduces the methods to perform on-chip THz spectroscopy measurements on a vdW heterostructure. This first requires an experimental setup to perform these measurements, as presented in Sect. 4.1. Section 4.2 then discusses the methods for preparing the on-chip THz spectroscopy measurements including calibrations. Section 4.3 briefly introduces the method of DC transport measurements, which can be performed simultaneously to on-chip THz spectroscopy. The on-chip THz spectroscopy measurement protocol is then discussed in Sect. 4.4 and the data processing of the obtained time-domain data is shown in Sect. 4.5.

4.1 Experimental Setup

The requirement of the experimental setup is to generate three sub-picosecond beams, that generate THz pulses at the THz antenna (α-silicon switch, see Sect. 3.6) and detect the reference and sample pulses at the two opposite detection switches as a function of time delay with respect to the generated THz beam (see Sect. 2.3). Thus, the timing between generator and detection beams has to be variable and very accurate. Additionally, chopping techniques are necessary to separate the small signals from the background noise.

To meet these requirements, the measurement setup was built by Dr. Benedikt Schulte around a Yb-based femtosecond laser (Spirit HE 1040-16) which provides pulses at a central wavelength of 1040 nm and full width half maximum of 310 fs at

G. Kipp, *Self-Cavity Electrodynamics of Van der Waals Heterostructures*, Springer Theses, https://doi.org/10.1007/978-3-032-31979-1_4

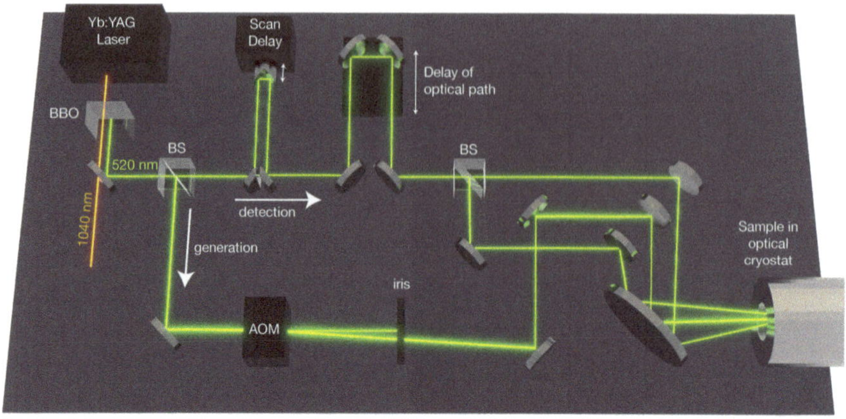

Fig. 4.1 Optical setup: 310 fs short pulses at a wavelength of 1040 nm are generated with a Yb:YAG laser. A thin Barium borate (BBO) crystal is used to frequency-double the pulses and achieve a wavelength of 520 nm, that is needed to operate the photoconductive switches on the sample. A beam splitter (BS) splits the green beam into two beams, one for generation and one for detection. The generation beam is sent through an acousto-optic modulator (AOM), used for high frequency chopping. The detection beam is delayed by a Scan Delay (shaking mirror) and by two mirrors on a translational stage that allow longer delays for coarser adjustments. The detection beam is sent through another beam splitter to obtain two detection beams for sample and reference detection switches

a repetition rate of 200 kHz. For more efficient operation of the photoconductive switches, the 1040 nm pulses were frequency-doubled in a thin Barium borate (BBO) crystal to generate green 520 nm light (see Fig. 4.1). This green light was split into two paths, one for THz generation and the other for THz detection. The beam for THz generation was intensity modulated by an acousto-optic modulator (AOM) that sent every second pulse onto the chip. The beam for detection was routed through two different optical delay lines: a translation stage and a fast optical delay line (scanDelay from APE). The latter was used for THz measurements to generate a time delay of up to 50 ps between generation and detection pulses at a scanning frequency of 2 Hz. After the delay lines, the detection beam was split into two for detecting the sample and reference side of the optoelectronic circuitry. All three beams were coupled in a microscope assembly and focused onto the different photoconductive switches. The microscope provided an optical image of the device and was used for active alignment stabilization (for more details see [2]).

The sample was mounted in a continuous-flow optical cryostat (ST500 from Janis) and all measurements were performed at a base temperature of 7–20 K. Currents generated by the photoconductive detector switches were collected by home-built transimpedance amplifiers from Dr. Toru Matsuyama (2×10^8 V A^{-1}, 100 kHz bandwidth) and the generated voltage signal was demodulated by a lock-in amplifier at the frequency of the intensity modulation of the THz generation (100 kHz). The lock-in output was digitized with a data acquisition card and synchronized to the fast optical delay line (2 Hz) to recover the THz signal [2].

4.2 Preparations and Calibrations

After the sample has been mounted in the optical cryostat and vacuum conditions are realized, a few preparations and calibrations have to be performed before measurements can take place. The following sections discuss the annealing of the photoconductive switches (Sect. 4.2.1) as well as aligning the laser beams and taking switch calibration curves (Sect. 4.2.2).

4.2.1 *Photoconductive Switch Annealing*

First, the photoconductive switches must be annealed to improve their performance. This is done by applying a bias voltage (~0.1–1 V) across a switch and shining a green laser beam (~0.005 J/cm^2 fluence) onto the switch at room temperature. Thereby, impurities diffuse from the switch contacts into the silicon, leading to defects and electron scattering [2], and a decrease of the initial switch resistance of ~100 MΩ to a final resistance of ~10 MΩ.

The switch resistance significantly increases again when the sample is cooled down (to several tens of GΩ at ~10 K), as the diffused defects from impurities are frozen out. A certain level of impurities is desirable, as they cause more scattering of carriers when the switch is excited by a short laser pulse, resulting in a shorter recombination time. This, in turn, produces a shorter current pulse and increases the bandwidth of the THz pulse (see Sect. 2.1). A drawback of having an excessive concentration of impurities is that it reduces the switch resistance, leading to an increased dark current in the switch's off-state, which in turn lowers the signal-to-noise ratio [3]. Thus, the right balance has to be found which is achievable with the silicon switch deposition recipe provided in Sect. 3.6 and the annealing procedure introduced in this section.

The sample is then cooled to a temperature of 7–20 K in a continuous-flow optical cryostat (ST500 from Janis) using liquid helium. It is crucial to maintain a slow cooling rate (around 1 K/min) while ensuring all contacts are grounded and the laser beams are blocked. Failure to do so risks damaging, or even destroying, the vdW heterostructure.

4.2.2 *Laser Beam Alignment and Calibrations*

The first step at low temperatures is to align the three laser beams onto the switches. This is done by applying a small voltage across the switch[1] (~0.1 V across the

[1] A higher bias voltage enables higher precision during the alignment. However, one should be particularly careful with the detector switch on the sample side, as only a thin hBN layer is

detector switches ~1 V across the THz antenna). Then, the photocurrent that flows through the switch can be measured upon switch illumination with the excitation beam (using a mechanical chopper and a Lock-In Amplifier). Alignment is achieved, when the photocurrent is maximized at a certain beam position.

Once all three laser beams are aligned, $I - V$ calibration curves are taken on all three switches. The calibration curves relate the THz electric field in the coplanar strip to the photocurrent measured at the detection switch.

Voltages in the range of ~ -0.05 to 0.05 V for the detection switches and ~ -10 to 10 V for the THz antenna are applied across the switches and the photocurrent is measured. The voltage range for the THz antenna is larger because 5–10 V are typically applied when generating THz pulses. In contrast, the range for the detection switches is kept small as the THz pulses themselves bias the switch in the on-chip THz spectroscopy measurements, and they have a typical peak amplitude of <0.5 V.

4.3 DC Transport Measurements

After the laser beams are aligned and the switch calibration curves are taken, measurements can be performed. For gate-tunable vdW heterostructures (see Fig. 1.2), DC transport measurements are first taken. To measure the resistance of a sensitive 2D material such as graphene, a small current of I=100 nA is applied from one contact to the other and the voltage drop U is detected as a function of the back gate voltage applied to the graphite. In case of graphene, the carrier density as a function of back gate voltage can be calculated with [4]:

$$n = \frac{\epsilon_0 \epsilon_{\mathrm{hBN}} (V_{\mathrm{gate}} - V_{\mathrm{cnp}})}{d_{\mathrm{hBN}} e}, \tag{4.1}$$

where ϵ_0 is the vacuum permittivity, ϵ_{hBN} is the dielectric constant of hBN, V_{gate} is the applied gate voltage, V_{cnp} is the voltage offset from the charge neutrality point that is defined as the applied gate voltage yielding the maximum graphene resistance, d_{hBN} is the distance of graphene and gate and e is the electron charge.

The DC transport measurement can help to identify the back gate voltage at which the graphene is electrically tuned to the charge neutrality point. To ensure that transport and on-chip THz measurements probe the same region of the device, careful attention must be paid to the current path defined by the contact geometry. For optimal comparability, it is advisable to use two wide edge contacts placed on opposite sides of the 2D material. This configuration enables resistance

---

separating the sample from the transmission lines across which the bias voltage for the beam-on-switch alignment is applied. To treat both detector switches equal, the bias voltage across both should be kept below 1 V, just to be save.

measurements that encompass the entire heterostructure, mirroring the spatial extent probed by on-chip THz spectroscopy, rather than isolating only subsections (see Fig. 3.8).

4.4 On-chip THz Spectroscopy Measurements

For on-chip THz spectroscopy measurements, time zero has first to be found both for the reference side and the sample side. This is achieved by generating THz pulses with the THz antenna (switch bias voltage is ~5–10 V) and then using the translation stage (see Sect. 4.1) to adjust the optical path between the generator beam and detector beams, which detect the photocurrent across each detector switch. The positions for both reference and sample side at which a maximum photocurrent is detected define time zero.

After both time zeros for sample and reference side are found, time traces are detected on the reference side and at different back gate voltages on the sample side using the fast optical delay line (Scan Delay, see Sect. 4.1) averaging each trace 5000–20,000 times.

4.5 Data Processing

Since the photocurrents at the generator switch are detected as a function of the shaking mirror's position in the Scan Delay (see Sect. 4.1), the mirror position x must be converted into a time axis t to express the detected photocurrents as a function of time. As the laser beams are propagating with the speed of light c, this can simply be achieved by $t = x/c$, delivering time-domain voltages on the sample line $V_{\mathrm{sam}}(t)$ and reference line $V_{\mathrm{ref}}(t)$. The Fourier transform of each measurement is taken to calculate the frequency-dependent voltages ($V_{\mathrm{sam}}(\omega)$ and $V_{\mathrm{ref}}(\omega)$). Thereby, the way that the lock-in amplifier (Model SR830 from Stanford Research systems) outputs the signal has to be considered [5]:

$$V_{\mathrm{output}} = \left(\frac{V_{\mathrm{actual\ signal}}}{\mathrm{sensitivity}} \right) \cdot 10\,\mathrm{V}, \tag{4.2}$$

where V_{output}, $V_{\mathrm{actual\ signal}}$ and sensitivity are given in Volts. The sensitivity refers to the sensitive voltage range of the lock-in amplifier during measurements, which is chosen wide enough to capture the signal's full amplitude, but narrow enough to maintain high resolution for detecting small variations.

The switch calibration curves (see Sect. 4.2.2) are then used to translate the recorded voltages into electric field strengths and the calibration curves are fitted with a linear function to extract the slope $m_{\mathrm{calibration}}$. To obtain the electric field strength, the detected voltage is divided by $m_{\text{calibration}}$, and by the distance between the coplanar strips W_2 to determine the field as a function of time or frequency:

$$E_{\text{THz}}(t) = \frac{V_{\text{output}}(t)}{m_{\text{calibration}}} \frac{\text{sensitivity}}{10\,\text{V}} \frac{1}{W_2}. \tag{4.3}$$

If the electric field strength between two different samples is to be compared, one must also consider the applied bias voltage and the switch calibration curves of the two THz antennae. For example, if one electric field was detected on a sample where the THz antenna was biased with 5 V and the signal shall be compared to a second sample, where the THz antenna was biased with 10 V, the electric field strength of the first device has to be multiplied by two to be comparable to the detected electric field strength of the second device.[2] The same theory applies for the switch calibration curves. A THz antenna, where a calibration curve shows a larger slope, will generate a larger THz signal into the coplanar strips. Thus, the slopes of the THz antenna calibration curves also have to be compared in order to analyze detected electric field strengths on two different samples.

To eliminate the influence of reflections from known sources on the processed signal, the amplitudes of the time-domain traces are windowed. This involves truncating the time-domain data at a specified delay and fitting it to an exponential decay function, which models a smooth transition to zero, thereby excluding signal reflections from the analysis. To eliminate high frequency noise, a binomial low-pass filter is applied at 1.5 THz.

4.6 Discussion

This chapter introduced the experimental setup and methods for performing on-chip THz spectroscopy on vdW heterostructures, including the processing of time-domain data collected from both the sample and reference measurement arms. By comparing the THz pulse that interacted with the vdW heterostructure on the sample side to the pulse detected on the empty reference transmission line arm (see Fig. 2.6), one can assess any distortion caused by the sample. The calibration of the photoconductive switches accounts for possible variations in the quality of the individual generation and detection switches, enabling precise extraction of electric field amplitudes from the measured voltage signals over time.

This approach has one subtle weak point: it relies on the perfectly symmetric launching of the sample and reference THz pulses at the antenna. Slight asymmetries in the positioning of the THz antenna in the transmission line geometry,

[2] This comparison is possible because the sample behaves linearly for the used THz field strengths of this work.

or an off-center excitation of the antenna by the laser pulse, can result in small discrepancies (such as the THz pulse launched to one side being slightly larger than the one launched to the other side). However, this issue can be largely mitigated by precise fabrication of a perfectly symmetric THz antenna and using large laser beams for switch illumination, with spot sizes significantly larger than the silicon patch width of the THz antenna [6].[3]

Once the calibrated sample and reference signals from the device have been obtained as a function of time, the next step is to extract optical information about the vdW heterostructure. The following section will therefore dive into the principles of optical conductivity and outline how it can be extracted from the measured time-domain on-chip THz spectroscopy data.

References

1. Kipp G, Bretscher HM, Schulte B, Herrmann D, Kusyak K, Day MW, Kesavan S, Matsuyama T, Li X, Langner SM, Hagelstein J, Sturm F, Potts AM, Eckhardt CJ, Huang Y, Watanabe K, Taniguchi T, Rubio A, Kennes DM, Sentef MA, Baudin E, Meier G, Michael MH, McIver JW (2025) Cavity electrodynamics of van der Waals heterostructures. Nat Phys 21:1926–1933. https://doi.org/10.1038/s41567-025-03064-8
2. Schulte BF (2022) Ultrafast Transport in Tunable Graphene Heterostructures. PhD Thesis. University of Hamburg
3. Miles RE, Harrison P, Lippens D (2001) Terahertz sources and systems. NATO science series II: mathematics, physics and chemistry (NAII, vol 27). Springer
4. Wang F, Zhang Y, Tian C, Girit C, Zettl A, Crommie M, Shen YR (2008) Gate-variable optical transitions in graphene. Science 320:206–209. https://doi.org/10.1126/science.1152793
5. Systems SR (2011) MODEL SR830 DSP Lock-In Amplifier Manual, Revision 2.5
6. Kusyak K, Schulte B, Matsuyama T, Kipp G, Bretscher HM, Day MW, Meier G, Potts AM, McIver JW (2025) Monolithic optoelectronic circuit design for on-chip terahertz applications. APL Photonics 10:076117. https://doi.org/10.1063/5.0278183

[3] In cases where the symmetry of the THz pulse launching is uncertain, a full 3D electromagnetic simulation can serve as an additional reference, as discussed in more detail in Sect. 6.1.

Chapter 5
Extracting the Near-Field Optical Conductivity

Abstract The goal of on-chip THz spectroscopy is to extract optical properties of van der Waals materials, such as their frequency-dependent optical conductivity. This chapter introduces the relevant electrical and optical conductivity models and demonstrates how the near-field optical conductivity can be determined from on-chip THz spectroscopy measurements.

This chapter shows how optical material properties, such as the optical conductivity, can be extracted from measured on-chip THz spectroscopy data. This requires the understanding of the difference between electrical and optical conductivity, as explained in Sect. 5.1. Next, it is helpful to know what optical conductivity spectra can be expected from a sample measured with on-chip THz spectroscopy. Section 5.2 therefore introduces two different optical conductivity models, the Drude model (Sect. 5.2.1) and the Drude-Lorentz model (Sect. 5.2.2). Finally, Sect. 5.3 shows how the near-field optical conductivity can be extracted from measured on-chip THz spectroscopy data.

5.1 Electrical and Optical Conductivity

The electrical (DC) conductivity σ_0 of a material is the reciprocal of the electrical resistivity ($\sigma = 1/\rho$), and is defined as the proportionality constant between the induced current density $\vec{j}$ and the inducing electric field $\vec{E}$ [3]:

$$\vec{j} = \sigma_0 \vec{E}. \tag{5.1}$$

G. Kipp, *Self-Cavity Electrodynamics of Van der Waals Heterostructures*, Springer Theses, https://doi.org/10.1007/978-3-032-31979-1_5

While the electrical conductivity usually refers to the static, time-independent limit, the optical conductivity describes how a material responds to AC electromagnetic waves as a function of frequency:

$$\vec{j}(\omega) = \sigma(\omega)\vec{E}(\omega) \tag{5.2}$$

and thus provides the ability to capture a series of characteristics of matter that the DC conductivity cannot. Specifically, the optical conductivity can capture

- **Dynamics of charge carriers:** At low frequencies, charge carriers dominate conductivity, but their contribution diminishes at higher frequencies due to scattering (as interpreted by the Drude model, see Sect. 5.2.1 below). Optical conductivity enables the direct extraction of scattering rates without needing additional parameters like carrier density or effective mass, which are typically required when estimating scattering rates from electrical conductivity measurements.
- **Interband dynamics:** Interband excitations in an insulator or semiconductor give rise to an absorption resonance at the gap energy, which can be seen as a peak in the real part of the optical conductivity, enabling the extraction of gap magnitudes [4].
- **Superfluid densities:** The superfluid density ρ_s quantifies the number of Cooper pairs or charges that are in the superfluid (superconducting) state and is encoded in the imaginary part of the optical conductivity (ρ_s is related to the $1/\omega$ divergence of $\sigma_2(\omega)$ for $\omega \rightarrow 0$). By tracking the superfluid density against different tuning parameters, such as carrier density or temperature, one can make predictions about the gap symmetry in different systems [5–7].
- **Resonant excitations of collective modes** such as phonons, plasmons or magnons, that can alter the frequency landscape of the optical conductivity from a Drude response to a Drude-Lorentz response, as described in more detail in the following sections.

5.2 Optical Conductivity Models

There are different optical conductivity models that describe how materials respond to electromagnetic fields as a function of frequency. Which model applies depends on the material's properties (if it is insulating, metallic, semiconducting, etc.) and the nature of the light-matter interaction. The two relevant conductivity models for this thesis are the Drude model and the Drude-Lorentz model, as introduced in the sections below.

5.2.1 Drude Model

Paul Drude developed a model around 1900 [8] that describes the transport properties of materials, and is most applicable in the case of simple metals. The model is able to explain the linear dependence of induced current and inducing electric field (see Eq. 5.1) [3].

Drude assumed that simple metals can be described as a classical free electron gas, where electrons are propagating with a drift velocity $\vec{v}$ and are continuously scattered at heavier, stationary crystal ions, leading to a scattering rate τ. The electrons are further accelerated by the inducing electric field $\vec{E}$ and are slowed down by the collisions, resulting in the following equation of motion:

$$m\ddot{x} = -e\vec{E} - m\frac{\dot{x}}{\tau}, \tag{5.3}$$

which can also be written as:

$$m\frac{d\vec{v}}{dt} = -e\vec{E} - m\frac{\vec{v}}{\tau}, \tag{5.4}$$

where the term $m\frac{\vec{v}}{\tau}$ describes the frictional force due to collisions. In the stationary case ($d\vec{v}/dt = 0$), one obtains:

$$\vec{v} = -\frac{e\tau}{m}\vec{E} = -\mu\vec{E}, \tag{5.5}$$

where μ is the electron mobility ($\mu = |\vec{v}|/\left|\vec{E}\right| = e\tau/m$), defining which drift velocity the electrons reach for a specific electrical field strength $\vec{E}$. The current density $\vec{j}$ can be defined as

$$\vec{j} = -en\vec{v}, \tag{5.6}$$

where n is the charge carrier density of the material, which can further be transformed into

$$\vec{j} = \frac{ne^2\tau}{m}\vec{E} = ne\mu\vec{E}, \tag{5.7}$$

using Eq. 5.5. The stationary (DC) electrical conductivity σ_0 is thus given by

$$\boxed{\sigma_0 = \frac{\vec{j}}{\vec{E}} = \frac{ne^2\tau}{m} = \frac{ne^2}{\Gamma m}} \tag{5.8}$$

where $\Gamma = 1/\tau$ [3].

While Drude assumed in his model that all electrons can contribute to the electrical conductivity, Arnold Sommerfeld noticed that electrons obey Fermi-Dirac statistics (not classical Maxwell-Boltzmann statistics) and follow the Pauli exclusion principle. Sommerfeld corrected the Drude model by assuming that only a small fraction of electrons, that have energies within an energy band of $\sim k_B T$ near the Fermi energy E_F, can participate in the conduction. Electrons with lower energies cannot contribute as the maximum energy transfer in a collision is $\sim k_B T \ll E_F$ and through the Pauli principle, there are no free states that these electrons could scatter to. This correction of the Drude model enables the calculation of the mean free path (path that an electron can propagate in the mean time τ between collisions) as:

$$l = v_F \tau, \tag{5.9}$$

where $v_F = \hbar \vec{k}_F / m$ is the Fermi velocity of the electrons [3].

The Drude model can also describe the current response in an oscillatory, time-dependent case ($d\vec{v}/dt \neq 0$). If the motion is assumed to be harmonic, then:

$$\begin{aligned} \vec{x}(\omega) &= \vec{x}_0 e^{-i\omega t} \\ \vec{E}(\omega) &= \vec{E}_0 e^{-i\omega t} \end{aligned} \tag{5.10}$$

which can be inserted into Eq. 5.3 to obtain:

$$-i\omega m \dot{x} = -e\vec{E}(\omega) - m\frac{\dot{x}}{\tau}. \tag{5.11}$$

Solving the above equation for $\dot{x} = \vec{v}$ gives:

$$\dot{x} = \frac{-e\vec{E}(\omega)}{m\left(\frac{1}{\tau} - i\omega\right)}. \tag{5.12}$$

Inserting this equation into Eq. 5.6 leads to:

$$\vec{j} = \frac{e^2 n \vec{E}(\omega)}{m\left(\frac{1}{\tau} - i\omega\right)} = \frac{e^2 n \tau}{m} \frac{1}{\left(1 - \frac{i\omega}{\tau}\right)} \vec{E}(\omega), \tag{5.13}$$

such that the complex, frequency-dependent optical conductivity defined by the Drude model is given by [6]:

$$\boxed{\sigma(\omega) = \frac{\sigma_0}{1 - i\omega\tau} = \frac{\sigma_0}{1 - \frac{i\omega}{\Gamma}}}. \tag{5.14}$$

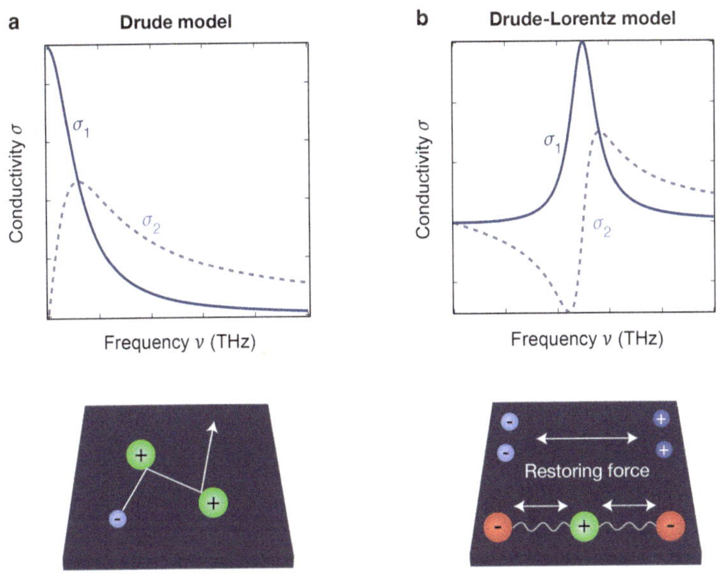

Fig. 5.1 Drude model and Drude-Lorentz model: Real and imaginary parts of the optical conductivity as a function of frequency as described by the (**a**) Drude model and (**b**) by the Drude-Lorentz model

An example plot of the real and imaginary parts of $\sigma(\nu)$ ($\nu = \omega/2\pi$) is given in Fig. 5.1a. The real part σ_1 has its maximum value at $\nu = 0$, reaching the value of the DC conductivity ($\sigma_1(\nu = 0) = \sigma_0$). The imaginary part σ_2 is small at low frequencies, indicating that the current is in phase with the applied electric field. The imaginary part peaks at the frequency at which $\sigma_1 = \sigma_0/2$, defining the scattering rate (linewidth) Γ, which can be used to calculate the scattering time $\tau = 1/\Gamma$. The larger the scattering rate, the broader and lower in amplitude σ_1 appears and the higher the frequency of the maximum value of σ_2.

While the Drude model captures key features of simple metals such as DC conductivity, scattering rates, mean free paths and intraband transitions, it cannot capture interband transitions, electron-electron interactions or the formation of collective modes that lead to defined features at finite frequencies. However, an extension to the Drude model, the so called Drude-Lorentz model gives intuition on these different finite-frequency effects.

5.2.2 *Drude-Lorentz Model*

The Drude-Lorentz model is an extension of the Drude model in which the electrons are additionally described to be bound and to experience a restoring force $-Kx$. The resonance frequency of the oscillating electrons can be defined as

$$\omega_0 = \sqrt{\frac{K}{m}}, \tag{5.15}$$

such that Eq. 5.3 can be extended to [3]

$$m\ddot{x} = -e\vec{E} - m\frac{\dot{x}}{\tau} - m\omega_0^2 x. \tag{5.16}$$

Equation 5.16 is simply the equation of motion for a driven damped harmonic oscillator, where $-e\vec{E}$ is the driving force through the inducing electric field, $-m\frac{\dot{x}}{\tau}$ is the frictional force due to electron scattering and $-m\omega_0^2 x$ is the restoring force.

Under the assumption of a harmonic motion (see Eq. 5.10), one finds

$$\vec{x}(\omega) = \frac{-\frac{e}{m}}{\omega_0^2 - \omega^2 - i\Gamma\omega}\vec{E}(\omega) \tag{5.17}$$

as the stationary solution for the displacement $x(\omega)$.

To find the Drude-Lorentz conductivity, one can again use Eq. 5.6, such that Eq. 5.17 can be transformed to

$$\vec{j} = -en\dot{\vec{x}} = -en(-i\omega)\vec{x}(\omega) = \frac{\frac{e^2 n}{m}\omega}{\Gamma\omega + i(\omega_0^2 - \omega^2)}\vec{E}(\omega). \tag{5.18}$$

With $\sigma_0 = ne^2/(\Gamma m)$, as given by Eq. 5.8, the Drude-Lorentz conductivity is then given by:

$$\boxed{\sigma(\omega) = \frac{\sigma_0\Gamma}{1 - \frac{i(\omega^2-\omega_0^2)}{\Gamma\omega}} = \sigma_0 \cdot \frac{\Gamma\omega}{\Gamma\omega + i(\omega_0^2 - \omega^2)}}. \tag{5.19}$$

Equation 5.19 shows that the Drude-Lorentz model becomes the Drude model in the limit of $\omega_0 \rightarrow 0$. Figure 5.1b plots the real and imaginary parts of the Drude-Lorentz model. The real part σ_1 peaks at the resonance frequency ω_0 of the oscillator, the same frequency at which σ_2 becomes zero. The full width half maximum of σ_1 defines the scattering rate $\Gamma = 1/\tau$.

Figure 5.2 shows plots of Eqs. 5.17 and 5.19 as a function of frequency for different scattering rates Γ, defined by the quality factors $Q = \omega_0/\Gamma$. Both equations just differ by a factor of ω in the conductivity formula. This factor defines if the resonance frequency shifts for $Q < 1$, as in the case for the displacement $x(\omega)$ (see Fig. 5.17a), or if the resonance frequency is fixed in frequency as in the case of the Drude-Lorentz conductivity $\sigma(\omega)$ (see Fig. 5.17b).

I would like to emphasize that the Drude-Lorentz model is comparable to a *driven* damped harmonic oscillator, but not to an *undriven* damped harmonic oscillator. In case of an undriven damped harmonic oscillator, $E = 0$ in Eq. 5.16, which leads to

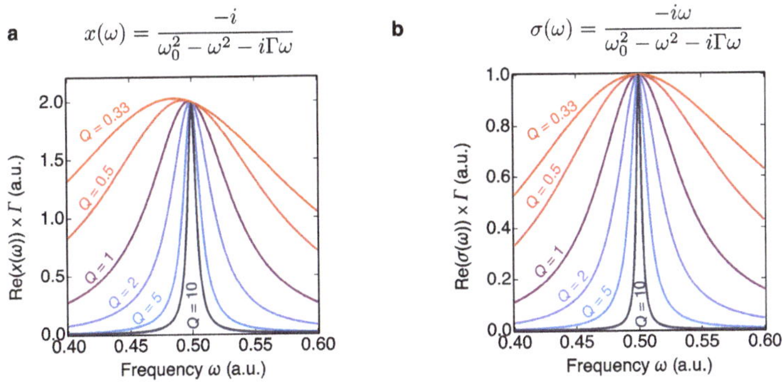

Fig. 5.2 Quality factor dependence of the Drude-Lorentz model: (**a**) Real part of the displacement $x(\omega)$ and (**b**) real part of the Drude-Lorentz conductivity $\sigma(\omega)$ as a function of frequency for different quality factors $Q = \omega_0/\Gamma$

the famous resonance condition:

$$\omega_{\text{undriven, damped}} = \sqrt{\omega_0^2 - \frac{\Gamma^2}{4}} = \omega_0\sqrt{1 - \left(\frac{1}{2Q}\right)^2}. \tag{5.20}$$

In contrast to the Drude-Lorentz model and to the case of a driven damped harmonic oscillator, there is no resonance ($\omega_{\text{undriven, damped}} = 0$) in the critically damped case (if $Q \approx 0.5$) of an undriven damped harmonic oscillator.

Having discussed two different frequency-dependent conductivity models, let's next see how $\sigma(\omega)$ can be extracted from the measured and processed on-chip THz spectroscopy data, from $V_{\text{sam}}(\omega)$ and $V_{\text{ref}}(\omega)$.

5.3 Extracting the Optical Conductivity From On-chip THz Spectroscopy Data

The goal of on-chip time-domain THz spectroscopy is to extract the complex, frequency-dependent conductivity of a sample of interest in the GHz-THz spectral range. In this section, the models used to extract the optical conductivity of the devices are described, derived and discussed. Based on these methods, Dr. Hope M. Bretscher implemented an analysis code in *Igor pro* that was used to analyze the data of this thesis. The time-domain traces from both the sample and reference sides, measured and rescaled based on the detected calibration curves (Sect. 4.5), are used as input for the analysis code. The first step in the process involves applying a Fourier transform to the time-domain data, converting the sample and reference signals into functions of frequency, $\omega = 2\pi\nu$. These signals as a function of

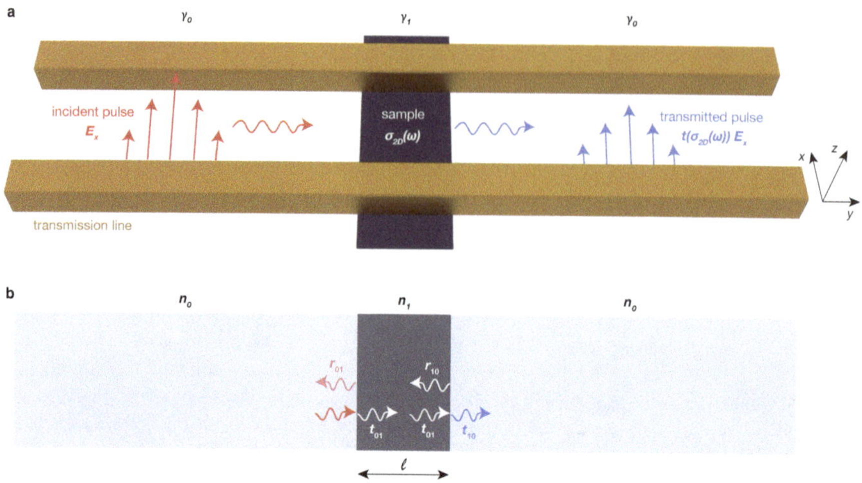

Fig. 5.3 Optics ansatz for on-chip THz spectroscopy: (**a**) An incident THz pulse E_x propagates in a coplanar strip transmission line with propagation constant γ_0. When it interacts with the sample with 2D conductivity $\sigma_{2D}(\omega)$, the propagation constant is γ_1. The transmitted pulse $t(\sigma_{2D}(\omega))E_x$ continues propagating with a propagation constant γ_0. (**b**) The scenario can be compared to an optics picture, where waves propagate in a dielectric with refractive index n_0 before interacting with the sample (refractive index n_1), where the wave gets reflected at both interfaces. The sample length l defines how much of the wave is reflected. See also Ref. [10]. (Figures adapted from [1] and [2] under a CC BY 4.0 license)

frequency are then converted into the near-field optical conductivity following the methods described below.

Importantly, the setup of on-chip THz spectroscopy is very comparable to a scenario in optics, where an electromagnetic wave propagates from a medium with a refractive index n_0 to a medium of length l with a refractive index n_1, before it continues propagating in a medium with n_0 (see Fig. 5.3). At the interfaces between the different media, the wave is partially reflected and partially transmitted. The transmission and reflection coefficients at the two interfaces can be obtained by using Fresnel's equations at normal incidence.

The reflection and transmission coefficients at the interface from medium with n_0 to a medium with n_1 are therefore [9]:

$$t_{01} = \frac{2n_0}{n_0 + n_1}, \tag{5.21}$$

$$r_{01} = \frac{n_0 - n_1}{n_0 + n_1}. \tag{5.22}$$

For the second interface between the sample medium with n_1 and the medium with n_0, the coefficients are similarly:

$$t_{10} = \frac{2n_1}{n_0 + n_1}, \tag{5.23}$$

$$r_{10} = \frac{n_1 - n_0}{n_0 + n_1}. \tag{5.24}$$

When a wave propagates on the sample side, including the section of the n_1 medium of length l, it experiences a phase shift Φ relative to the reference side, given by [11]:

$$\Phi(\omega, l) = e^{\frac{i\omega l}{c}n}. \tag{5.25}$$

Thus, the transmitted electromagnetic wave $E_t(\omega)$ that propagated across both interfaces on the sample side can be described by:

$$E_t(\omega) = t_{01}t_{10}E_0(\omega)e^{\frac{i\omega l}{c}n_1}. \tag{5.26}$$

Another effect that also has to be considered is that the wave can reflect several times within the sample (medium with n_1), as in a Fabry-Pérot interferometer. The internal resonance enhancement factor of a Fabry-Pérot resonator is defined as [12]:

$$F = \frac{1}{1 + r_{01}r_{10}e^{2i\frac{\omega l}{c}n_1}}. \tag{5.27}$$

The overall transmitted electromagnetic wave is therefore on the sample side:

$$E_{\text{sam}}(\omega) = \frac{t_{01}t_{10}E_0(\omega)e^{\frac{i\omega l}{c}n_1}}{1 + r_{01}r_{10}e^{2i\frac{\omega l}{c}n_1}}, \tag{5.28}$$

and on the reference side:

$$E_{\text{ref}}(\omega) = E_0(\omega)e^{\frac{i\omega l}{c}n_0}. \tag{5.29}$$

It is helpful to translate the derived equations in the optics picture to the language that is typically used to describe circuits, namely Telegrapher's equations (see Sect. 2.2.2). The goal is to transform Eqs. 5.28 and 5.29 into frequency dependent voltages $V_{\text{sam}}(\omega)$ and $V_{\text{ref}}(\omega)$, that are functions of propagation constants γ and can be measured with on-chip THz spectroscopy. With the ratios of $V_{\text{sam}}(\omega)$ and $V_{\text{ref}}(\omega)$, the near-field optical conductivity can then be calculated.

To isolate the propagation constant of the sample from that of the coplanar strips, a measurement scheme and analysis framework analogous to that of free-space THz can be derived [10, 13, 14]. In a dissipationless transmission line, which is a valid approximation for the empty coplanar strip transmission lines used in this experiment, the propagation constant is given by $\gamma_0 = i\omega\sqrt{LC}$ (see Eq. 2.9 for $R \to 0$ and $G \to 0$).

When a sample is placed beneath the coplanar strip transmission lines such as in the geometry of the experiment, it modifies the complex propagation constant, denoted by γ_{sam}.

The propagation constant γ (Telegrapher's equations) and refractive index n (optics) are then related by

$$\gamma = ik = \frac{i\omega n}{c}. \tag{5.30}$$

$V_{\text{sam}}(\omega)$ and $V_{\text{ref}}(\omega)$ can therefore be written as:

$$V_{\text{sam}}(\omega) = V_0(\omega)\frac{t_{01}t_{10}e^{\gamma_{\text{sam}}l}}{1 + r_{01}r_{10}e^{2\gamma_{\text{sam}}l}}, \tag{5.31}$$

$$V_{\text{ref}}(\omega) = V_0(\omega)e^{\gamma_0 l}. \tag{5.32}$$

By using Fresnel's equations (Eqs. 5.21–5.24) and replacing the refractive indexes by propagation constants, the fraction of the measured $V_{\text{sam}}(\omega)$ and $V_{\text{ref}}(\omega)$ can be calculated:

$$\frac{V_{\text{sam}}(\omega)}{V_{\text{ref}}(\omega)} = \frac{\frac{2\gamma_0}{\gamma_0+\gamma_{\text{sam}}}\frac{2\gamma_{\text{sam}}}{\gamma_0+\gamma_{\text{sam}}}e^{(\gamma_{\text{sam}}-\gamma_0)l}}{1 + \frac{\gamma_0-\gamma_{\text{sam}}}{\gamma_0+\gamma_{\text{sam}}}\frac{\gamma_{\text{sam}}-\gamma_0}{\gamma_0+\gamma_{\text{sam}}}e^{2\gamma_{\text{sam}}l}}, \tag{5.33}$$

which can be further transformed to [10, 13]:

$$\frac{V_{\text{sam}}(\omega)}{V_{\text{ref}}(\omega)} = \frac{2\frac{\gamma_{\text{sam}}}{\gamma_0}e^{-\gamma_0 l}}{(2\frac{\gamma_{\text{sam}}}{\gamma_0})(\cosh(\gamma_{\text{sam}}l)) - (1 + (\frac{\gamma_{\text{sam}}}{\gamma_0})^2)\sinh(\gamma_{\text{sam}}l)}. \tag{5.34}$$

Here, $\gamma_0(\omega)$ is the propagation constant of the empty coplanar strips simulated using full 3D electromagnetic simulations and l is the length of the sample. While this can be analytically simplified using a thin film assumption [10], numerically solving Eq. 5.34 provides more accurate results and was used for all the calculations of conductivities shown throughout this study. Numerical solving was performed using the *findroots* solver in IgorPro, which uses a Jenkins-Traub algorithm for solving the zeros of a polynomial [15].

As we will see later, the changes in the propagation constant that occur when a sample is present are due to the excitation of a current in the material between the coplanar strips. Thus, the γ_{sam}, calculated above, can be related to the longitudinal optical conductivity of the material $\sigma_{\text{eff}}(\omega)$ (the complex conductivity and the complex dielectric constant are related by $\epsilon = (1 + i\sigma/(\epsilon_0\omega))$, which in turn is related to the dielectric constant (ϵ_{sam}) of a material through (see Eq. 5.30 and $\epsilon(\omega) = n(\omega)^2$) [4]:

$$\gamma_{\rm sam}^2(\omega) = -\frac{\omega^2}{c^2}\epsilon_{\rm sam}(\omega) = -\frac{\omega^2}{c^2}(1+\frac{i\sigma_{\rm eff}(\omega)}{\epsilon_0\omega}). \tag{5.35}$$

Thus, the near-field optical conductivity $\sigma_{\rm eff}$[1] can be extracted from measured sample and reference signals using Eqs. 5.34 and 5.35.

5.4 Discussion

This chapter introduced the concept of the frequency-dependent optical conductivity and explored two distinct models: the Drude model and the Drude-Lorentz model. The Drude model describes the optical conductivity of free carriers (like electrons in metals) as a response, where the real part peaks at zero frequency and then decays towards higher frequencies representing dissipative transport without restoring forces. In contrast, the Drude-Lorentz model considers bound electrons by introducing a restoring force. This results in a peak at finite frequency, meaning that there is a resonance within the optical conductivity spectrum. When measuring metallic samples, one of the two discussed models should apply.

Section 5.3 outlined the method for extracting the optical conductivity from on-chip THz spectroscopy data measured on both the sample and reference sides of the circuit architecture that enables measurements of metallic samples such as graphite flakes (Sect. 2.3). As graphite is commonly used as an electrostatic gate in gate-tunable vdW heterostructures (see Fig. 1.2), capturing the THz optical conductivity of graphite flakes (presented in Chap. 6) is essential for understanding gate-induced contributions to the THz response of gate-tunable vdW heterostructures (discussed in Chap. 8).

References

1. Kipp G, Bretscher HM, Schulte B, Herrmann D, Kusyak K, Day MW, Kesavan S, Matsuyama T, Li X, Langner SM, Hagelstein J, Sturm F, Potts AM, Eckhardt CJ, Huang Y, Watanabe K, Taniguchi T, Rubio A, Kennes DM, Sentef MA, Baudin E, Meier G, Michael MH, McIver JW (2025) Cavity electrodynamics of van der Waals heterostructures. Nat Phys 21:1926–1933. https://doi.org/10.1038/s41567-025-03064-8
2. Michael MH, Kipp G, Potts AM, Day MW, Matsuyama T, Meier G, Bretscher HM, McIver JW (2025) Resolving self-cavity effects in two-dimensional quantum materials. arXiv preprint arXiv:2505.12799
3. Gross R, Marx A (2014) Festkörperphysik. De Gruyter Oldenbourg, München. https://doi.org/10.1524/9783110358704
4. Dressel M, Gruüner G (2002) Electrodynamics of solids: optical properties of electrons in matter. Cambridge University Press

[1] $\sigma_{\rm eff}$ is later renamed to $\sigma_{\rm cavity}$ based on the findings of Chap. 6.

5. Basov DN, Timusk T (2005) Electrodynamics of high-T_c superconductors. Rev Mod Phys 77:721–779. https://doi.org/10.1103/RevModPhys.77.721
6. Armitage N (2009) Electrodynamics of correlated electron systems. arXiv preprint arXiv:0908.1126
7. Sigrist M, Ueda K (1991) Phenomenological theory of unconventional superconductivity. Rev Mod Phys 63:239–311. https://doi.org/10.1103/RevModPhys.63.239
8. Drude P (1900) Zur Elektronentheorie der Metalle, I. Teil, II. Teil, und Berichtigung. Ann Phys 1:556–613
9. Born EWM (1970) Principles of optics. Pergamon Press Ltd.
10. Gallagher P, Yang C-S, Lyu T, Tian F, Kou R, Zhang H, Watanabe K, Taniguchi T, Wang F (2019) Quantum-critical conductivity of the Dirac fluid in graphene. Science 364:158–162. https://doi.org/10.1126/science.aat8687
11. Schulte BF (2022) Ultrafast transport in tunable graphene heterostructures. PhD Thesis. University of Hamburg
12. Ismail N, Kores CC, Geskus D, Pollnau M (2016) Fabry-Pérot resonator: spectral line shapes, generic and related Airy distributions, linewidths, finesses, and performance at low or frequency-dependent reflectivity. Opt Express 24:16366–16389. https://doi.org/10.1364/OE.24.016366
13. Neu J, Schmuttenmaer CA (2018) Tutorial: an introduction to terahertz time domain spectroscopy (THz-TDS). J Appl Phys 124. https://doi.org/10.1063/1.5047659
14. Nuss MC, Orenstein J (2007) Terahertz time-domain spectroscopy. In: Millimeter and submillimeter wave spectroscopy of solids, pp. 7–50. https://doi.org/10.1007/BFb0103419
15. Jenkins MA (1975) Algorithm 493: Zeros of a real polynomial [c2]. ACM Trans Math Softw 1:178–189

Chapter 6
THz Electrodynamics of Graphite Heterostructures

Abstract This chapter discusses the progress from the first on-chip THz measurement of a graphite heterostructure to a detailed understanding of tunable van der Waals material "self-cavities". By combining experimental THz spectra from a series of different graphite sample geometries, full 3D electromagnetic simulations, and analytical theory development, the observed resonances are identified as 2D plasmonic standing waves naturally confined by the graphite edges. The chapter further demonstrates how sample geometry and the interplay of screened and unscreened plasmons enable tunability, establishing design principles for accessing either Drude-like or self-cavity responses in van der Waals heterostructures.

This chapter documents the progress from the first graphite heterostructure probed using on-chip THz spectroscopy to a detailed and full understanding of tunable 2D material cavities ("self-cavities").

The chapter begins with the extraction of the near-field THz conductivity spectrum from experimental data obtained from a typical graphite heterostructure (see Sect. 6.1). The spectrum reveals a distinct resonance within the spectral sensitivity range, which is in sharp contrast to the expected Drude response (Sect. 5.2.1), commonly observed in far-field THz spectroscopy of large metallic samples [3, 4], and previously reported in on-chip THz spectroscopy experiments [5–7]. To therefore investigate the origin of this resonance and how it can be tuned, a series of graphite—hBN heterostrutures of different dimensions and geometries were measured and analyzed, as shown in Sect. 6.2.

Full 3D electromagnetic simulations advance the understanding of the experimental data by verifying the near-field conductivity obtained from graphite heterostructures and computing current density profiles (see Sect. 6.3). Based on the experimental and simulated results, an analytical theory was developed that identifies 2D plasmonic standing waves in the graphite as the origin for the detected

G. Kipp, *Self-Cavity Electrodynamics of Van der Waals Heterostructures*, Springer Theses, https://doi.org/10.1007/978-3-032-31979-1_6

resonances, where the edges of the graphite layer naturally act as a plasmonic cavity (see Sect. 6.4).

Based on the analytical theory, the cavity tunability through geometry is explored in Sect. 6.5, including the effect of the graphite regions extending beyond the coplanar strip metal traces (Sect. 6.5.1) and the cavity length (Sect. 6.5.2). As both the graphite regions beneath the metal traces as well as in between and outside of them contribute to the overall THz signal, the hybridization of so called screened and unscreened 2D plasmons is discussed in Sect. 6.5.3.

The chapter concludes by discussing design principles for vdW heterostructures comprising one conductive 2D material to either access the Drude response regime or the self-cavity regime (see Sect. 6.6).

6.1 Near-field THz Spectrum of A Bare Graphite Heterostructure

A typical graphite heterostructure embedded in on-chip THz circuitry is depicted in Fig. 6.1. A 16 nm thick graphite flake of roughly $13 \times 13\ \mu m^2$ is covered by a 26 nm thick hBN flake and placed on a sapphire substrate. THz pulses are generated by the THz antenna as described in Sect. 2.3 and are launched simultaneously to the left in an empty coplanar strip and to the right, where the graphite heterostructure is embedded (see Fig. 6.1a).

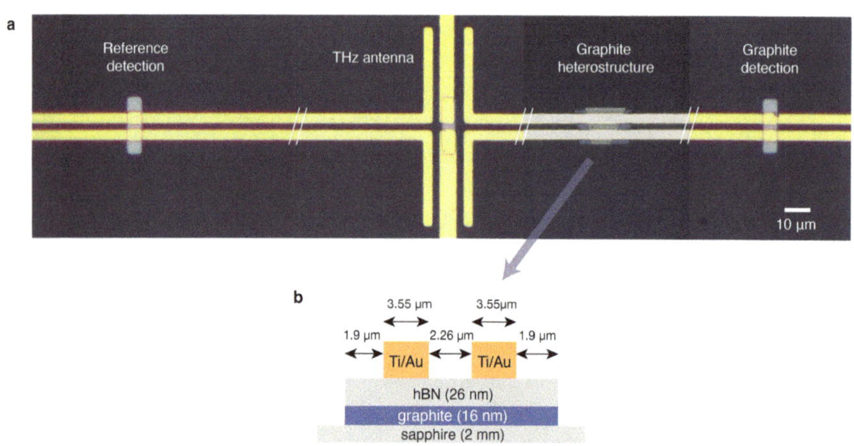

Fig. 6.1 Optical micrograph of a graphite heterostructure in on-chip THz circuitry. (**a**) Stitched overview of the ultrafast circuitry with THz antenna, coplanar strip, detection switches and graphite heterostructure. While the heterostructure is roughly 10 μ m long, the distance between detection switches and pulse launcher is 2 mm. (**b**) Cross-section of the graphite heterostructure with dimensions

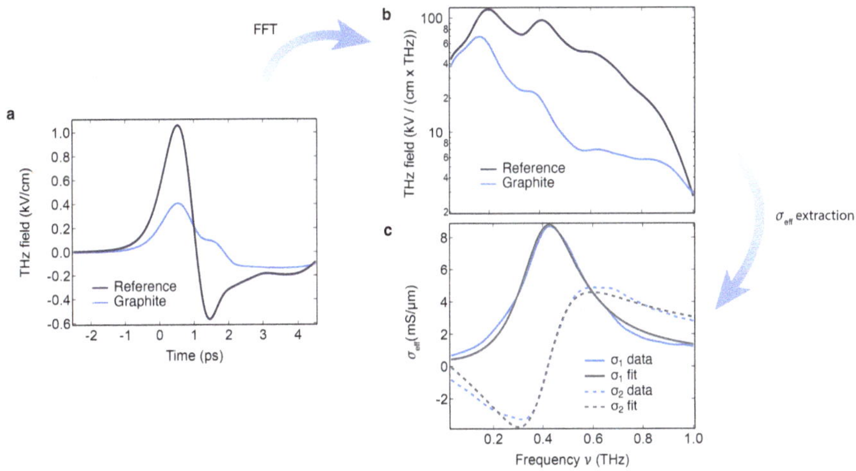

Fig. 6.2 Extracting the optical conductivity of a graphite heterostructure. (**a**) Measured time transients on the graphite heterostructure and reference side of the device shown in Fig. 6.1. The detected graphite heterostructure signal is significantly modified as compared to the reference trace. (**b**) Real part of the Fourier transforms of the measured transients shown in (**a**). (**c**) Real and imaginary parts of the optical conductivity obtained from the data in (**b**). The grey traces are fits to the data using a Drude Lorentz model (see Sect. 5.2.2). (Figure adapted from [1] under a CC BY 4.0 license)

Time-domain transients are detected on the reference and graphite detection switches from the device depicted in Fig. 6.1. The electric field amplitude of the signal can be determined from the measured voltage as a function of time with Eq. 4.3 (sensitivity = 10 mV, $m_{\text{calibration}} = 0.0309$, $W_2 = 2.26\,\mu\text{m}$), and is shown in Fig. 6.2a. The time-domain signals are then Fourier transformed (see Fig. 6.2b) to calculate the transmission coefficients (the transmission coefficients equal V_{sam}/W_2 and V_{ref}/W_2, whereby V_{sam} and V_{ref} are defined in Sect. 5.3), which are then used to numerically calculate real and imaginary parts of the effective near-field conductivity (see Fig. 6.2c) using the procedures outlined in Sect. 5.3.

The conductivity data shown in Fig. 6.2c was found to be independent of the windowing procedure. However, there is some uncertainty in the overall conductivity amplitude due to variations in the reference and sample trace amplitudes. In the case of the present graphite heterostructure, a systematic uncertainty of approximately 10 % in the amplitude of the calculated conductivities was identified through full 3D electromagnetic simulations, which were used to extract the transmission ratio of the sample compared to an empty circuit. Importantly, this uncertainty has negligible impact on the resonance lineshape.

The data looks like a Lorentz oscillator response and were fit using the Drude-Lorentz model (see Sect. 5.2.2 for model and Fig. 6.2c for fit). A linewidth of (0.29 ± 0.01) THz and a resonance frequency of (0.43 ± 0.01) THz were extracted from the fits, giving a quality factor Q of 1.48 ± 0.01, where the uncertainty comes from the covariance of the fits. In the case of negligible inhomogeneous broadening, the obtained linewidth can be related to the mean free scattering time (τ) or scattering rate (τ^{-1}) by $\Gamma = \tau^{-1}$. Thus, a clear resonance from the graphite heterostructure is visible, centered within the spectral sensitivity range.

6.2 Results from Different Device Designs

The geometric parameters of all devices are provided in Appendix A.1.
To investigate the origin of the resonance detected in the graphite heterostructure (see Fig. 6.2c), a series of different graphite heterostructures were fabricated in collaboration with Dorothee Herrmann as part of her Master's thesis research [8], and studied using on-chip THz spectroscopy. Based on the observations in Fig. 1.3, it is highly plausible that the resonance arises from the excitation of plasmons, collective oscillations of free charge carriers, within the graphite layer. Given that the penetration depth of THz light into graphite (see Table 2.2) is significantly greater than the typical thickness of graphite flakes, the plasmon could exhibit a two-dimensional (2D) character. Such 2D plasmons (see Fig. 1.3), in contrast to their 3D counterparts, are known for their unique dispersion relations (see Sect. 6.4.2) and deeply sub-wavelength confinement [9]. However, existing literature predominantly reports on either fully screened (metal in close proximity) or unscreened (no metal nearby) 2D plasmons. The graphite flake in this study is partially screened by metal traces and partially unscreened, so if the resonance indeed originates from a 2D plasmon excitation, one must also ask in which region(s) of the heterostructures this excitation occurs.

To find the origin of the resonance, a series of devices was explored in which the graphite layer, electrically isolated from the coplanar strips, does not span the entire width of the coplanar strip but was limited either to the region covered by the metal strips or to the coplanar strip gap. This initial data series helps determine whether the resonance is excited solely within sections of the graphite, defined by the edges of the coplanar strip's metal traces, or across the entire graphite from edge to edge, outside the coplanar strip.

6.2.1 *Series of Partial Coplanar Strip Coverage Graphite Devices*

To determine if the measured resonance (see Fig. 6.2c) arises specifically from regions of the graphite flake confined by the sharp dielectric boundaries of the metal traces in the coplanar strip, a series of test devices is characterized. In these devices, narrow graphite or gated graphene flakes are positioned either under one or both metal traces (see Fig. 6.3) for comparison.

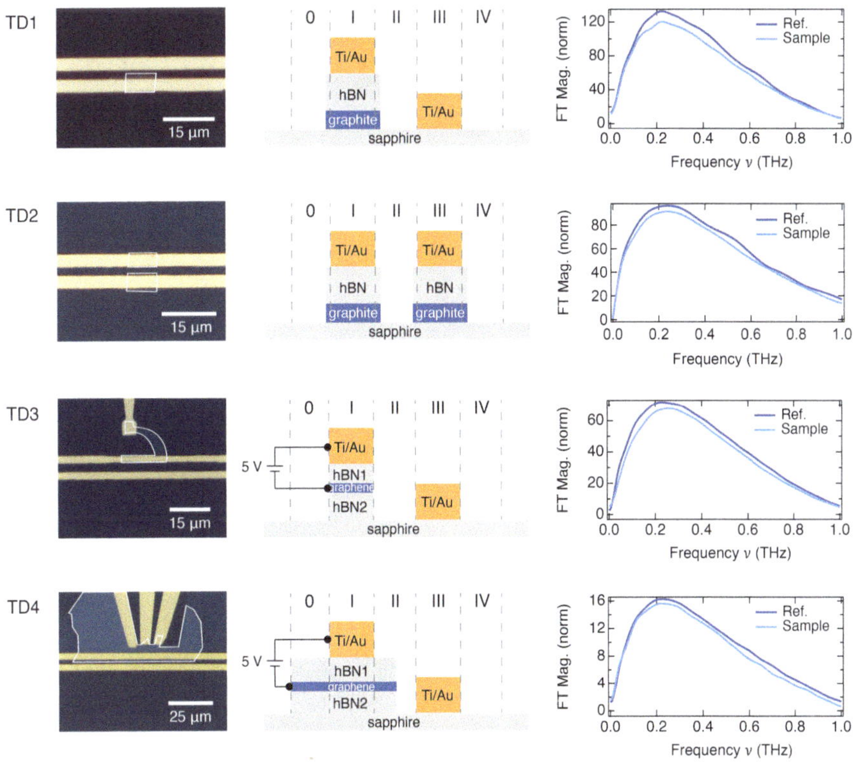

Fig. 6.3 Plasmon excitation test devices. Left panel: Optical micrograph of device, middle panel: Device cross-section, right panel: Fourier transform magnitudes of sample and reference signal. *Note*: The measurements for Device TD2 exhibited an unusually high noise level, and the signal measured on the sample side was unexpectedly larger than that on the reference side. To facilitate comparison, the Fourier transform magnitude of the signal from the sample side was rescaled by a factor of 0.75 to match the magnitude of the signal detected on the reference side. Device parameters are given in Table A.1

The use of monolayer graphene in some devices (see TD3 and TD4 in Fig. 6.3) was chosen for ease of fabrication and should not cause confusion. As will be seen later, a gated graphene layer at a typical carrier density can have the same number of carriers as a thin graphite flake. The disadvantage of using graphene layers as test devices, however, is that they must be contacted to apply a voltage between one of the coplanar strip traces (serving as a topgate) and the graphene layer in order to induce carriers into the graphene (see Fig. 1.2).

However, for all four test devices shown in Fig. 6.3, the Fourier spectra detected on both the graphite heterostructure side and the reference side are nearly identical (see right panels of Fig. 6.3 and compare to Fig. 6.2b). This similarity suggests that the THz pulse either did not interact with the material or did so only weakly, below the detection sensitivity. Consequently, the resonance observed in Fig. 6.2c cannot

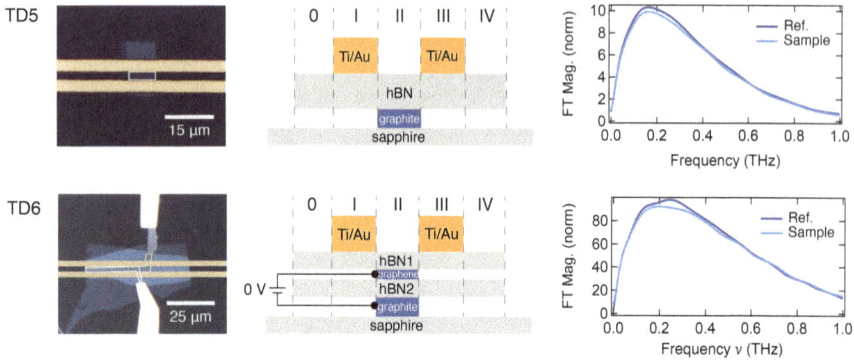

Fig. 6.4 Plasmon excitation test devices. Left panel: Optical micrograph of device, middle panel: device cross-section, right panel: FT magnitudes of sample and reference signal. Device parameters are given in Table A.1

arise from a collective mode confined exclusively to the graphite regions beneath the metal traces.

The next possibility to examine is whether the detected collective oscillation in the full coplanar strip width graphite heterostructure is only excited in the coplanar strip gap. Two samples were built: One with a narrow graphite flake located in the coplanar strip gap (see TD5 in Fig. 6.4), and one with a narrow graphene and graphite layer stack in the coplanar strip gap. The graphene in device TD6 could generally be electrostatically gated using the graphite flake, but is assumed to be transparent to the THz pulse if 0 V are applied between the graphene and the graphite flake. As shown in the right panel of Fig. 6.4, again no significant difference between the Fourier transforms detected on graphite heterostructure and reference side is visible for TD5 and TD6.

These results indicate that the resonance can only be seen when the graphite bridges the two metal traces of the coplanar strip. Therefore, it is most likely that the resonance detected on the device discussed in Figs. 6.1 and 6.2 stems from a collective excitation that oscillates from one outer coplanar strip metal edge to the other or even from graphite flake edge to graphite flake edge, if the graphite flake is broader than the coplanar strip, which is investigated in the following section.

6.2.2 Series of Full Coplanar Strip Coverage Graphite Devices

The series of graphite devices discussed in the previous section showed that the resonance detected from the graphite device shown in Figs. 6.1 and 6.2 must be an excitation of a collective oscillation in multiple or all coplanar strip regions. Thus, a series of graphite devices covering the full coplanar strip with varying dimensions is investigated in the following.

As shown in Fig. 6.5 (for all device geometry parameters see Table A.2 in the Appendix), the resonance frequency of graphite heterostructure can be tuned through geometry. Additionally, while the most-prominent lowest-frequency resonance is located at frequencies below 0.5 THz, a second resonance appears at frequencies between 0.5 THz and 0.8 THz (if the first resonance is located below 0.3 THz), but this mode appears with lower amplitude and broader linewidth. All spectra can be fit by a sum of one or two Drude-Lorentz models (see Sect. 5.2.2) to extract resonance frequencies ν_0, linewidths Γ and resulting quality factors $Q = \nu_0/\Gamma$. Typical quality factors range from 0.5 to 1.5, where the higher frequency resonances always have broader linewidths, but often still a higher quality factor (see Table A.3 in the Appendix).

The bare graphite devices are sorted from lowest detected resonance frequencies (BG1 in Fig. 6.5) to highest detected resonance frequencies (BG7 in Fig. 6.5), and the corresponding device geometry, where the most significant geometry change is marked, is shown in the center panel (for all device parameters see Table A.2). The lowest resonance frequencies are detected in devices with rather thin graphite and hBN flakes and with the widest coplanar strip widths (see BG1). Decreasing the coplanar strip width leads to a slight increase of the resonance frequencies (see BG2 and BG3). If the length of the device is significantly increased from the typical 10 μm to 38 μm, no significant change of the THz spectrum is observed when compared to the spectra obtained from the shorter devices.[1] Increasing the graphite thickness from 6 nm (BG5) to 16 nm (BG6E) leads to a significant increase of the resonance frequency from around 0.2 THz to 0.4 THz for the detected resonance that is lowest in frequency. When increasing the hBN thickness instead of the graphite thickness (see BG7), also a high resonance frequency is observed. The second higher frequency resonances detected for devices BG1–BG5 presumably also exist in devices BG6E and BG7, but are pushed outside the spectral sensitivity range.

The bare graphite devices BG1 to BG3 suggest that increasing the width of the coplanar strips may shift the resonance frequencies to lower values. To explore this trend further, device BGXL was fabricated (see Fig. 6.6), incorporating coplanar strip widths that are 3 to 5 times larger than those of the devices shown in Fig. 6.5. The THz spectrum of BGXL is notably broad, exhibiting a series of fixed-frequency oscillations. These oscillations remain unaffected by the data windowing procedure, suggesting that the broad spectrum arises from the superposition of multiple low-frequency resonances. On one hand, this finding further verifies the previous observation that wider coplanar strip widths lead to lower frequencies of the resonances. On the other hand, the spectrum detected from device BGXL suggests that in principle significantly more than two resonances can be excited in graphite heterostructures, which are either within the spectral sensitivity range (as for device BGXL) or outside the spectral sensitivity range (as for devices BG1 to

[1] Device BG4 was built by Sivasruthi Kesavan as part of her Master's thesis [10].

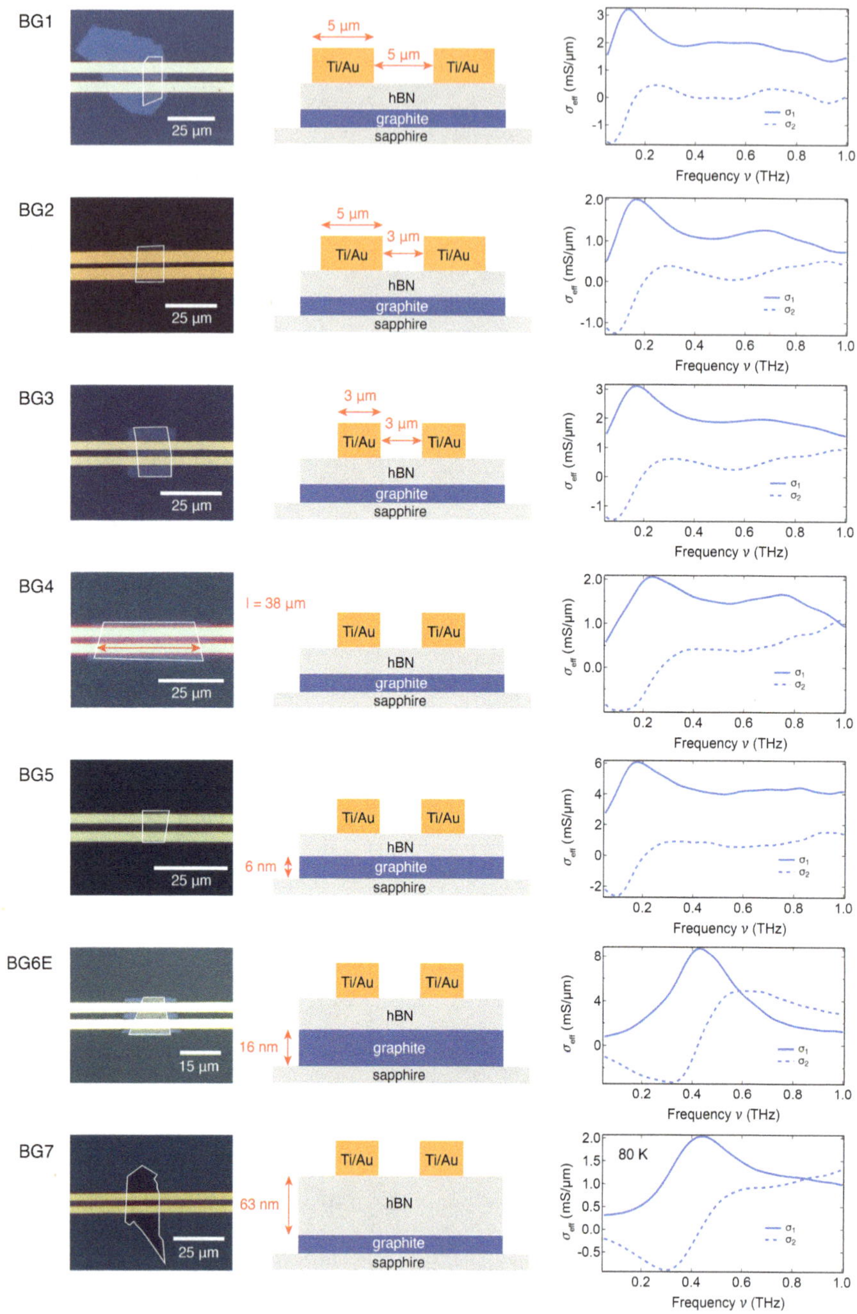

Fig. 6.5 Conductivity spectra of graphite devices with different dimensions

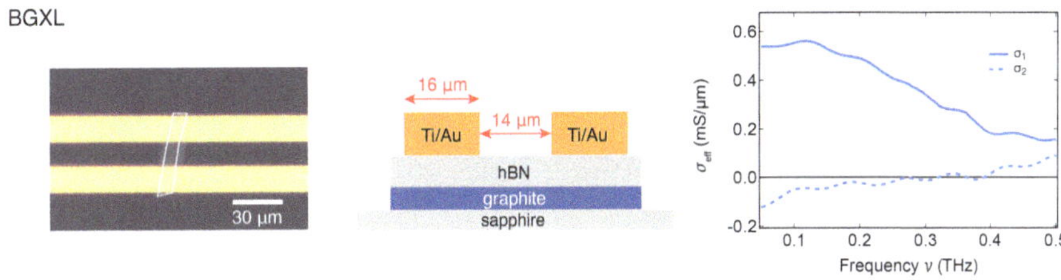

Fig. 6.6 Conductivity spectrum of a graphite device with particularly wide coplanar strip widths. The oscillations of the broad spectrum are found to be fixed in frequency, independent of the data processing or windowing, indicating that the spectrum consists of many resonances with frequency separations that are comparable to the linewidths of each resonance. Device parameters are given in Table A.2. (Figure adapted from [1] under a CC BY 4.0 license)

BG7). In summary, a series of resonances can be observed in bare graphite devices, which can be tuned by adjusting the geometry of the coplanar strip and the graphite heterostructure.

6.2.3 *Effect of Graphite Regions Extending the Coplanar Strip*

Next, it is investigated whether the graphite regions extending beyond the coplanar strip width influence the observed resonances. A series of devices is explored that have very wide regions extending the coplanar strip, which are then etched into graphite heterostructures with well defined shapes that do not extend significantly beyond the metal traces. THz spectra obtained before and after etching are compared to determine the contribution of the graphite regions outside the coplanar strip (Fig. 6.7).

The effective near-field conductivity spectrum of a graphite heterostructure with a wide, inhomogeneous (almost triangular) graphite region extending beyond the coplanar strip exhibits a broad spectrum (see Device BG8 in Fig. 6.7), rather than individual resonances. Additionally, when a wide graphite flake with relatively uniform width along the THz pulse propagation is incorporated into the on-chip THz spectroscopy circuit, such that one side extends the coplanar strip by several multiples of its width (see Devices BG9 and BG6 in Fig. 6.7), the near-field conductivity spectrum reveals two closely spaced resonances. In contrast, when the extension of the graphite flake beyond the coplanar strip is reduced to approximately 1 μ m, only one resonance remains (compare the data for Devices BG9E and BG6E in Fig. 6.7). This suggests that the region of the graphite flake extending beyond the coplanar strip contributes to the near-field conductivity spectrum, implying that the collective excitation arises from charge oscillations *between the edges* of the graphite flake.

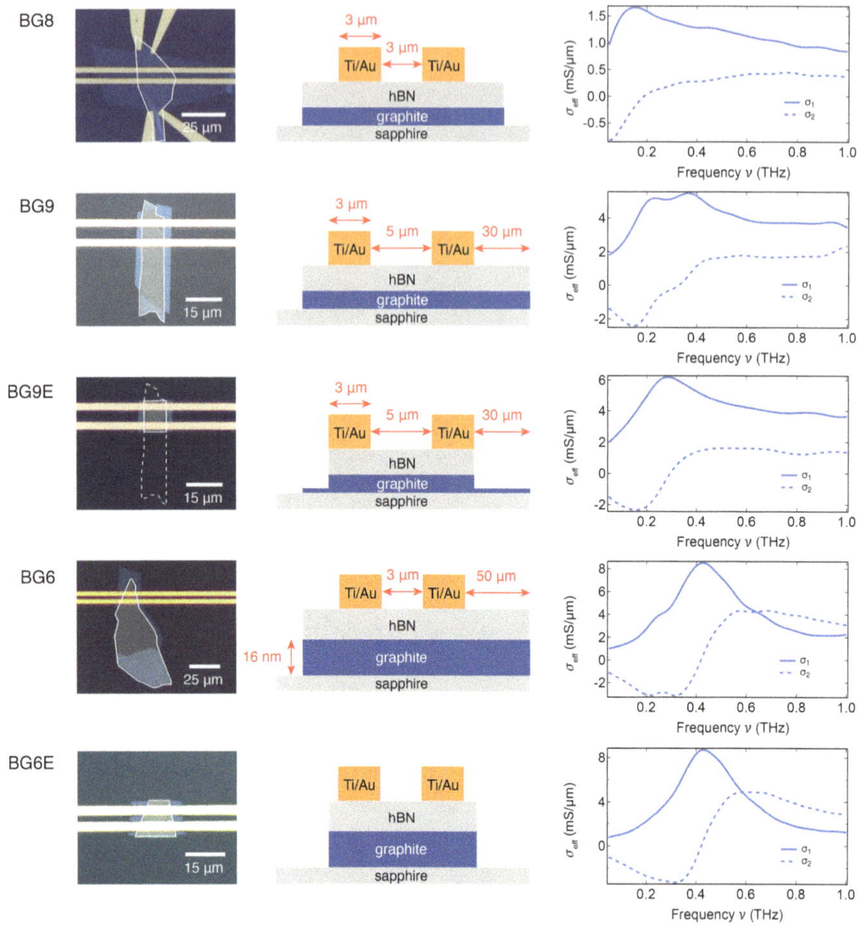

Fig. 6.7 Conductivity spectra of graphite devices with different graphite widths extending the coplanar strip. BG9 and BG9E (and BG6 and BG6E) is the same device but where only the graphite region extending the coplanar strip was etched (E). Device parameters are given in Table A.2

6.3 Full 3D Electromagnetic Simulations for Bare Graphite Heterostructures

To gain a better understanding of the experimental findings, 3D electromagnetic simulations were carried out in *CST Studio Suite* by Dr. Toru Matsuyama to investigate and verify the effective near-field conductivity obtained from graphite heterostructures, and to compute current density profiles in the range between 0.3 GHz and 1 THz [11] (for more details about the method, see supplementary information of Ref. [1]).

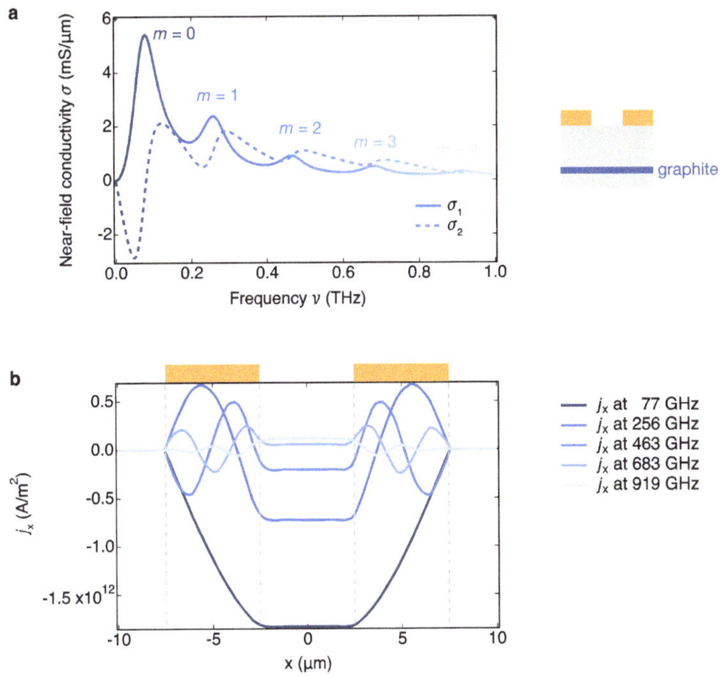

Fig. 6.8 Full 3D electromagnetic simulation of a graphite heterostructure. (**a**) Frequency-dependent real and imaginary parts of the near-field conductivity for a particularly thin and wide graphite heterostructure, designed to obtain five resonances within the spectral sensitivity range. The individual resonances are labeled with $m = 0, 1, 2, 3, 4$ indicating the number of nodes the simulated current density profiles show beneath each metal trace. (**b**) Current density profiles as a function of x for the five plasmonic resonances obtained in the conductivity spectrum in (**a**). (Figure adapted from [1] under a CC BY 4.0 license)

An example simulation of a thin and wide graphite sample, which exhibits five resonances within the spectral sensitivity range, is shown in Fig. 6.8. The graphite heterostructure discussed here consists of a 5 nm graphite, a 7 nm hBN, and a coplanar strip with $W_1 = W_2 = 5\,\mu$m, whereby W_1 defines the widths of the metal traces of the coplanar strip and W_2 is the width of the coplanar strip gap, and the graphite ends with the coplanar strip width such that $W_0 = 0\,\mu$m. The simulation was performed assuming a device length $l = 10\,\mu$m, a graphite 3D plasma frequency of $\nu_{3D} = 118$ THz (0.49 eV) and a scattering rate of $\tau^{-1} =$ 0.08 THz. The graphite 2D plasma frequency $\nu_{3D} = 118$ THz was found to match the experimental data of the bare graphite heterostructures best and is in good agreement with literature [12].

The simulated bare graphite conductivity spectrum of real and imaginary parts (σ_1 and σ_2) is shown in Fig. 6.8a. The current density profiles (j_x) as a function of x for the five resonances that appear in the spectrum are illustrated in Fig. 6.8b. The profiles show that the resonances indeed stem from standing waves of current density, namely 2D plasmons, where the standing wave of each resonance has a different wavelength and thus momentum. While the current density is continuous across the device, oscillations occur in the screened regions of the sample, those covered by the metallic coplanar strips. In contrast, in the unscreened regions between the metal strips, the current density is primarily flat.

When looking at the wavelength of the current density oscillations in the screened regions, one can see that the lowest frequency mode is approximately a quarter wavelength mode, meaning that the wavelength of the plasmonic standing wave in the screened region is about four times as long as the width of the metal strip, W_1. In the limit where the conductor separation $W_2 \to 0$, the two screened regions would together admit a half wavelength of standing current. The higher frequency modes are then approximately three-quarter, five-quarter, seven-quarter and nine-quarter wavelength modes, defined relative to the individual widths of each coplanar metal strip (W_1). These modes can be labeled with $m = 0, 1, 2, 3, 4$, corresponding to the number of current density nodes in the screened regions, below each metal strip.

The electric field distribution of the graphite heterostructure at the resonance frequency of the $m = 0$ (= 77 GHz) was simulated for different phases and is shown in Fig. 6.9. A phase range of $0° - 360°$ defines a full plasmon oscillation. As shown in Fig. 6.9, most of the electric field is confined in between the two metal strips, pointing in-plane and forming an odd mode. The in-plane orientation of the field in this mode is sensitive to the optical conductivity of a 2D sample. It can be seen that this mode excites a current in the graphite, which further leads to modulations of the electric field confined between the gold and graphite.

In summary, the resonances observed in the bare graphite devices (BG1–BG9) result from the excitation of 2D plasmonic standing waves, predominantly confined beneath the metal traces. The THz pulse's electric field induces currents in the graphite, which reflect at dielectric discontinuities, particularly at the sample edges, forming standing waves of current density with varying wavelengths at different resonance frequencies (see Fig. 6.8b). This behavior resembles standing waves on guitar strings, where different wavelengths produce distinct tones, or light waves in optical cavities that are confined in between two parallel mirrors (see Fig. 6.10). Since the edges of the graphite flakes embedded in the on-chip THz spectroscopy circuitry reflect the excited currents, functioning as mirrors without the need for actual mirrors like in optical cavities, these micropatterned graphite heterostructures are henceforth referred to as graphite ***self-cavities*** for THz light. Accordingly, the calculated effective near-field conductivity of these self-cavities is referred to as ***cavity conductivity*** moving forward.

Next, an analytical theory is developed to describe the plasmonic self-cavity modes of graphite heterostructures and their conductivity.

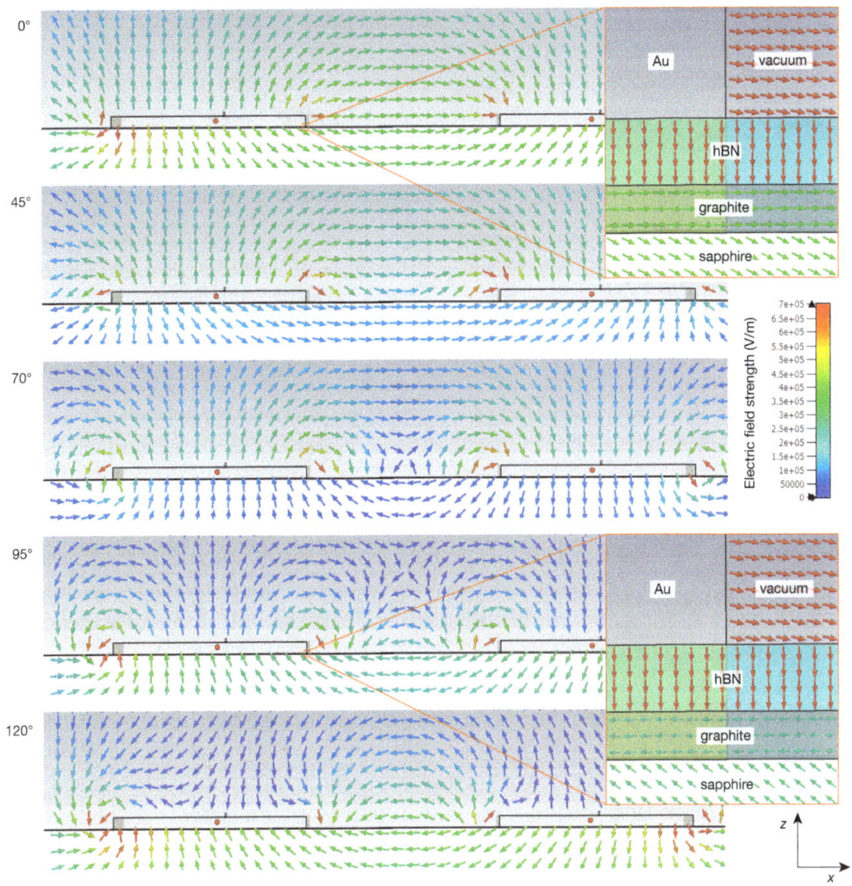

Fig. 6.9 Electric field distributions for the 77 GHz mode in Fig. 6.8. Electric field distributions for the 77 GHz mode at different phases (0° – 120°). Right panels in orange boxes show zoomed in selections of the electric field distributions in the graphite heterostructure. (Figure adapted from [1] under a CC BY 4.0 license)

6.4 Theoretical Description of Graphite Self-cavities

This section outlines the analytical framework, developed in close collaboration with Dr. Marios H. Michael, for modeling vdW heterostructures comprising a single conductive 2D material (such as a graphite flake) integrated into on-chip THz spectroscopy. The theory is based on Maxwell's equations and is a purely classical model.

This section is structured as follows: Sect. 6.4.1 begins with a discussion of Maxwell's equations, including their solution in the near-field limit and the behavior of electromagnetic waves at interfaces between different media. Based on these foundations, the dispersion relation of 2D plasmons is derived. In typical on-chip

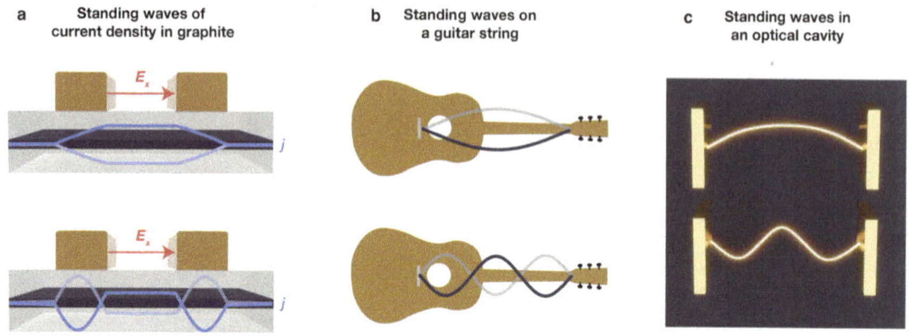

Fig. 6.10 Comparison of graphite devices to guitars and optical cavities. (**a**) Standing waves of current density confined between the edges of the graphite flake embedded in coplanar strips can be compared to (**b**) standing waves of guitar strings or (**c**) standing waves of light between two parallel mirrors in an optical cavity

THz spectroscopy circuitry, graphite regions beneath the gold traces are "screened", while those in the coplanar strip gap remain "unscreened". The distinct dispersion relations of screened and unscreened graphite plasmons are analyzed in Sect. 6.4.2.

Both the experimental results in Sect. 6.2 and the full 3D electromagnetic simulations in Sect. 6.3 demonstrate that 2D plasmonic standing waves are excited in both screened and unscreened regions of the graphite. To accurately model the electrodynamics of the full integrated graphite heterostructure, the dispersion relations of these plasmons serve as key inputs to the analytical theory. While the dispersion relation of unscreened 2D plasmons is straightforward, the screened case requires approximation. To find accurate solutions of the screened 2D plasmon dispersion relation, the Newton method is employed (Sect. 6.4.3).

Using these precise dispersion relations, boundary conditions are defined in Sect. 6.4.4 to solve for the overall cavity conductivity of the entire integrated heterostructure (see Sect. 6.4.5).

The overall approximations made in the analytical theory are discussed in Sect. 6.4.6. To compare the calculated cavity conductivity amplitudes from the analytical theory with experimental data and full 3D electromagnetic simulations, a "filling" factor must be considered, as discussed in Sect. 6.4.7. Section 6.4.8 offers a comparison of the analytical theory with simulations and experimental results.

Finally, the impact of the scattering rate on cavity resonance frequencies and the suitability of the Drude-Lorentz model for fitting the cavity conductivity spectra are discussed in Sect. 6.4.9.

6.4.1 Maxwell's Equations and Light Propagation at Interfaces

The analytical theory describes the interaction of the THz pulse (light) with the embedded bare graphite heterostructure (matter), so that there is no way around

Maxwell's equations that fully describe the interaction of electromagnetic waves with a material. Maxwell's equations are briefly summarized in this section from Ref. [13].

6.4.1.1 Maxwell's Equations in a Homogeneous Medium

When an electromagnetic wave, such as a THz pulse, is travelling through a homogeneous medium, the four Maxwell equations in SI units are:

$$\vec{\nabla} \times \vec{E} + \frac{\partial \vec{B}}{\partial t} = 0 \tag{6.1}$$

$$\vec{\nabla} \cdot \vec{B} = 0 \tag{6.2}$$

$$\vec{\nabla} \times \vec{B} - \frac{1}{c^2} \frac{\partial \vec{E}}{\partial t} = \frac{1}{\epsilon_0 c^2} \vec{j} \tag{6.3}$$

$$\vec{\nabla} \cdot \vec{E} = \frac{\rho}{\epsilon_0}, \tag{6.4}$$

where $\vec{E}$ is the electric field strength, $\vec{B}$ is the is the magnetic induction, $c = 1/\sqrt{\mu_0 \epsilon_0}$ is the speed of light in free space, j is the current density and ρ is the charge density.

Equations 6.1 and 6.2 can be satisfied by a vector potential $\vec{A}$ and a scalar potential V:

$$\vec{B} = \vec{\nabla} \times \vec{A} \tag{6.5}$$

$$\vec{E} + \frac{1}{c} \frac{\partial A}{\partial t} = -\vec{\nabla} V. \tag{6.6}$$

If there is no vector potential or if A is constant in time, Eq. 6.6 simplifies to:

$$\vec{E} = -\vec{\nabla} V. \tag{6.7}$$

Another helpful equation is the continuity equation for electric charge:

$$\frac{\partial \rho}{\partial t} + \vec{\nabla} \cdot \vec{j} = 0, \tag{6.8}$$

which expresses the conservation of charges: The divergence of j equals the negative rate of change of ρ in time.

In free space, there are no free currents or external charges, so $j = 0$ and $\rho = 0$. When taking the curl of Eq. 6.1 and by using the general vector identity $\vec{\nabla} \times (\vec{\nabla} \times \vec{E}) = \vec{\nabla}(\vec{\nabla} \cdot \vec{E}) - \vec{\nabla}^2 \vec{E}$, where the first term $\vec{\nabla}(\vec{\nabla} \cdot \vec{E})$ is zero (Eq. 6.4), one obtains:

$$-\vec{\nabla}^2\vec{E} + \frac{\partial}{\partial t}\vec{\nabla}\times\vec{B} = 0. \tag{6.9}$$

Now, by using Eq. 6.3 with $j = 0$, the above equation can be transformed into the wave equation without dissipation:

$$\vec{\nabla}^2\vec{E} - \frac{1}{c^2}\frac{\partial^2\vec{E}}{\partial t^2} = 0. \tag{6.10}$$

The wave equation can easily be satisfied by a harmonic propagating wave

$$\vec{E}(\vec{r},t) = \vec{E}_0 e^{-i\omega t} e^{i\vec{q}\vec{r}}, \tag{6.11}$$

that is characterized by the wave vector q.

Next, the influence of a homogeneous medium (rather than free space) on the propagation of the electromagnetic wave has to be discussed. In that case, charge and current densities are no longer zero as assumed before. Instead, the electromagnetic wave might induce electric dipoles, magnetic moments, polarization charges, field fluctuations or currents in the medium, that change the electric and magnetic fields.

Maxwell's equations in a medium are not derived here, but can be found in [13]. However, some consequences will be discussed. First, the presence of an electric field in matter leads to a motion of the electrons in the materials, where Ohm's law applies:

$$\vec{j} = \sigma\vec{E}, \tag{6.12}$$

where $\sigma = \sigma_1 + i\sigma_2$ is the complex optical conductivity. Second, the wave equation for the electric field (assuming no additional external charges) transforms to:

$$\vec{\nabla}^2\vec{E} - \frac{\epsilon_1\mu_1}{c^2}\frac{\partial^2\vec{E}}{\partial t^2} - \frac{4\pi\mu_1\sigma_1}{c^2}\frac{\partial\vec{E}}{\partial t} = 0, \tag{6.13}$$

with the same solution of a propagating wave as in the free space case:

$$\vec{E}(\vec{r},t) = \vec{E}_0 e^{-i\omega t} e^{i\vec{q}\vec{r}}. \tag{6.14}$$

Then, from Eqs. 6.13 and 6.14, the dispersion relation between wavevector $\vec{q}$ and frequency ω can be found:

$$\vec{q} = \frac{\omega}{c}\sqrt{\epsilon_1\mu_1 + i\frac{4\pi\mu_1\sigma_1}{\omega}}\vec{n}_q, \tag{6.15}$$

where c is the speed of light in vacuum, ϵ_1 is the real part of the dielectric permittivity, μ_1 is the real part of the magnetic permeability, σ_1 is the real part of the conductivity and $\vec{n}_q$ is a unit vector in $\vec{q}$-direction. In the case of no magnetization ($\mu_1 = 1$), the real part of Eq. 6.15 is:

$$q = \sqrt{\epsilon_1}\frac{\omega}{c}. \tag{6.16}$$

6.4.1.2 Eigenmodes of Maxwell's Equations in the Near-Field Limit

To derive the dispersion relations of screened and unscreened plasmons from Maxwell's equations, one must consider their form in the near-field limit. An electromagnetic wave interacting with a material can be either s- or p-polarized: in the s-polarized case, the electric field is perpendicular to the plane of incidence (parallel to the surface and along the y-axis), while in the p-polarized case, it lies within the plane of incidence (x-z-plane). Since s-polarized light lacks an electric field component normal to the surface, it cannot excite plasmon modes. Therefore, only p-polarized fields are relevant for plasmon coupling, and without loss of generality, the electric field can be assumed to lie entirely within the x-z-plane in the following:

$$\vec{E}(\vec{r}, t) = \vec{E}_0 e^{-i\omega t} e^{i\vec{q}\vec{r}} = \vec{E}_0 e^{-i\omega t} e^{iq_x x} e^{iq_z z}. \tag{6.17}$$

A 2D plasmon is a collective oscillation of charge carriers confined to a plane, generating electromagnetic fields that extend into the surrounding 3D space. In the sub-wavelength regime ($q_x \ll \sqrt{\epsilon_i}\omega/c$), the out-of-plane wave vector is given by

$$q_x^2 + q_y^2 = \epsilon_i \frac{\omega^2}{c^2} \tag{6.18}$$

$$q_z \overset{\epsilon_i\omega^2/c^2 \ll q_x}{=} \pm i q_x. \tag{6.19}$$

Since the wave vector is imaginary, the fields are evanescent, meaning the plasmon's field is tightly confined near the 2D layer.

Thus, Eq. 6.17 can be transformed to:

$$\boxed{\vec{E}(\vec{r}, t) = \vec{E}_0 e^{-i\omega t} e^{iqx} e^{qz}.} \tag{6.20}$$

The above equation serves as a starting point for the following derivations.

6.4.1.3 Introducing an Interface

As already introduced in Sect. 5.3, an electromagnetic wave that propagates from one medium with a reflective index n_0 to another medium with refractive index n_1 is

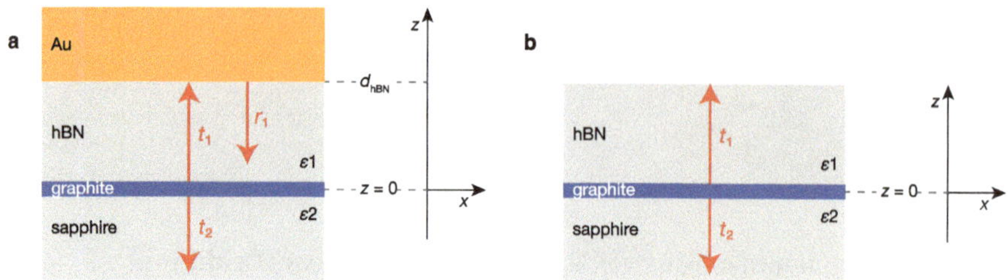

Fig. 6.11 Screened and unscreened graphite layer. Transmission and reflection coefficients of the electric field in a graphite heterostructure that is screened by a metallic gold layer (**a**) or unscreened (**b**). (Figure adapted from [2] under a CC BY 4.0 license)

partially reflected at the interface. This leads to a partial transmission (described by a transmission coefficient t) and partial reflection (described by a reflection coefficient r) of the electromagnetic wave, which is well described by Fresnel's equations (see Sect. 5.3).

A current oscillating within the plane and propagating along the x-direction generates an evanescent electric field that decays exponentially in the z-direction. When a metal trace is placed above the 2D material, this evanescent field reflects off the metal, producing a counter-propagating evanescent wave that decays in the negative z-direction. Thus, the electric field in x-direction can be written as a superposition of reflected and transmitted waves:

$$E_x = te^{-qz}E_0e^{iqx-i\omega t} + re^{qz}E_0e^{iqx-i\omega t}. \tag{6.21}$$

This approach enables the derivation of the dispersion relations for both screened and unscreened 2D plasmons, as presented in the following section.

6.4.2 *Screened and Unscreened 2D Plasmons*

In this section, the dispersion relations of screened and unscreened 2D plasmons are derived from Maxwell's equations, as known from literature (see Ref. [14]). To derive the dispersion relations of 2D modes, the near-field solutions in the bulk discussed in the previous section need to be supplemented by boundary conditions imposed on the material, such as the presence ("screened") or absence ("unscreened") of a nearby metal layer.

In the case of the screened geometry, the electric field is split into two distinct regions (see Fig. 6.11a):

$$E_x = \begin{cases} \left(t_1e^{-kz} + r_1e^{qz}\right)E_0e^{iqx-i\omega t}, & 0 < z < d_{\mathrm{hBN}} \\ \left(t_2e^{kz}\right)E_0e^{iqx-i\omega t}, & z < 0 \end{cases} \tag{6.22}$$

where $z = 0$ defines the out-of-plane position of the graphite and $z < 0$ defined the region in the sapphire. From the electric field along the x-direction (in-plane), the electric field along the z-direction (out-of-plane) can be found using Gauss's law: $\vec{\nabla} \cdot \vec{E} = 0$:[2]

$$E_z = \begin{cases} \left(\frac{iq}{k} t_1 e^{-kz} - \frac{iq}{k} r_1 e^{qz}\right) E_0 e^{iqx - i\omega t} & , 0 < z < d_{\mathrm{hBN}} \\ -\left(\frac{iq}{k} t_2 e^{kz}\right) E_0 e^{iqx - i\omega t} & , z < 0. \end{cases} \tag{6.23}$$

We can also write down the magnetic field along the y-direction that is given by the Maxwell's equation $\partial_x B_y = \frac{\epsilon_i}{c^2} \partial_t E_z$:

$$B_y = \frac{-i\omega}{c^2} \begin{cases} \epsilon_1 \left(\frac{1}{k} t_1 e^{-kz} - \frac{1}{k} r_1 e^{qz}\right) E_0 e^{iqx - i\omega t} & , 0 < z < d_{\mathrm{hBN}} \\ -\epsilon_2 \left(\frac{1}{k} t_2 e^{kz}\right) E_0 e^{iqx - i\omega t} & , z < 0. \end{cases} \tag{6.24}$$

Now, the density distribution in the graphite is given by (using Eq. 6.4):

$$\begin{aligned} \rho_{2\mathrm{D}} &= -\epsilon_0 \left(\epsilon_1 E_z(z = 0^+) - \epsilon_2 E_z(z = 0^-)\right) \\ &= -\frac{iq\epsilon_0}{k} E_0 \left(\epsilon_1 t_1 - \epsilon_1 r_1 + \epsilon_2 t_2\right) e^{iqx - i\omega t}, \end{aligned} \tag{6.25}$$

as $\vec{\nabla} \cdot \vec{E} = \rho/(\epsilon_0 \epsilon_{1,2})$ in a material. ϵ_1 is ϵ_{hBN} and ϵ_2 is $\epsilon_{\mathrm{sapphire}}$ in our case.

Next, the current density $j_{2\mathrm{D}}$ in the graphite layer can be calculated from the charge density $\rho_{2\mathrm{D}}$ by using the continuity equation (Eq. 6.8):

$$\frac{\partial \rho_{2\mathrm{D}}}{\partial t} = i\omega \frac{iq\epsilon_0}{k} E_0 (\epsilon_1 t_1 - \epsilon_1 r_1 + \epsilon_2 t_2) e^{iqx - i\omega t}, \tag{6.26}$$

$$j_{2\mathrm{D}} = -\int \frac{\partial \rho_{2\mathrm{D}}}{\partial t} dx = -\frac{i\omega\epsilon_0}{q} E_0 (\epsilon_1 t_1 - \epsilon_1 r_1 + \epsilon_2 t_2). \tag{6.27}$$

Another way to obtain the above Eq. 6.27 is to use Eq. 6.3, and calculate the magnetic field difference:

$$B_{\mathrm{y,\, z<0}} - B_{\mathrm{y,\, 0<z<d_{hBN}}} = -\frac{1}{\epsilon_0 c^2} j_{2\mathrm{D}}, \tag{6.28}$$

$$B_{\mathrm{y,\, z<0}} - B_{\mathrm{y,\, 0<z<d_{hBN}}} = \frac{-i\omega}{c^2 q} E_0(-\epsilon_2 t_2) - \frac{-i\omega}{c^2 q} E_0(\epsilon_1 t_1 - \epsilon_1 r_1) \tag{6.29}$$

[2] Gauss's law can be transformed to (assuming $E_y = 0$) $\frac{\partial E_z}{\partial z} = -\frac{\partial E_x}{\partial x}$, such that $E_z = -\int \frac{\partial E_x}{\partial x} dz$.

$$= \frac{i\omega}{c^2 q} E_0(\epsilon_1 t_1 - \epsilon_1 r_1 + \epsilon_2 t_2), \tag{6.30}$$

$$j_{2D} = -\epsilon_0 c^2 \left(B_{\mathrm{y,\,z<0}} - B_{\mathrm{y,\,0<z<d_{hBN}}}\right) = -\frac{i\omega\epsilon_0}{q} E_0(\epsilon_1 t_1 - \epsilon_1 r_1 + \epsilon_2 t_2). \tag{6.31}$$

At the metal interface, the electric field parallel to the surface vanishes:

$$E_x(z = d_{\mathrm{hBN}}) = t_1 e^{-q d_{\mathrm{hBN}}} + r_1 e^{q d_{\mathrm{hBN}}} = 0, \tag{6.32}$$

$$r_1 = -t_1 e^{-2q d_{\mathrm{hBN}}}, \tag{6.33}$$

while at $z = 0$, the electric field parallel to the interface can be matched (Eq. 6.22):

$$\boxed{t_1 + r_1 = t_2}, \tag{6.34}$$

which can be transformed to (by inserting 6.33 for r_1):

$$t_1 \left(1 - e^{-2q d_{\mathrm{hBN}}}\right) = t_2. \tag{6.35}$$

Now, Eqs. 6.27 and 6.33 (and $j_{2D} = \sigma_{2D} E$ with $E = E_0 t_2$) can be used to obtain:

$$j_{2D} = \sigma_{2D} E_0 t_2 = \frac{-i}{q} \omega\epsilon_0 E_0 \left(\epsilon_1 t_1 + \epsilon_1 t_1 e^{-2q d_{\mathrm{hBN}}} + \epsilon_2 t_2\right), \tag{6.36}$$

which can be transformed to:

$$\sigma_{2D} t_2 = \frac{-i}{q} \omega\epsilon_0 \left(\epsilon_1 t_1 \left(1 + e^{-2q d_{\mathrm{hBN}}}\right) + \epsilon_2 t_2\right). \tag{6.37}$$

By then inserting Eq. 6.35 into the above equation, one obtains:

$$\sigma_{2D} t_2 = \frac{-i}{q} \omega\epsilon_0 \left(\epsilon_1 t_2 \frac{1 + e^{-2q d_{\mathrm{hBN}}}}{1 - e^{-2q d_{\mathrm{hBN}}}} + \epsilon_2 t_2\right), \tag{6.38}$$

$$\sigma_{2D} = \frac{-i}{q} \omega\epsilon_0 \left(\epsilon_1 \coth(q d_{\mathrm{hBN}}) + \epsilon_2\right). \tag{6.39}$$

Using the fact that $q d_{\mathrm{hBN}} \ll 1$ (leading to $\coth(q d_{\mathrm{hBN}}) \approx 1/(q d_{\mathrm{hBN}})$), the equation can again be transformed to:

$$\sigma_{2D} = \frac{-i}{q} \omega\epsilon_0 \left(\frac{\epsilon_1}{q d_{\mathrm{hBN}}} + \epsilon_2\right). \tag{6.40}$$

This leads to the dispersion relation of a screened 2D plasmon as:

$$\omega_{sc} = \frac{-\sigma_{2D} q}{i\epsilon_0 \left(\frac{\epsilon_1}{q d_{hBN}} + \epsilon_2\right)}, \tag{6.41}$$

that can be simplified with $q d_{hBN} \ll 1$ to

$$\omega_{sc} = \frac{-\sigma_{2D} q^2 d_{hBN}}{i\epsilon_0 \epsilon_1}, \tag{6.42}$$

and using the two-dimensional Drude-conductivity for graphite (see Appendix Sect. A.2),

$$\sigma_{2D,\,\text{graphite}} = \frac{\epsilon_0 \omega_{3D}^2 d_{gr}}{-i\omega + \Gamma}, \tag{6.43}$$

one obtains the plasmon wave vector q for screened 2D graphite plasmons as:

$$\boxed{q_{\text{screened, graphite}} = \sqrt{\frac{i\epsilon_1 \Gamma \omega + \epsilon_1 \omega^2}{\omega_{3D}^2 d_{gr} d_{hBN}}}}. \tag{6.44}$$

In case of no dissipation ($\Gamma = 0$), the resonance frequency of a screened 2D plasmon is:

$$\boxed{\omega_{\text{screened, graphite}} = \sqrt{\frac{\omega_{3D}^2 q^2 d_{gr} d_{hBN}}{\epsilon_1}}}. \tag{6.45}$$

In case of a graphene layer, instead of a graphite layer, the 2D conductivity is (the Fermi energy of graphene is defined as $E_F = \text{sign}(n)\hbar v_F \sqrt{\pi |n|}$, where $v_F = 1 \cdot 10^6\ \text{m s}^{-1}$ is the Fermi velocity, and n is the charge carrier density)

$$\sigma_{2D,\,\text{graphene}} = \frac{e^2 E_f}{\hbar^2 \pi} \frac{1}{-i\omega + \Gamma} = \frac{e^2 v_F \sqrt{n}}{\hbar \sqrt{\pi}} \frac{1}{-i\omega + \Gamma}, \tag{6.46}$$

such that the dispersion relation of a screened 2D graphene plasmon is:

$$\boxed{q_{\text{screened, graphene}} = \sqrt{\frac{\epsilon_0 \hbar \sqrt{\pi}(i\epsilon_1 \Gamma \omega + \epsilon_1 \omega^2)}{e^2 v_F \sqrt{n} d_{hBN}}}}, \tag{6.47}$$

$$\boxed{\omega_{\text{screened, graphene}} = \sqrt{\frac{e^2 v_F \sqrt{n}}{\epsilon_0 \hbar \sqrt{\pi}} \frac{q^2 d_{hBN}}{\epsilon_1}}}. \tag{6.48}$$

In the opposite limit of $d_{\text{hBN}} \to \infty$, we recover the **unscreened limit** (see Fig. 6.11b) and Eq. 6.39 becomes:

$$\sigma_{\text{2D}} = \frac{-i}{q}\omega\epsilon_0\left(\epsilon_1 + \epsilon_2\right), \tag{6.49}$$

as the coth(qd_{hBN}) in Eq. 6.39 becomes 1 for large d_{hBN}.

Inserting the Drude conductivity either for graphite (Eq. 6.43) or for graphene (Eq. 6.46) for σ_{2D}, one obtains the plasmon wave vector for unscreened 2D graphene/graphite plasmons:

$$\boxed{q_{\text{unscreened, graphite}} = \frac{(\epsilon_1 + \epsilon_2)\omega(i\Gamma + \omega)}{d_{\text{gr}}\omega_{3D}^2}}, \tag{6.50}$$

$$\boxed{q_{\text{unscreened, graphene}} = \frac{\epsilon_0\hbar\sqrt{\pi}(\epsilon_1 + \epsilon_2)\omega(i\Gamma + \omega)}{e^2 v_F\sqrt{n}}}. \tag{6.51}$$

These terms can be reformulated in the case of no dissipation to the well-known formula for unscreened 2D plasmons [14, 15]:

$$\boxed{\omega_{\text{unscreened, graphite}} = \sqrt{\frac{\omega_{3D}^2 q d_{\text{gr}}}{\epsilon_1 + \epsilon_2}}}, \tag{6.52}$$

$$\boxed{\omega_{\text{unscreened, graphene}} = \sqrt{\frac{e^2 v_F\sqrt{n}}{\epsilon_0\hbar\sqrt{\pi}}\frac{q}{\epsilon_1 + \epsilon_2}}}. \tag{6.53}$$

The dispersion relations for screened and unscreened graphene plasmons are introduced here for completeness, and are discussed in detail later in Chap. 9. Two important points are:

1. A thin graphite flake of ~5 nm thickness and with $\nu_{3D} = \omega_{3D}/(2\pi) = 118$ THz has the same 2D conductivity as a graphene layer that is doped to $n = 3.14 \cdot 10^{12}$ cm^{-2}, a typical value for gated graphene.
2. Equations 6.48 and 6.53 show, that the resonance frequency of both screened and unscreened 2D plasmons in graphene scale with $\omega \propto n^{1/4}$. In the remaining sections of this Chapter, the focus lies on graphite heterostructures.

The dispersion relations for $\nu(q)$ with $\nu = \omega/(2\pi)$ of screened (Eq. 6.45) and unscreened (Eq. 6.52) 2D plasmons are plotted in Fig. 6.12 using typical graphite heterostructure parameters ($d_{\text{gr}} = 10$ nm, $d_{\text{hBN}} = 20$ nm, $\epsilon_1 = 3.7$ $(= \epsilon_{\text{hBN}})$, $\epsilon_2 = 10$ $(= \epsilon_{\text{sapphire}})$ and $\nu_{3D} = 118$ THz). The resonance frequencies of unscreened 2D plasmons are significantly higher than of screened 2D plasmons at

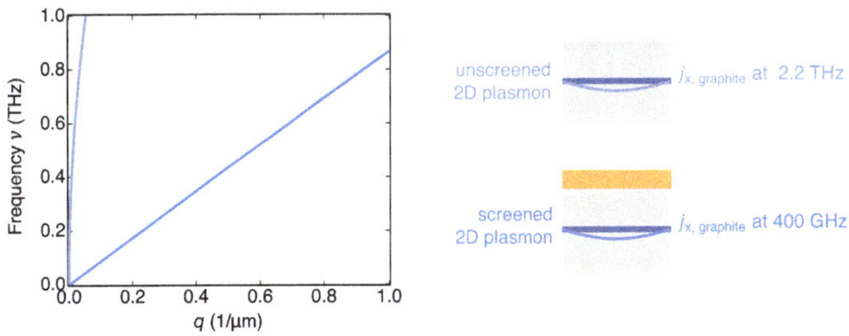

Fig. 6.12 Dispersion relations for screened and unscreened 2D plasmons in graphite. The resonance frequencies of unscreened 2D plasmons ($\nu \propto \sqrt{q}$) are much higher than of screened plasmons ($\nu \propto q$) at a given q. For example, a standing wave of current density can form with a wavelength four times the width $W = 3.4\,\mu$ m of the screened/unscreened region ($q = 0.46\,\mu\text{m}^{-1}$). The resonance of the screened plasmon occurs at 400 GHz, which falls within the spectral sensitivity range of on-chip THz spectroscopy, whereas the resonance of the unscreened plasmon would be at 2.2 THz, far outside the bandwidth

a given q (see Fig. 6.12). This has multiple implications. First, it highlights that the dielectric environment has a major influence on plasmonic oscillations in a 2D material. Secondly, it explains why test devices TD5 and TD6, where the graphite layers were unscreened (see Fig. 6.4), did not show a resonance within the spectral sensitivity range. As the graphene or graphite layers of test devices TD1 to TD4 (see Fig. 6.3) were screened by the metal trace, a plasmonic resonance within the spectral sensitivity range would be expected (see Table A.1 in the Appendix). However, the absence of the resonances in the detected spectra indicates a more complex plasmon excitation mechanism, which is explored in greater detail in the following sections.

6.4.3 *The Newton Method: More Precise Dispersion Relations*

As discussed in the previous section, the dispersion relation for the unscreened case is solved exactly (see Eq. 6.50) without any approximations. However, for the screened case, an approximation is made, assuming that $qd_{\text{hBN}} \ll 1$, which allows us to approximate $\coth(qd_{\text{hBN}})$ as $1/(qd_{\text{hBN}})$. This leads to a simplified dispersion relation for screened 2D plasmons, as shown in Eq. 6.44. In the analytical theory of on-chip THz cavities, quantifying the validity of this approximation is essential for achieving maximum accuracy.

A way to estimate the error between the approximated dispersion relation and the actual solution is to use Newton's method [16]. Newton's method is an approximation algorithm for the numerical solution of non-linear equations, which allows one to find the zero points of a continuously differentiable function $f(q) = 0$, based on an initial guess q_0. The basic idea of this approach is to linearize the function at q_0 (determining the tangent) and to use the q_1-value at which the tangent crosses zero as an improved approximation of the zero of the function:

$$q_{\mathrm{improved},1} = q_0 - \frac{f(q_0)}{f'(q_0)}, \tag{6.54}$$

where $f'(q_0)$ is the derivative of $f(q_0)$. By repeating the procedure with the novel $q_{\mathrm{improved},1}$-value, the approximation can be further improved. In principle, the approach can be repeated several times until the error becomes sufficiently small [16].

For finding the most accurate q of screened 2D plasmons, the function $f(q)$ is (Eq. 6.39):

$$f(q) = \sigma_{\mathrm{2D}} - \frac{i}{q}\omega\epsilon_0\left(\epsilon_1 \coth(q d_{\mathrm{hBN}}) + \epsilon_2\right), \tag{6.55}$$

and our guess is the approximated q_{screened} from Eq. 6.44:

$$q_0 = \sqrt{\frac{i\epsilon_1\Gamma\omega + \epsilon_1\omega^2}{\omega_{3D}^2 d_{\mathrm{gr}} d_{\mathrm{hBN}}}}. \tag{6.56}$$

The error of the approximated q_{screened} based on the first correction is then given by $-\frac{f(q_0)}{f'(q_0)q}$ and is plotted for a range of hBN layer thicknesses and frequencies in Fig. 6.13. The error is increasing linearly with increasing frequency and is about 1 % at 1 THz (see Fig. 6.13a). Furthermore, the error is increasing proportional to $\sqrt{d_{\mathrm{hBN}}}$ with increasing hBN layer thickness and is about 2.5 % for a heterostructure with a 100 nm thick hBN layer (see Fig. 6.13b).

The correction of the approximated q_{screened} can be further improved by applying Newton's method again to the first correction:

$$q_{\mathrm{improved},2} = q_{\mathrm{improved},1} - \frac{f(q_{\mathrm{improved},1})}{f'(q_{\mathrm{improved},1})}, \tag{6.57}$$

however, the second correction is small as seen in Fig. 6.13, such that the first correction is sufficient to continue with a very precise plasmon dispersion relation for screened 2D plasmons.

6.4.4 Boundary Conditions for Bare Graphite Cavities

With the precise dispersion relations for screened and unscreened 2D plasmons now established, the next step is to incorporate them into a model describing the THz electrodynamics of an entire graphite heterostructure embedded in on-chip THz spectroscopy circuitry. A cross-section of a typical graphite heterostructure is shown

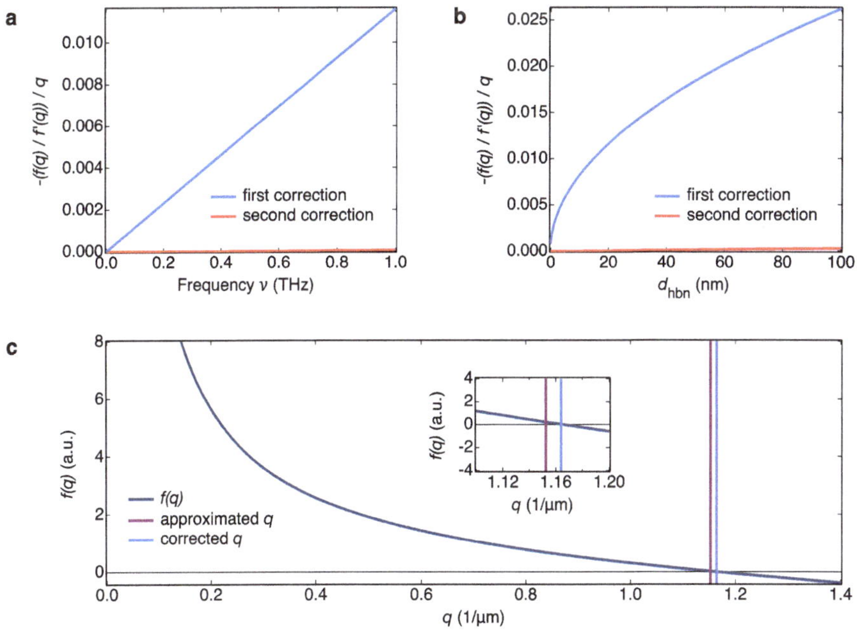

Fig. 6.13 Correcting the approximated q_{screened} with Newton's method. (**a**) Errors of the approximated q_{screened} after the first and second correction as a function of frequency. The error obtained after the first correction scales linear in frequency. The error of the first correction (= second correction) is negligible, showing that the first correction is sufficient. (**b**) Errors of the approximated q_{screened} after the first and second correction as a function of hBN thickness. The error scales with $\propto \sqrt{d_{\text{hBN}}}$. The error after the second correction (red) is also negligible. **c** Function $f(q)$ as a function of q. Solutions of q can be obtained for $f(q) = 0$. The approximated q (purple) is a good guess, but the value obtained after the first correction (blue) is more precise

in Fig. 6.14. As discussed in Sect. 6.4.1.2, plasmons can be excited by p-polarized near-field electric fields in the x-z plane. In the case of a coplanar strip geometry, where a THz pulse (*quasi*-TEM mode, both s- and p-polarized with respect to the 2D material) propagates along the y-direction with its electric field primarily aligned in the x-direction between the metal traces and evanescent components in the z-direction above and below, 2D plasmons can be efficiently launched in the embedded graphite layer.

The resulting current in the graphite along the x-axis can be divided into five distinct regions (labeled 0 to IV in Fig. 6.14), corresponding to either "screened"(regions I and III) or "unscreened" (regions 0, II, and IV) segments. For analytical simplicity, the heterostructure is assumed to have mirror symmetry in the x-axis.

To capture the response of the entire graphite heterostructure, with broken translational symmetry due to the finite size and different momenta q_{screened} and $q_{\text{unscreened}}$, all five different regions have to be considered. This is done by expressing the current in the graphite layer in terms of transmission and reflection

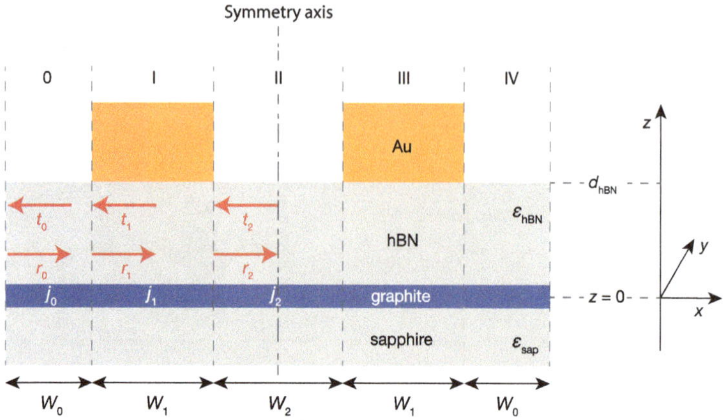

Fig. 6.14 Cross-section of a graphite heterostructure embedded in THz circuitry. The graphite—hBN heterostructure is placed on a sapphire substrate and embedded in a coplanar strip geometry, which provides an inhomogeneous dielectric environment for the 2D plasmon standing waves that form in the graphite layer. The analytical theory developed in this work captures the total cavity conductivity by solving Maxwell's equations for transmitted $t_{0,1,2}$ and reflected $r_{0,1,2}$ waves at the boundaries (dashed grey lines) between screened (I and III) and unscreened (0, II and IV) regions. (Figure adapted from [1] under a CC BY 4.0 license)

coefficients in each region as follows:

$$j_{\text{graphite}}(x,t) = \sigma_{2D}(\omega)\begin{cases} r_0 E_0 e^{-iq_{\text{unscreened}}(\nu)x} + t_0 E_0 e^{iq_{\text{unscreened}}(\nu)x}, & \text{case } a \\ r_1 E_0 e^{-iq_{\text{screened}}(\nu)x} + t_1 E_0 e^{iq_{\text{screened}}(\nu)x}, & \text{case } b \\ r_2 E_0 e^{-iq_{\text{unscreened}}(\nu)x} + t_2 E_0 e^{iq_{\text{unscreened}}(\nu)x} + E_0, & \text{case } c, \end{cases} \quad (6.58)$$

with case $a: -W_0 - W_1 - W_2/2 < x < -W_1 - W_2/2$, case $b: -W_1 - W_2/2 < x < -W_2/2$ and case $c: -W_2/2 < x < W_2/2$, where due to symmetry, regions III and IV are the symmetry counterparts of 0 and I and omitted in the above expression. Moreover, the symmetry around the middle of the coplanar strip also leads to $t_2 = r_2$. In addition to hosting the eigenmodes of the graphite layer, region II also supports a current directly excited by the probe pulse, given by $\sigma_{2D}(\omega)E_0$. The coefficients are determined by matching current density j and electrostatic potential V of the plasma modes at the boundaries of each region at $x = -W_1 - W_2/2$ and at $x = -W_2/2$ [17]. Maxwell's equations are used to relate the amplitudes of reflected and transmitted waves of current density to the reflected and transmitted amplitudes of the electrostatic potential V ($j = \sigma E$ and $E = -\vec{\nabla} V$ such that $V \propto \int j$). This is shown schematically in Fig. 6.14, with the grey dashed lines marking the boundaries. At the edge of the graphite, the boundary conditions $j_0(x = -W_0 - W_1 - W_2/2) = 0$ are imposed. By then defining the coordinate systems for the cases a (region 0), b (region I) and c (region II) in a way that $x = 0$ at the (left) graphite edge for case a, $x = 0$ at the left metal edge for case b and

$x = 0$ at the center of the sample for case c, six boundary equations can be obtained in total:[3]

$$\boxed{\begin{aligned}
&(1) j_{1,\text{graphite}}(W_1, t) = j_{2,\text{graphite}}(-W_2/2, t)\\
&(2) V_{1,\text{graphite}}(W_1, t) = V_{2,\text{graphite}}(-W_2/2, t)\\
&(3) r_2 = t_2\\
&(4) j_{1,\text{graphite}}(0, t) = j_{0,\text{graphite}}(W_0, t)\\
&(5) V_{1,\text{graphite}}(0, t) = V_{0,\text{graphite}}(W_0, t)\\
&(6) j_{0,\text{graphite}}(0, t) = 0,
\end{aligned}} \tag{6.59}$$

where E_{ext} is the external electric field, that excites the plasmons in the graphite layer in the coplanar strip gap (region II).

Finally, by solving the six boundary equations above, the optical near-field conductivity, called cavity conductivity, can be extracted as shown in the next section.

6.4.5 Extracting the Cavity Conductivity

This section reveals how the overall experimentally measured cavity conductivity emerges from, and can be calculated using, the theoretical principles and boundary conditions developed in the preceding sections.

The cavity conductivity detected with on-chip THz spectroscopy from a bare graphite heterostructure is proportional to the total current density ($\sigma = j_{tot}/E$) and can be obtained by a RPA (random phase approximation) conductivity approach. The principle of this approach is that the electric field of the THz pulse induces a current j in the material that follows the Drude model (Fig. 6.15a):

$$j = \sigma_{\text{graphite}}(\omega) E_{\text{ext}}. \tag{6.60}$$

This excited current leads to a charge displacement in x-direction (Fig. 6.15b). However, since the material is sub-wavelength in size, the charges cannot travel beyond the sample's edges, which serve as strict boundaries, forcing the current density to be $j = 0$ at these limits. The assembled charge accumulation, in turn, leads to an active restoring force (Fig. 6.15c). This restoring force has the effect that an additional electric field is created by the charge accumulation that points in

[3] The offset of the different x-axes amounts to a rescaling of the r and t coefficients in the different regions. The RPA conductivity is later defined such that $x = 0$ at the sample center and the equations are solved for r_2, such that the rescaling of the other r and t coefficients does not contribute to the overall solution.

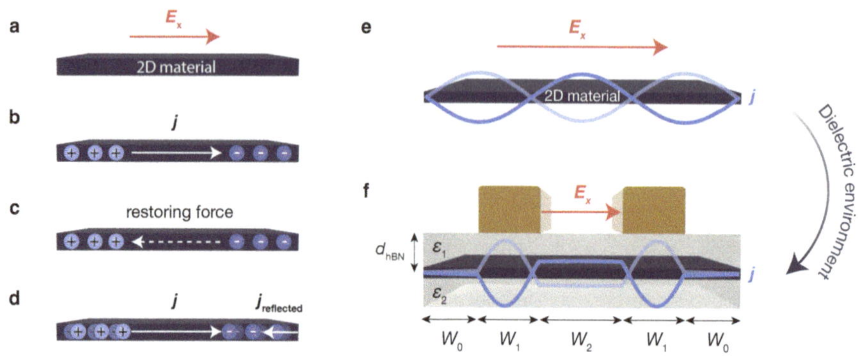

Fig. 6.15 Excitation of standing waves of current density in a 2D material. (**a**) If the electric field of the THz pulse interacts with a 2D material, it (**b**) excites a current in the 2D material that pushes electrons towards one sample edge and creates holes towards the other sample edge. (**c**) This charge accumulation leads to an active restoring force, such that (**d**) the current is back-reflected from the sample edge. This reflected current now propagates in opposite direction as the initial excited current by the THz pulse, leading to (**e**) standing waves of current density in the 2D material, where $j = 0$ at the sample edges. (**f**) The current density oscillation is not only dependent on the width, thickness and conductivity of the flake, but furthermore highly defined by the dielectric environment. (Figure adapted from [2] under a CC BY 4.0 license)

opposite direction as the electric field of the THz pulse (Fig. 6.15d), which leads to a reflected current $j_{\text{reflected}}$. Thus, the total current in the graphite flake amounts to:

$$\begin{aligned} j_{\text{tot,graphite}} &= \sigma_{\text{graphite}}(\omega) E_{\text{ext}} + j_{\text{reflected}} \\ &= \sigma_{\text{graphite}}(\omega) E_{\text{ext}} + \alpha(\omega) E_{\text{ext}} \\ &= \left(\sigma_{\text{graphite}}(\omega) + \alpha(\omega)\right) E_{\text{ext}} \\ &= \sigma_{\text{RPA,graphite}} E_{\text{ext}} \end{aligned} \tag{6.61}$$

where $\sigma_{\text{RPA,graphite}}$ is the RPA conductivity of the graphite layer.

In such a way, standing wave patterns of current density can form (Fig. 6.15e), and while the edges of the 2D material itself already provide the setup for standing waves of current density for THz light with a wavelength that is much longer than typical sample dimensions, the dielectric environment majorly influences the current density distribution (see Fig. 6.15f).

In Eq. 6.61, $\alpha(\omega)$ denotes the response function characterizing the transmitted and reflected waves within the coplanar strip gap region (where the photoconductive detector switch measures the transmitted THz field) of the graphite layer, and is defined as:

$$\alpha(\omega) = \sigma_{\text{graphite}}(\omega) \left(t_2 e^{i q_{\text{un}} x} + r_2 e^{-i q_{\text{un}} x}\right) \tag{6.62}$$

For $r_2 = t_2$ (see boundary equations (3)) and in the limit of small q, this term simplifies to:

$$\begin{aligned} \alpha(\omega) &= \sigma_{\text{graphite}}(\omega) 2r_2 \cos(q_{\text{un}} x) \\ &= 2\sigma_{\text{graphite}}(\omega) r_2, \end{aligned} \tag{6.63}$$

such that the total current j_{tot} in the graphite (or other 2D material) layer is:

$$j_{\text{tot,graphite}} = \sigma_{\text{graphite}}(\omega)\,(1 + 2r_2)\,E_{\text{ext}}, \tag{6.64}$$

such that the 2D cavity conductivity, that can be detected with on-chip THz spectroscopy, is:

$$\boxed{\sigma_{\text{2D,cavity}} = \sigma_{\text{graphite}}(\omega)\,(2r_2(\omega) + 1)}, \tag{6.65}$$

where r_2 can be obtained by solving the six boundary equations given above:

$$\begin{aligned} r_2 &= -\frac{r_{2,1}}{r_{2,2} + r_{2,3} + r_{2,4}} \\ r_{2,1} &= e^{\frac{i q_{\text{un}} W_2}{2}} q_{\text{un}} \Big(\left(1 + e^{2 i q_{\text{un}} W_0}\right) \left(1 + e^{2 i q_{\text{sc}} W_1}\right) q_{\text{sc}} \\ &\quad + \left(-1 + e^{2 i q_{\text{un}} W_0}\right) \left(-1 + e^{2 i q_{\text{sc}} W_1}\right) q_{\text{un}} \Big) \\ r_{2,2} &= \left(1 + e^{2 i q_{\text{un}} W_0}\right) \left(-1 + e^{2 i q_{\text{sc}} W_1}\right) \left(-1 + e^{i q_{\text{un}} W_2}\right) q_{\text{sc}}^2 \\ r_{2,3} &= 2\left(1 + e^{2 i q_{\text{sc}} W_1}\right) \left(1 + e^{i q_{\text{un}} (2W_0 + W_2)}\right) q_{\text{sc}} q_{\text{un}} \\ r_{2,4} &= \left(-1 + e^{2 i q_{\text{un}} W_0}\right) \left(-1 + e^{2 i q_{\text{sc}} W_1}\right) \left(1 + e^{i q_{\text{un}} W_2}\right) q_{\text{un}}^2 \end{aligned} \tag{6.66}$$

and the 2D conductivity of graphite is given by:

$$\sigma_{\text{graphite}}(\omega) = \sigma_{\text{graphite}}(2\pi\nu) = \frac{\nu_{\text{3D}}^2 d_{\text{gr}}}{-i\nu + \Gamma}. \tag{6.67}$$

These terms can be simplified to obtain for the cavity conductivity:

$$\boxed{\sigma_{\text{c}} = \sigma_{\text{g}} \left(1 + \frac{1}{-\cos\left(\frac{q_{\text{un}} W_2}{2}\right) + \frac{q_{\text{sc}}(q_{\text{un}} \cos(q_{\text{sc}} W_1) \sin(q_{\text{un}} W_0) + q_{\text{sc}} \cos(q_{\text{un}} W_0) \sin(q_{\text{sc}} W_1)) \sin(q_{\text{un}} W_2/2)}{q_{\text{un}}(q_{\text{sc}} \cos(q_{\text{un}} W_0) \cos(q_{\text{sc}} W_1) - q_{\text{un}} \sin(q_{\text{un}} W_0) \sin(q_{\text{sc}} W_1))}} \right)}, \tag{6.68}$$

with the cavity conductivity σ_{c} and the graphite Drude-conductivity σ_{g}, that are both functions of ω, as the screened and unscreened plasmon wave vectors q_{sc} and q_{un}

are functions of σ_g, that is a function of ω. Now, this term can be further simplified in the special case of setting $W_0 = 0$:

$$\boxed{\sigma_{cavity}(W_0=0) = \sigma_g\left(1 + \frac{q_{un}}{-q_{un}\cos(q_{un}W_2/2) + q_{sc}\sin(q_{un}W_2/2)\tan(q_{sc}W_1)}\right)}. \tag{6.69}$$

Thus, the full effective conductivity experienced by the probe, σ_{cavity} depends both on the bare graphite conductivity $\sigma_{graphite}(\omega)$, but also on reflected and transmitted waves from boundaries of the on-chip cavity. The influence of the boundaries is given by the reflection coefficient $r_2(\omega)$ and is responsible for the appearance of plasmonic resonances.

6.4.6 Approximations of the Analytical Theory

The analytical theory makes three distinct approximations:

1. Radiative corrections of the cavity modes are neglected: Oscillating charges, such as electrons, can emit electromagnetic waves at a frequency matching their oscillation frequency. However, this effect is expected to be small in the deeply subwavelength region and is neglected in the analytical theory.
2. The refraction at the dielectric boundaries between screened and unscreened regions is assumed to be purely local, meaning that the electric fields are suppressed quickly. This approximation is found to be valid for screened systems [17], allowing to match the potential and current at boundaries, which is further confirmed with the full 3D electromagnetic simulations.
3. The efficiency of the plasmonic excitation of the THz electric field is assumed to be independent of the coplanar strip geometry or the separation between the 2D material and the coplanar strip (hBN thickness). However, the coplanar strip geometry and hBN thickness do influence the cavity conductivity amplitudes of the results, as discussed in the following Sect. 6.4.7.

6.4.7 Filling Factor

As discussed in Sect. 5.3, the propagation constant γ is expected to change when the THz pulse interacts with a sample beneath the coplanar strip. The full 3D electromagnetic simulations and the developed analytical theory show that these changes of the propagation constant are due to the excitation of a current in the material between the coplanar strip, that are highly dependent on the dielectric environment of the 2D material (see Fig. 6.15). Thus, γ_{sam} can be related to the longitudinal, effective optical conductivity $\sigma_{cavity}(\nu)$, such that Eq. 5.35 transforms to:

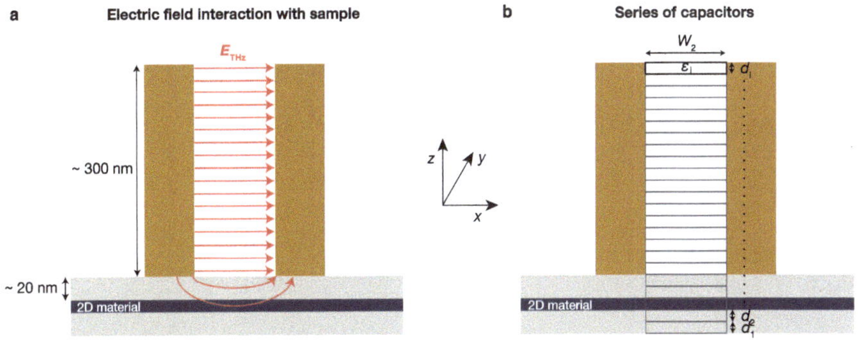

Fig. 6.16 Visualization of the filling factor C_{filling}. (**a**) The electric field of the THz pulse is mostly confined in x-direction between the metal traces with a thickness of ~300 nm, but starts bending beneath the metal traces, where the electric field interacts with a 2D material that is electrically isolated from the metal traces by a tens of nanometers thick hBN layer. (**b**) To describe the interaction of the THz pulse with the 2D material, the region between the metal traces can be considered as a stack of capacitors of height d_i (and length l_y, which is not shown in the sketch) and capacitor plate separation of W_2 with a dielectric permittivity ϵ_i. By calculating the effective capacitance of the cavity, one finds that a multiplicative "filling" factor C_{filling} has to be considered in every on-chip THz experiment. (Figure adapted from [2] under a CC BY 4.0 license)

$$\gamma_{\text{sam}}^2(\nu) = -\frac{(2\pi)^2\nu^2}{c^2}\epsilon(\nu) = -\frac{(2\pi)^2\nu^2}{c^2}(1 + \frac{i\sigma_{\text{cavity}}(\nu)}{2\pi\epsilon_0\nu}). \tag{6.70}$$

However, the degree to which the THz field is modified by a sample is governed by both the geometry of the coplanar strip and the thickness of the insulating layer that separates the strip from the 2D material. For instance, placing a material in direct contact with the coplanar strip leads to strong attenuation of the transmitted THz pulse, while the same material, when separated by a thick hBN layer, produces only a modest transmission change.

This behavior is fundamentally linked to the confinement and distribution of the THz electric field. In a coplanar strip, the field is primarily concentrated in the x-direction between the two metal traces, but field lines also extend beneath the traces with a z-component, causing them to bend (see Fig. 2.4f). The extent of interaction between the field and the 2D material is thus defined by the overlap between these bent field lines and the conductive plane of the 2D material (see Fig. 6.16a).

Crucially, this overlap is determined by the coplanar strip geometry: the thickness of the metal traces (fixed at ~300 nm in all measured and simulated devices), the gap W_2 between the traces, and the vertical separation from the 2D layer all influence how much the field bends and penetrates the material.

To describe the electric field interaction with the 2D material, the setup consisting of the coplanar strip and the electrically protected 2D material can be imagined as a stack of capacitors with plate area $d_i l_y$, where l_y is the length of the self-cavity, capacitor plate separation W_2 and the dielectric constant ϵ_i^{3D} (see Fig. 6.16b). The overall effective capacitance is then

$$C_{\text{eff}} = \sum_i \frac{\epsilon_i^{3D} d_i l_y}{W_2}. \tag{6.71}$$

By defining

$$\sum_i d_i = l_z, \tag{6.72}$$

Eq. 6.71 can be rewritten:

$$C_{\text{eff}} = \frac{l_y l_z}{W_2} \sum_i \frac{d_i}{l_z} \epsilon_i^{3D} = \frac{A}{W_2} \sum_i \frac{d_i}{l_z} \epsilon_i^{3D} = \frac{A}{W_2} \epsilon_{\text{eff}}. \tag{6.73}$$

Thus, the effective capacitance, and thereby the effective conductivity, is defined by an effective dielectric constant:

$$\epsilon_{\text{eff}} = \sum_i \frac{d_i}{l_z} \epsilon_i^{3D} = \sum_i \frac{1}{l_z} \epsilon_i^{2D} = \sum_i C_{\text{filling}} \epsilon_i^{2D}, \tag{6.74}$$

where ϵ_i^{2D} is the two-dimensional dielectric constant of a 2D material, similar to the 2D conductivity of a 2D material (see Sect. 6.4.2). The cavity conductivity in Eq. 6.70 is thus defined as:

$$\sigma_{\text{cavity}} = C_{\text{filling}} \sigma_{\text{analytical theory}}. \tag{6.75}$$

The multiplicative "filling" factor is a free parameter in the analytical theory and depends on details of the coplanar strip itself. For example, the filling factor decreases with increasing hBN thickness, as shown in Fig. 6.17. This trend is expected from Eq. 6.74, as the effective capacitor thickness l_z increases with increasing hBN thickness, leading to a reduced filling factor C_{filling}. This effect is still relatively small over the range of a few tens of nanometers, but becomes prominent if the hBN thickness differs by about 100 nm.

In summary, the filling factor emerges as a fundamental parameter in on-chip THz spectroscopy, as it is defined solely by the coplanar strip geometry and the thickness of the insulating layers, independent of the 2D material's conductivity. This makes it possible to detect conductivity changes in a stacked heterostructure when, for example, the carrier density in graphene is varied (see Fig. 6.18). However, to reliably extract parameters such as the Drude weight or spectral weight from measured conductivity spectra, the filling factor must always be taken into account. Since a precise theoretical model for the filling factor is not yet available, each measured heterostructure must be individually simulated using a full 3D electromagnetic solver. The filling factor is then determined by comparing the results of the 3D simulation with those derived from the analytical theory for the

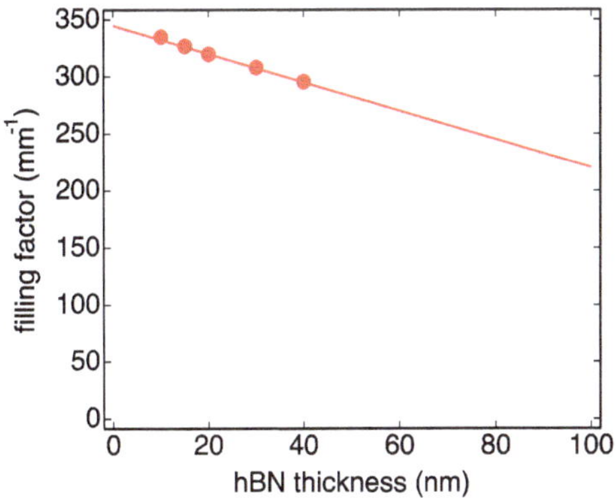

Fig. 6.17 Filling factor tunability through geometry. The filling factor as a function of hBN thickness is extracted from full 3D electromagnetic simulations and analytical theory calculations of the same cavity geometries. The data was fit with a linear function. Cavity geometry: $W_0 = 10.5\,\mu\,\text{m}$, $W_1 = 3\,\mu\,\text{m}$ and $W_2 = 3\,\mu\,\text{m}$ and $d_{\text{gr}} = 5\,\text{nm}$. (Figure adapted from [2] under a CC BY 4.0 license)

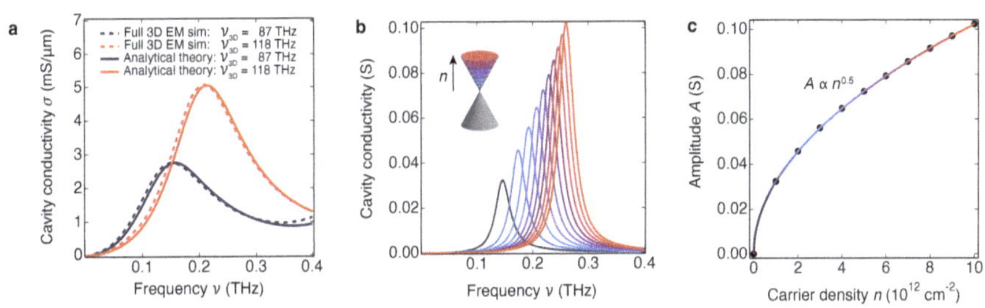

Fig. 6.18 Independence of the filling factor on the carrier density. (**a**) Cavity conductivity from the analytical theory in comparison to full 3D electromagnetic simulations of a 2D material in a fixed dielectric environment with two different 3D plasma frequencies ν_{3D}. Both analytical theory curves were multiplied by a filling factor of $366\,\text{mm}^{-1}$. The good agreement of theory and 3D simulation for both plasma frequencies verifies that the filling factor is independent of the carrier density of the material. Cavity parameters: $W_0 = 0.275\,\mu\,\text{m}$, $W_1 = 3.45\,\mu\,\text{m}$ and $W_2 = 2.55\,\mu\,\text{m}$ and $d_{\text{hBN}} = 18\,\text{nm}$. (**b**) As the filling factor is independent of carrier density, this allows to compare conductivity amplitudes of a cavity, such as the calculated graphene self-cavity here, as a function of carrier density. (**c**) The measured amplitudes as a function of carrier density in (**b**) can be fit to $n^{0.5}$, as expected for the 2D conductivity of graphene (see Eq. 6.46). (Parts of the figure adapted from [2] under a CC BY 4.0 license)

same cavity geometry. Thus, the cavity conductivity, $\sigma_{\text{cavity}}(\nu)$, reflects not only the intrinsic properties of the material but also the geometric confinement of the THz field, and without properly accounting for this factor, accurate interpretation of experimental data is not possible.

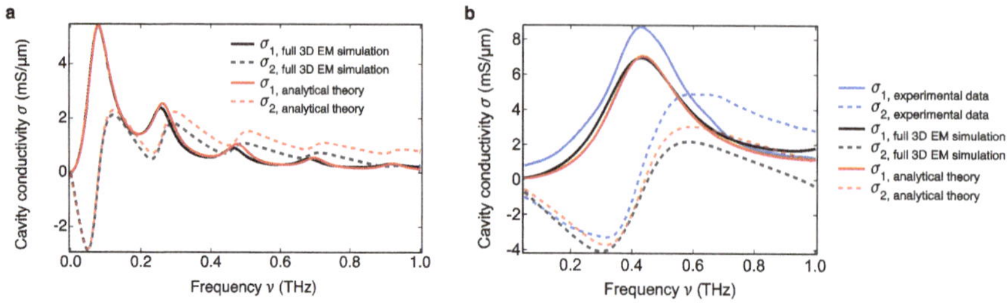

Fig. 6.19 Comparison of analytical theory to full 3D electromagnetic simulations and experimental data. (**a**) Real and imaginary parts of the cavity conductivity of a thin graphite heterostructure that hosts 5 resonances below 1 THz obtained (as shown in Fig. 6.8) from a full 3D electromagnetic simulation (black) and the analytical theory (red). The results of the analytical theory agree with the full 3D electromagnetic simulation. $C_{\text{filling}} = 200\,\text{mm}^{-1}$. (**b**) Comparison of the cavity conductivity of a thick graphite heterostructure (device BG6E, see Fig. 6.2.2) that hosts only one resonance below 1 THz obtained experimentally (blue), from a full 3D electromagnetic simulation (black) and the analytical theory (red). $C_{\text{filling}} = 360\,\text{mm}^{-1}$. (Figure adapted from [1] under a CC BY 4.0 license)

6.4.8 Comparison to Full 3D Electromagnetic Simulations and Experimental Data

The cavity conductivity obtained from the analytical theory and from full 3D electromagnetic simulations for the same cavity as discussed in Sect. 6.3 shows comparable results (see Fig. 6.19a). The analytical theory matches further with full 3D electromagnetic simulations and experimental data obtained from Device BG1 (see Figs. 6.2 and 6.19b), providing strong evidence for the validity and accuracy of the analytical theory.

The full 3D electromagnetic simulations were performed on a cavity with the exact dimensions of the graphite and hBN flake in the vdW heterostructure, including deviations from the shape of a perfect rectangle, assuming a scattering rate of $\tau^{-1} = 0.29$ THz and a 3D plasma frequency of $\nu_{3\text{D}} = 121$ THz (0.5 eV).

The analytical theory calculations were performed assuming the same scattering rate and 3D plasma frequency as for the full 3D electromagnetic simulations, but simplifying the cavity to be perfectly symmetric with $W_0 = 1.9\,\mu$ m. As mentioned in Sect. 6.4.7, in order to compute the effective conductivity experienced by the probe pulse, the 2D cavity conductivity needs to be scaled by a filling factor. A filling factor of $200\,\text{mm}^{-1}$ was found to match the amplitudes of the theoretically calculated cavity conductivity with the cavity conductivity obtained from the full 3D electromagnetic simulations of the cavity shown in Fig. 6.19a and a filling factor of $360\,\text{mm}^{-1}$ was found for the cavity discussed in Fig. 6.19b.

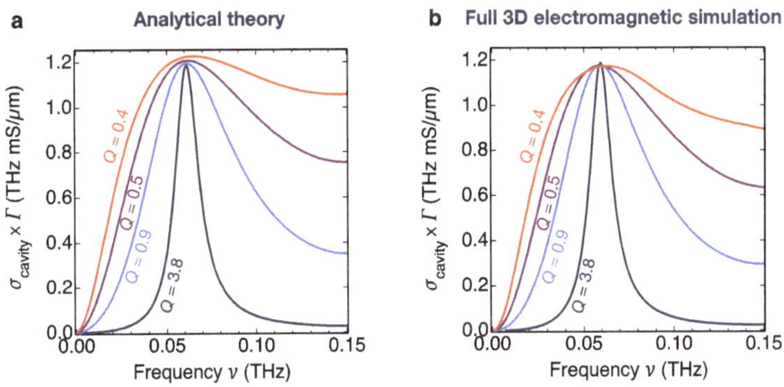

Fig. 6.20 Impact of scattering rate on resonance frequencies. **a** The $m = 0$ mode of a graphite self-cavity is fixed in resonance frequency with varying quality factors $Q = \nu_0/\Gamma$ (Γ in THz), as shown by the analytical theory and by (**b**) full 3D electromagnetic simulations of the same cavity. Cavity parameters: $W_0 = 0\,\mu\text{m}$, $W_1 = 16.2\,\mu\text{m}$ and $W_2 = 13.9\,\mu\text{m}$, $d_{gr} = 6.8\,\text{nm}$ and $d_{hBN} = 31.5\,\text{nm}$

6.4.9 *Impact of Scattering Rate on Self-cavity Resonance Frequencies*

Having derived and validated the analytical theory for self-cavities in 2D materials embedded in on-chip THz spectroscopy, the question arises if and how the scattering rate thus linewidth of the resonance impacts the resonance frequency of a cavity mode. As shown in Sect. 5.2.2 in Fig. 5.2, the frequency of the resonance appearing in the conductivity spectrum of a Drude-Lorentz model is found to be independent of the linewidth, while the frequency of the resonance appearing in the displacement x of the oscillator is red-shifted for quality factors $Q < 1$ [18].

The real part of the cavity conductivity as given by the analytical theory can be computed for a given heterostructure as a function of frequency for various different scattering rates. As shown in Fig. 6.20a, the resonance frequency does not shift with increasing scattering rates. This finding is supported by full 3D electromagnetic simulations, as shown in Fig. 6.20b.

Thus, the cavity conductivity follows the same trend as a driven damped oscillator, which is the Drude-Lorentz model. This can be further validated by the analytical theory for the special case of a mostly screened self-cavity.

If one assumes that the graphite heterostructure is embedded in a coplanar strip geometry with $W_0 = 0$, the cavity conductivity simplifies to (see Eq. 6.69):

$$\sigma_{\text{cavity}} = \sigma_{2D}\left(1 + \frac{q_{\text{un}}}{-q_{\text{un}}\cos\left(\frac{q_{\text{un}}W_2}{2}\right) + q_{\text{sc}}\sin\left(\frac{q_{\text{un}}W_2}{2}\right)\tan(q_{\text{sc}}W_1)}\right), \tag{6.76}$$

which can be further expanded for small W_2 and $q_{sc} \to \pi/(2W_1)$ to:

$$\boxed{\sigma_{\text{cavity}} = \frac{d_{\text{gr}}\omega_{3D}^2}{(\Gamma - i\omega)} \frac{W_2\sqrt{\frac{i\epsilon_2(\Gamma-i\omega)\omega}{d_{\text{hBN}}d_{\text{gr}}\omega_{3D}^2}}}{W_1\left(-\pi + 2W_1\sqrt{\frac{i\epsilon_1(\Gamma-i\omega)\omega}{d_{\text{hBN}}d_{\text{gr}}\omega_{3D}^2}}\right)}}. \tag{6.77}$$

Interestingly, the cavity conductivity does not simplify to a typical Drude-Lorentz form. Instead, it becomes a product of two terms: a Drude model component (the first fraction in Eq. 6.77) and an oscillator-like term (the second fraction in Eq. 6.77). Plotting only the second term reveals a red shift in resonance frequency with increasing scattering rates, but when plotting the product of the two terms, the resonance frequency remains unaffected by the scattering rate.

This overall behavior can be explained by the RPA-like conductivity of the self-cavities (see Sect. 6.4.5): the cavity conductivity represents the time convolution of a Drude response (the "driving force" of the damped oscillator) and a near-field damped oscillator. As this overall behavior mimics a Drude-Lorentz response, all data in this thesis is fit using a sum of multiple Drude-Lorentz models (see Eq. 5.19).

The discovery that the resonance frequencies of plasmonic self-cavity modes remain unaffected by the scattering rate carries several important implications. Firstly, linewidths in calculated self-cavity conductivity spectra can be narrowed to enhance features such as potential avoided crossings, without altering the resonance frequencies of the modes. More importantly, frequency shifts in the cavity conductivity spectrum can be directly attributed to changes in carrier density or external factors, making this especially valuable when investigating superconducting phases, where scattering is expected to decrease significantly below the critical temperature.

Having demonstrated that the scattering rate does not influence the resonance frequency of the self-cavity modes, let's next investigate how the geometry of the micropatterned vdW heterostructure impacts the cavity conductivity spectrum.

6.5 Cavity Tunability Through Geometry and Dielectric Environment

As demonstrated in previous sections, micron-sized 2D materials such as graphite flakes inherently form THz plasmonic self-cavities that are highly sensitive to their dielectric surroundings (see Fig. 6.15e, f). The analytical theory elucidates how both the dielectric environment and cavity geometry shape the resonance frequencies of distinct cavity modes. Key parameters, such as graphite thickness (d_{gr}), hBN thickness (d_{hBN}), coplanar strip width (W_1), gap width (W_2), and the surrounding dielectric constants, enable systematic tuning of the self-cavity's resonance frequencies. Figure 6.21a illustrates, via analytical theory calculations,

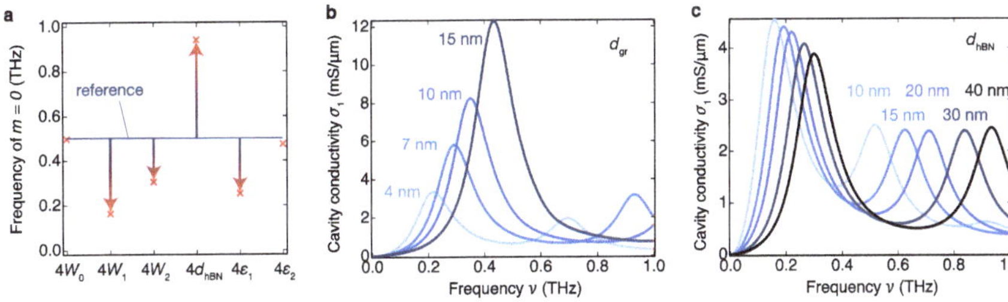

Fig. 6.21 Effect of cavity geometry and dielectric environment on self-cavity modes. (**a**) The standing waves of current density of the self-cavity modes in 2D materials are highly sensitive to the dielectric environment, such that the dielectric environment can be used to systematically tune the resonance frequencies of the cavity. For example, if individual parameters of a cavity with $W_0 = 3\,\mu$ m, $W_1 = 3\,\mu$ m and $W_2 = 4\,\mu$ m, $d_{gr} = 15.33$ nm, $d_{hBN} = 35$ nm are increased by a factor of four, the original resonance frequency of the cavity, $\nu_0 = 0.5$ THz, is changed as indicated by the arrows. (**b**, **c**) Full 3D electromagnetic simulations of the real part of the cavity conductivity as a function of frequency for different graphite and hBN thicknesses. Cavity geometry: $W_0 = 10.5\,\mu$ m, $W_1 = 3\,\mu$ m and $W_2 = 3\,\mu$ m, $d_{hBN} = 25$ nm (in panel **b**) and $d_{gr} = 5$ nm (in panel **c**)

how the $m = 0$ mode of a cavity initially resonating at 0.5 THz shifts in response to a fourfold increase in each parameter.

As shown in Fig. 6.21, the resonance frequency of the bare cavity mode can be increased by increasing the graphite thickness, by increasing the hBN thickness, by decreasing the coplanar strip metal width, by decreasing the coplanar strip gap width or by decreasing the relative permittivity of the insulating spacer ϵ_1 (assuming that a different insulator as hBN could be used). The coplanar strip gap width has the less-significant effect on the resonance frequency as this unscreened region hosts plasmons far outside the spectral sensitivity range in this cavity geometry. Larger cavity conductivity amplitudes can be most effectively achieved by increasing the graphite thickness or decreasing the thickness of the hBN layer. Thus, the resonance frequency and with that the wavelength and momentum of the confined plasmonic resonance is determined by the geometry and dielectric environment of the heterostructure [19–22]. The theoretical trends agree well with the experimentally measured spectra shown in Fig. 6.5.

In the following sections, other parameters such as the width W_0 of the 2D material extending the coplanar strip, the cavity length l, and the ratio of W_1 and W_2 that can tune the amount of hybridization between the screened and unscreened regions are discussed.

6.5.1 *Effect of Graphite Regions Extending the Coplanar Strip*

The width W_0 of the graphite regions extending the coplanar strip also has an impact on the cavity conductivity, particularly of the higher frequency modes $m = 1, 2, \ldots$,

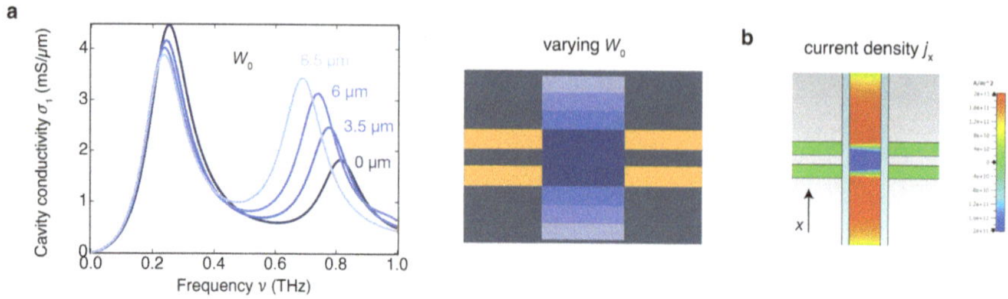

Fig. 6.22 Effect of W_0 on the cavity conductivity. (**a**) Full 3D electromagnetic simulations of σ_1 as a function of ν for varying W_0. The resonance frequency is decreasing for increasing W_0. Particularly the higher frequency mode is effected. Cavity geometry: $W_1 = W_2 = 3\,\mu$ m, $d_{\text{gr}} = 5$ nm, $d_{\text{hBN}} = 25$ nm, $\Gamma = 1.5/(2\pi)$ THz and $\nu_{3D} = 121$ THz. (**b**) Current density profile of a bare graphite cavity widely extending the coplanar strip width. For large W_0s, unscreened plasmons form in the regions outside the coplanar strips with resonance frequencies below 1 THz, leading to additional resonances. Cavity geometry: $W_0 = 20\,\mu$ m, $W_1 = 3.45\,\mu$ m, $W_2 = 2.55\,\mu$ m, $d_{\text{gr}} = 6$ nm, $d_{\text{hBN}} = 18$ nm, $\Gamma = 0.5/(2\pi)$ THz and $\nu_{3D} = 118$ THz

as seen in the full 3D electromagnetic simulations by Dr. Toru Matsuyama in Fig. 6.22. While the $m = 0$ mode remains mainly unaffected, the higher frequency modes $m = 1, 2, \ldots$ have decreasing resonance frequencies with increasing W_0 due to current oscillations in the W_0 regions (Fig. 6.22a).

If the W_0 regions become particularly large ($\sim 20\,\mu$ m for a typical bare graphite cavity as shown in Fig. 6.22b), unscreened plasmons can form in the W_0 regions that lead to additional resonances below 1 THz. This effect has also been observed experimentally in graphite samples with particularly wide wings as shown in Fig. 6.7. Cavity BG9 in Fig. 6.7 has two resonances, while after etching of the wide wings (cavity BG9E), only one resonance remains. The same trend is observed for cavities BG6 and BG6E.

In case that the cavity is inhomogeneous along the y-direction, particularly the higher frequency modes broaden due to their sensitivity to the W_0 region as shown in the full 3D electromagnetic simulations in Fig. 6.23. A good approximation for accounting for the inhomogeneity is a slice model of the cavity (see Fig. 6.23a). The overall conductivity of an inhomogeneous cavity is approximately the sum of the cavity conductivity of the slices. For largely inhomogeneous cavities such as for the experimentally measured cavity BG8 in Fig. 6.7, this results in a particularly broad spectrum where it becomes difficult to recognise individual resonances.

Thus, the experimentally and theoretically observed effects of W_0 on the cavity conductivity highlight that it is strongly desirable to use perfectly rectangular graphite flakes, particularly when being interested in cavity modes $\neq m = 0$, that have widths not much larger than the coplanar strip width. Such flakes can be obtained by atomic force lithography as described in Sect. 3.2.

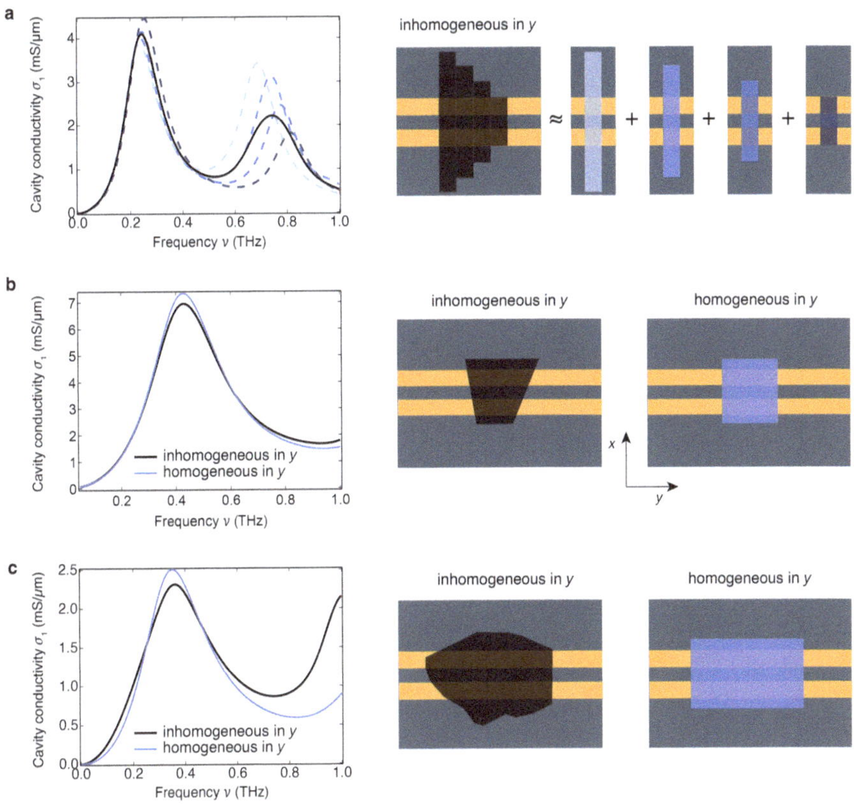

Fig. 6.23 Effect of W_0 inhomogeneity along the y-direction on the cavity conductivity. (**a**) Full 3D electromagnetic simulations of σ_1 as a function of ν for a cavity with varying W_0 along the y-direction. The cavity can be divided into four slices and the overall cavity conductivity (black) is approximately the sum of the conductivity of the four slices (blue). As mainly the higher frequency mode is effected by the width of W_0, the mode broadens due to inhomogeneous broadening. Cavity geometry is the same as in Fig. 6.22a. (**b**, **c**) Full 3D electromagnetic simulations of inhomogeneous cavities versus homogeneous cavities along the y-direction. (**c**) is a simulation of device BG6E and (**d**) of the Device GG2 discussed in Section 8.4. *Note*: The graphite flake simulated here is only 6.7 nm thick, while the actual graphite flake of Device GG2 is 10 nm thick. This leads to a lower resonance frequency of the $m = 0$ mode in (**c**), in comparison to what has been experimentally observed and calculated with the analytical theory in Section 8.4

6.5.2 Dependence of the Cavity Length on the Cavity Conductivity per Micron

Let's next discuss the dependence of the 2D material length along the y-direction (in propagation direction of the THz pulse) on the cavity conductivity. The transmitted signal varies with the sample length, as described by Equation (5.34) in Sect. 5.3, where longer samples cause more of the THz pulse to be reflected. However, the

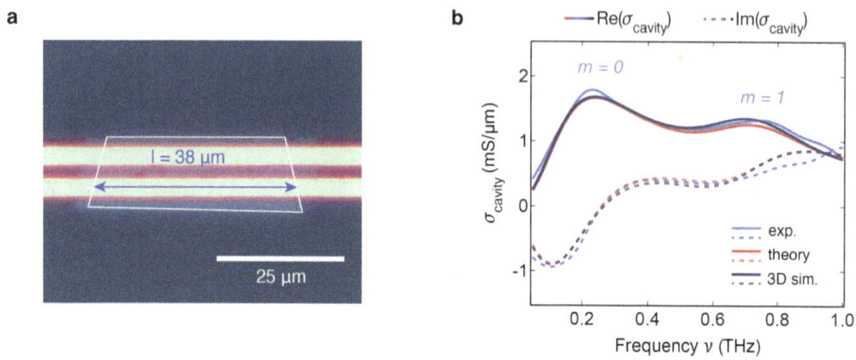

Fig. 6.24 Data set on a long bare graphite cavity. (**a**) Optical micrograph of the cavity. The cavity dimensions are: $l = 38\,\mu$ m, $W_0 = 1\,\mu$ m, $W_1 = 3.33\,\mu$ m, $W_2 = 2.56\,\mu$ m, $d_{gr} = 5$ nm and $d_{hBN} = 24$ nm. (**b**) Real and imaginary parts of the cavity conductivity obtained experimentally (blue), from the analytical theory (red) and from full 3D electromagnetic simulations (black). The two resonances stem from the excitation of the $m = 0$ and $m = 1$ self-cavity modes within the spectral sensitivity range. $C_{filling} = 360\,\text{mm}^{-1}$. (Figure adapted from [1] under a CC BY 4.0 license)

effective cavity conductivity remains independent of the sample length. This is because the THz pulse in the transmission lines is millimeters long and fairly uniform along the sample's y-direction over the length scales comparable to the sample size, while the electric field within the metal traces of the transmission line is almost entirely oriented along the x-direction. Under these two assumptions, the THz pulse launches a spatially patterned current only along the x-direction without any current along the y-direction. Consequently, the effective cavity conductivity per micron depends primarily on the sample's geometry along the x-direction, but is unaffected by its length along the transmission lines.

Further evidence of the independence of the cavity length on the cavity conductivity per micron is provided by the additional data set shown in Fig. 6.24. A long bare graphite cavity ($l = 38\,\mu$ m) was measured. Analytical theory and full 3D electromagnetic simulations match with the experimental findings independent of the increased cavity length.

6.5.3 *Hybridization of Screened and Unscreened 2D Plasmons*

In previous sections, it was shown that all five regions of an integrated graphite heterostructure (see cross-section in Fig. 6.14) contribute to the overall cavity conductivity through currents that are reflected and transmitted between screened and unscreened regions, vanishing at the graphite edges. Current oscillations are more pronounced in the screened regions, as unscreened plasmon resonances occur outside the spectral range of on-chip THz spectroscopy. This section investigates the interaction between screened and unscreened regions by analyzing the cavity conductivity as a function of effective plasmon wavevector q_{eff} (Sect. 6.5.3.1) and examining current (Sect. 6.5.3.2) and charge density profiles (Sect. 6.5.3.3) for various coplanar strip configurations.

6.5.3.1 Tunability Through Coplanar Strip Dimensions

A good way to see how "screened" or "unscreened" the self-cavity plasmons in the graphite actually behave is to plot the cavity conductivity as a function of an effective q_{eff}. As shown in Sect. 6.4, it is not really possible to define an actual q of the cavity modes, as both q_{screened} and $q_{\text{unscreened}}$ contribute to the overall mode. However, it is possible to tailor a way around it by taking a specific cavity, such as a cavity with $W_0 = W_1 = W_2$, and then defining the overall cavity width with $W = 2W_0+2W_1+W_2$, as shown in Fig. 6.25a. This allows one to define an effective $q_{\text{eff}} = 2\pi/W$ and by changing W while keeping $W_0 = W_1 = W_2$, it is possible to tune q_{eff}. The cavity conductivity of such a cavity as a function of q_{eff} is shown in Fig. 6.25b.

As shown in Fig. 6.25b, the resonances appearing in the cavity conductivity spectrum as a function of q_{eff} and frequency ν are neither purely linear, as would be expected for purely screened plasmons, nor scale as $\nu \propto \sqrt{q}$, as would be expected for purely unscreened plasmons. Instead, the resonances form avoided crossings. However, this effect becomes most pronounced at frequencies beyond the spectral sensitivity range, since the resonance frequencies of unscreened plasmons

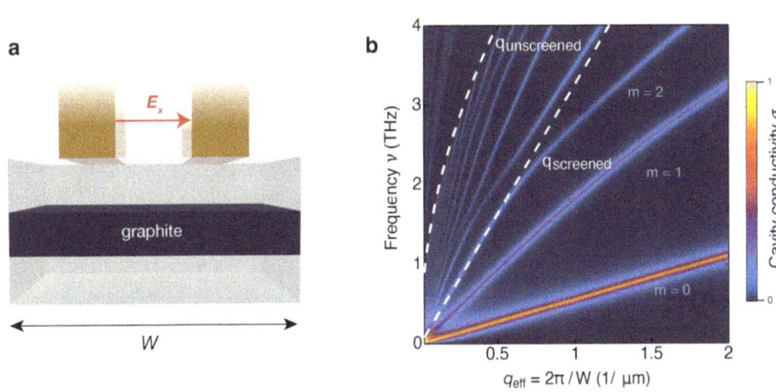

Fig. 6.25 Hybridization of screened and unscreened plasmons. (**a**) By varying the overall width W of a given cavity ($W_0 = W_1 = W_2$), $q_{\text{eff}} = 2\pi/W$ can be tuned. (**b**) Cavity conductivity as a function of q_{eff} and ν. Dashed lines show the dispersion relation for purely screened and purely unscreened plasmons. The resonances of the cavity form avoided crossings around the dashed lines, which is due to the hybridization of plasmons in the screened and unscreened regions of the bare graphite cavity. The cavity dimensions are: $W_0 = 0.2\,\mu$m, $W_1 = 3.27\,\mu$m, $W_2 = 2.9\,\mu$m, $d_{\text{gr}} = 6.3$ nm and $d_{\text{hBN}} = 17.9$ nm

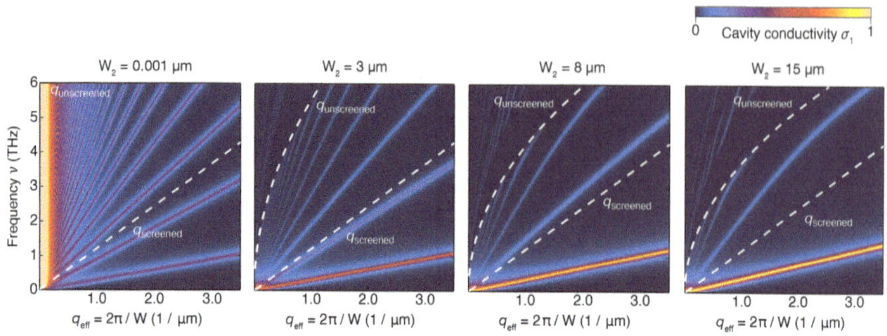

Fig. 6.26 Hybridization strength tunability through coplanar strip gap width. Cavity conductivity as a function of q_{eff} and ν for a cavity geometry with $W_0 = 0$ and varying coplanar strip gap widths W_2 ($d_{\text{gr}} = 10\,\text{nm}$, $d_{\text{hBN}} = 10\,\text{nm}$ and $\omega_{3\text{D}} = 118\,\text{THz}$). The resonances scale linear ($\nu \propto q$) for small coplanar strip gaps (W_2= 0.001 μ m), but start to form avoided crossings due to the hybridization of screened and unscreened plasmons for larger W_2. The hybridization becomes stronger with increasing W_2

lie outside this range (see Sect. 6.4.2), and the observed self-cavity resonances still exhibit approximately linear scaling below 1 THz.

Another verification of this approach is to see what happens if the graphite is designed to be as wide as the coplanar strip ($W_0 = 0$) and if the coplanar strip gap width W_2 is varied. As shown in Fig. 6.26, all self-cavity modes scale linear in frequency as a function of q_{eff} if the coplanar strip gap is very small (0.001 μ m), but become continuously more hybridized starting at lower frequencies with increasing W_2. Also, the amplitudes of the different modes is equal at a given q_{eff} for small W_2, but decreases with increasing mode node number if W_2 becomes larger, such that the $m = 0$ mode has the largest amplitude for example for a $W_0 = W_1 = W_2 = 3\,\mu\,\text{m}$ geometry. This is also verified by the experimental data discussed in Sect. 6.2.2.

The tuning of q_{eff} is only possible in theory. In experiment, q_{eff} is set for a given cavity geometry and cannot be tuned. For example, for a cavity with $W_0 = W_1 = W_2 = 3\,\mu\,\text{m}$, the effective plasmon wavevector is $q_{\text{eff}} \approx 0.42\,\mu\text{m}^{-1}$.

So, interestingly it is the case that screened and unscreened plasmons in the graphite cavities hybridize although they are either resonant to very different frequencies at a given q or have very different qs at a specific frequency. How is this possible? Typically, both q and resonance frequency have to match to allow hybridization of different modes, to satisfy momentum and energy conservation principles. To investigate this further, let's next look at current and charge density profiles of the plasmons for varying coplanar strip gap widths.

6.5.3.2 Current Density Profiles

Current density profiles can be computed with the analytical theory for a cavity with $d_{\text{gr}} = d_{\text{hBN}} = 10\,\text{nm}$, $W_0 = 0$ and $W_1 = 3\,\mu\,\text{m}$ (same parameters as in Fig. 6.26)

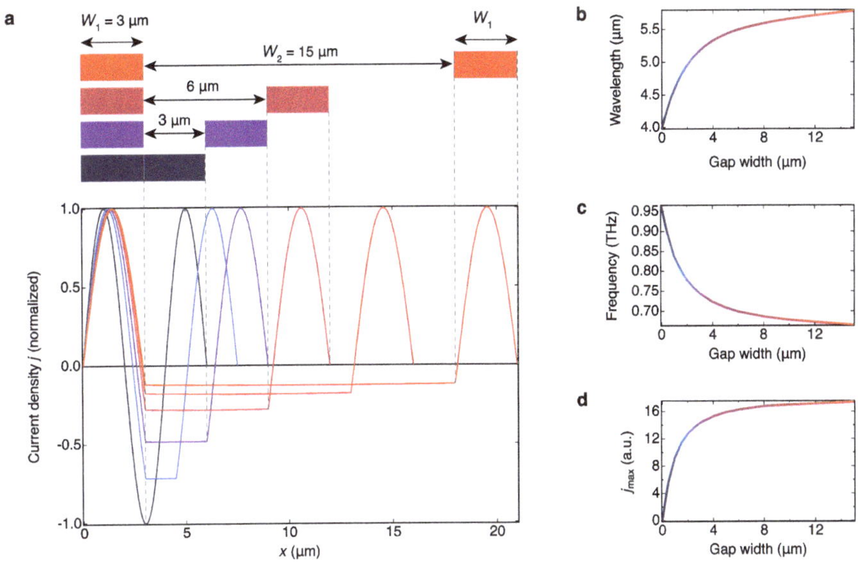

Fig. 6.27 Current density profiles for varying coplanar strip gap widths W_2. (**a**) Normalized current density amplitudes of the $m = 1$ mode as a function of x for cavities with different W_2. (**b**) The detected wavelength of the current oscillation in the W_1 region increases with increasing gap width W_2, which leads to a (**c**) decreasing resonance frequency of the overall self-cavity resonance. (**d**) The maximum current density (unnormalized) increases with increasing gap width W_2, as more charges are present and can oscillate in self-cavities with increased width, leading to a larger current. The filling factor is not considered here. Cavity parameters: $d_{gr} = d_{hBN} = 10\,\text{nm}$, $W_0 = 0$ and $W_1 = 3\,\mu\text{m}$

for different coplanar strip gap widths W_2 as shown in Fig. 6.27. Computed are the $m = 1$ modes (rather than the $m = 0$ modes) to visualize the oscillations of the currents in the screened regions better.

If the gap width is small, only 0.001 μm wide, the current density almost follows a perfect sinusoidal curve from cavity edge to cavity edge with a wavelength of $\lambda = \frac{4}{3} W_1 = 4\,\mu\text{m}$ (see darkest blue traces in Fig. 6.27a, b). The wavelength of the current density oscillation is then increasing with increasing gap width W_2, although the current density in the coplanar strip gap is relatively flat. The increase of wavelength leads to a decreasing resonance frequency (see Fig. 6.27c) (and an increasing maximum current density as shown in Fig. 6.27d). This is because an increasing wavelength leads to a decreasing plasmon wave vector ($q = 2\pi/\lambda$) and with that smaller resonance frequencies.

So, the computed current density profiles clearly show that the coplanar strip gap region W_2 effects the current density profiles and wavelength of the screened regions beneath the metal traces. Next, the charge density profiles and its Fourier transforms are computed to understand the underlying coupling mechanism between screened and unscreened plasmons better.

6.5.3.3 Charge Density Profiles

Charge densities can be computed from the analytical theory (see Sect. A.5 of the Appendix) and are calculated for the same cavities as shown in Fig. 6.27.

As shown in Fig. 6.28, the charge density is not continuous across the cavity in contrast to the current density profiles. In addition, the charge densities are antisymmetric and maximum at the edges in both the screened regions (see for $x = 0\,\mu$ m to $3\,\mu$ m in the center panel of Fig. 6.28) and in the unscreened coplanar strip gap (see bottom panel of Fig. 6.28). Furthermore, there are more charges accumulating at the boundaries with increasing coplanar strip gap width.

The charge density profiles of the individual screened (W_1) and unscreened (W_2) regions of the heterostructure can be Fourier transformed to reveal the charge distribution as a function of wave vector q. As illustrated in Fig. 6.29, the resulting spectra for both regions display broad, continuous features rather than discrete peaks. The observed nodes in the spectra arise from padding the current density traces with a series of zeros, which was done to improve the resolution of the Fourier transform in q-space. This observation matches with the expectation that an (infinite) long oscillation in real space would lead to a discrete feature in momentum space, while a short pulse with only a few oscillations in real space results in a broad momentum distribution, when being Fourier transformed.

Therefore, both screened and unscreened plasmons within distinct regions of a 2D material embedded in THz circuitry display broad momentum distributions. This facilitates efficient hybridization at the dielectric interfaces defined by the coplanar strip geometry, provided that the resonance frequencies of the screened and unscreened plasmons are closely matched. The degree of hybridization between the two regions is thus determined by the extent to which their plasmon resonance frequencies align. Since unscreened plasmons resonate at a higher frequency than their screened counterparts for the same wave vector q, increasing the size of the unscreened region, W_2, leads to a reduction in q, which in turn lowers the resonance frequency of the unscreened plasmon. This adjustment enables the plasmons in both regions to become resonant for specific values of W_1 and W_2, facilitating coupling. For example, assuming that $W_1 = 3\,\mu$ m, $q_{\text{screened}} = 2\pi/(4W_1)$, $q_{\text{unscreened}} = 2\pi/(4W_2)$ and $d_{\text{hBN}} = 20\,\text{nm}$, W_2 has to be $77\,\mu$ m in order to let the lowest frequency plasmon resonance of the screened W_1 region have the same resonance frequency as the lowest frequency plasmon resonance of the unscreened W_2 region. However, the W_2 region only has to be $8.6\,\mu$ m in order to let the third lowest plasmon resonance of the screened W_1 region have the same resonance frequency as the lowest frequency plasmon resonance of the unscreened W_2 region. Therefore, the resonance frequencies of the two lowest in frequency modes still scale mostly linear with q_{eff} in a typical $W_1 = W_2 = 3\,\mu$ m coplanar strip geometry, as shown in Fig. 6.26.

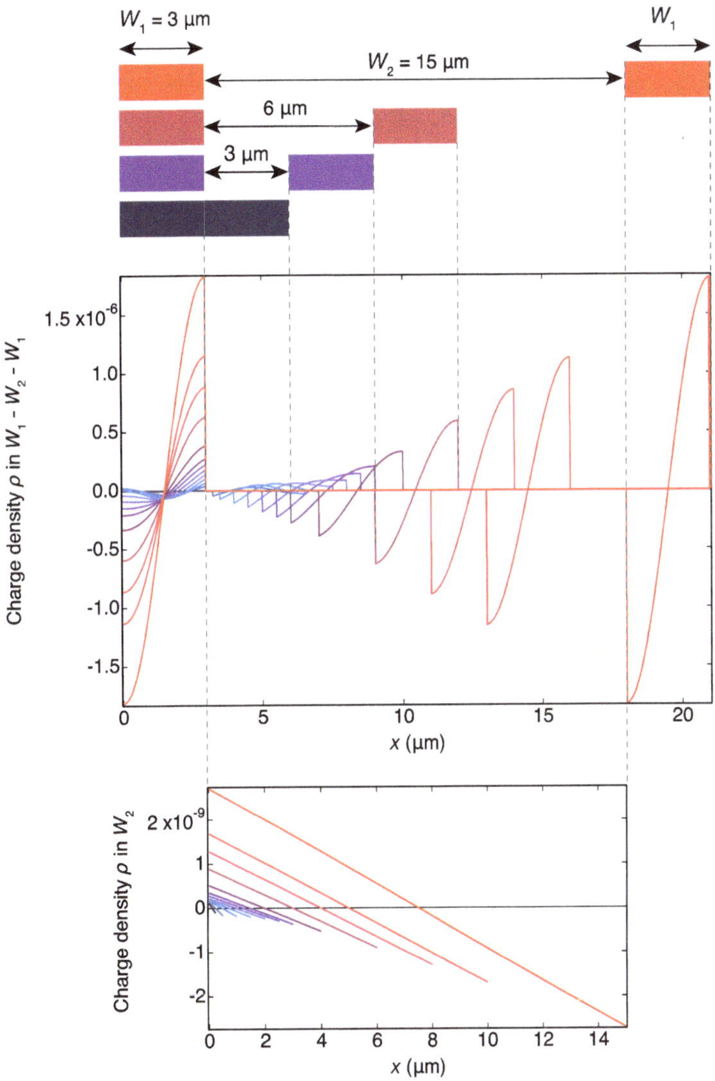

Fig. 6.28 Charge density profiles for varying coplanar strip gap widths W_2. Charges accumulate at the dielectric boundaries of the self-cavity, both at the graphite edges and at the boundaries between screened and unscreened regions. The charge density profile in the coplanar strip gap (bottom panel) is approximately linear, with roughly the same slope for all coplanar strip gap calculations. However, the larger W_2, the larger the number of charges at the dielectric boundaries, as more charges are present that can oscillate, and the larger the charge density amplitude in the screened regions beneath the metal traces. The cavity geometries are the same as in Fig. 6.27

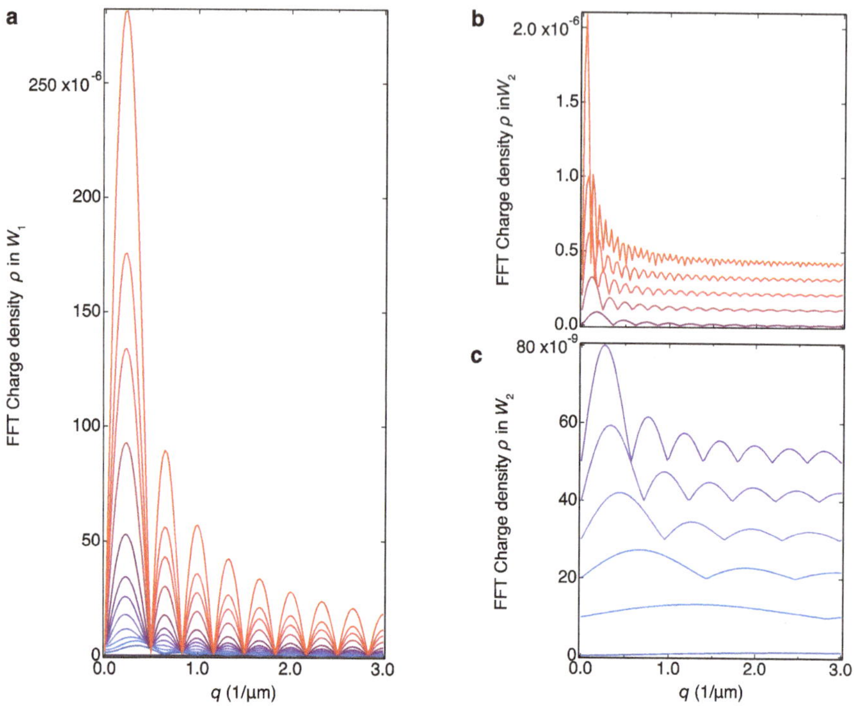

Fig. 6.29 Fourier Transforms of charge density profiles. Charge density distribution for different coplanar strip gap widths (see Fig. 6.28) as a function of plasmon wave vector q for (**a**) the screened W_1 region and (**b**) the unscreened coplanar strip gap region (W_2) of the heterostructure. The nodes in the spectra arise as the charge density profiles of the individual W_1 and W_2 regions shown in Fig. 6.28 were padded with zeros over an extended x-range. Thus, the momentum distributions of the charge densities of both screened and unscreened regions are broad, rather than discrete. This allows the hybridization of screened and unscreened 2D plasmons of different wavelengths at the dielectric boundaries

6.6 Detecting a Drude Response—Or Not?

In this section, I would like to discuss under which conditions cavity effects become significant, and when the measured conductivity can be reliably interpreted as a conventional Drude response, as typically expected for large metallic samples. This distinction is important for the accurate extraction of intrinsic material properties, and for understanding light-matter interactions in vdW materials. While the analysis here focuses on finite-size effects in semimetals, these results are generalisable to other phases and formations of polaritons with other collective modes.

Other works have suggested that it is possible to detect a Drude response of gated monolayer graphene [5] or a gated monolayer semiconductor WSe_2 [5, 6] and extracted parameters such as Drude weight or scattering rates. In the framework of this thesis, a sample was built with similar dimensions as in [5], replacing the gated

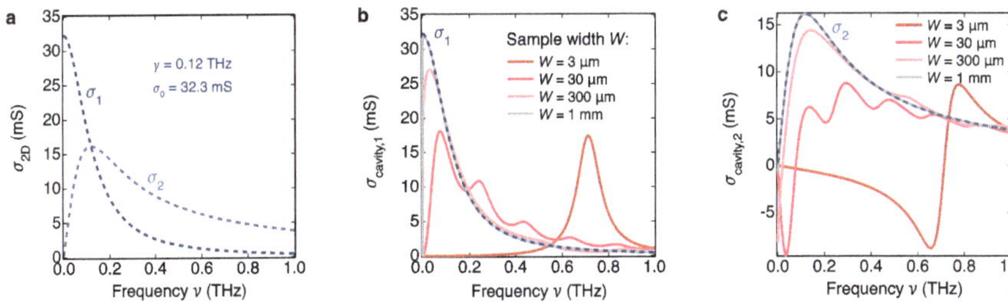

Fig. 6.30 Intrinsic Drude-like 2D conductivity vs cavity conductivity as a function of sample width. (**a**) Intrinsic 2D conductivity of a typical graphite flake. Real and imaginary parts follow perfectly the Drude-model, which allows to extract the scattering rate Γ and DC conductivity σ_0. (**b**) Real and (**c**) imaginary parts of the cavity conductivity for different sample widths W ($W = 2 \cdot W_1 + W_2$, $W_1 = W_2$, $d_{\mathrm{hBN}} = 25\,\mathrm{nm}$). In the limit of very large samples ($\sim$1 mm, $W \gg \lambda_{\mathrm{THz}}$), the cavity conductivity recovers σ_{2D}, with the small difference that σ_{cavity} is zero at $\nu = 0$ due to the capacitive sensing of on-chip THz spectroscopy. (Figure adapted from [1] under a CC BY 4.0 license)

graphene layer by a thin graphite layer that achieves a similar 2D conductivity (see Fig. 6.6). The obtained conductivity spectrum from this device appeared to be broad, but showed a series of oscillations at low frequencies with peaks that were found to be robust to the analysis procedure, indicating that the oscillations are due to low-frequency cavity modes. Thus, cavity effects potentially also have to be considered for graphene/graphite samples with sizes $\sim$45 µ m.

The analytical theory can be used to calculate the measured cavity conductivity obtained with on-chip THz spectroscopy (σ_{cavity}), as a function of sample size (W), and intrinsic 2D conductivity of a 2D material, σ_{2D}, to determine under which conditions discrete cavity modes appear in the spectrum and when a Drude response can be detected. Figure 6.30a shows σ_{2D} (as defined by a Drude lineshape, $\sigma_{2D} = \frac{\epsilon_0 \omega_{3D}^2 d_{gr}}{-i\omega+\Gamma}$) of a graphite flake of typical thickness. The value of $\sigma_{2D,1}(\nu = 0)$ corresponds to the DC conductivity, 32.3 mS, and a scattering rate of 0.12 THz. This conductivity is used as an input for the analytical theory, which calculates the corresponding cavity conductivity for different sample widths W ($W = 2W_1 + W_2$, $W_1 = W_2$, $d_{\mathrm{hBN}} = 25\,\mathrm{nm}$), as shown in Fig. 6.30b, c. For graphite samples of typical dimensions achievable using the current generation of dry-transfer vdW heterostructure fabrication techniques, the cavity conductivity is noticeably different from the intrinsic 2D conductivity, even for an unrealistically large sample width of 300 µ m. When the sample lateral dimensions become larger than the THz diffraction limit, for example, 1 mm, a response comparable to σ_{2D} is recovered, with only a small discrepancy occurring at frequencies typically lower than the spectral sensitivity range, which is due to the capacitive sensing nature of on-chip THz spectroscopy: The 2D material of interest is always isolated from the circuitry by an insulating layer such as a hBN flake, preventing the detection of the DC ($\nu = 0$) contribution.

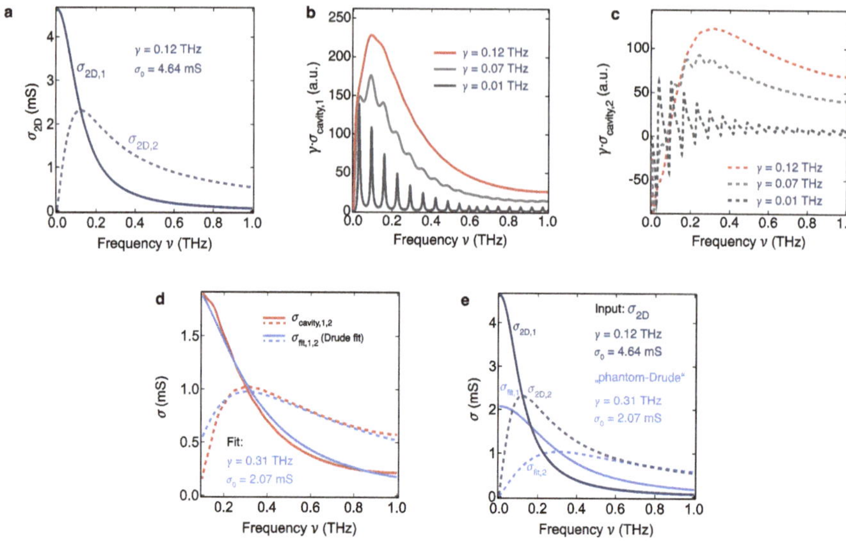

Fig. 6.31 Intrinsic 2D conductivity vs cavity conductivity for a sample with low σ_{2D}. (**a**) Real and imaginary parts of the intrinsic 2D conductivity of a 2D material with DC conductivity $\sigma_0 = 4.64\,\text{mS}$, which served as input to the analytical theory. (**b**) Real and (**c**) imaginary parts of the calculated cavity conductivity σ_{cavity} for two different scattering rates γ, normalized through a multiplication with the scattering rate to allow seeing the shape of the different linewidth spectra. A series of many low-frequency cavity modes form a broad peak for $\gamma = 0.12\,\text{THz}$. (**c**) Typically, the spectral resolution in few-mm long on-chip THz spectroscopy circuits is limited to 0.1 THz at low frequencies. In that case, the conductivity spectrum for $\gamma = 0.12\,\text{THz}$ could be mistaken for a Drude response ("phantom Drude" response). (**d**) "Phantom Drude" response in comparison to the intrinsic 2D conductivity to the analytical theory. The extracted DC conductivity from the Drude-fit is off by a factor of two and the extracted scattering rate is off by a factor of three. Cavity parameters: $d_{\text{hBN}} = 20\,\text{nm}$, $W_0 = W2 = 3\,\mu\,\text{m}$, $W_2 = 10\,\mu\,\text{m}$, $\omega_{3D} = 100\,\text{THz}$ and $d_{\text{2D material}} = 1\,\text{nm}$. Note: All panels do not include the scaling factor C_{filling}. However, as C_{filling} is the same for all curves in all shown panels, only the overall amplitudes change, not the relative features. (Figure adapted from [1] under a CC BY 4.0 license)

There are also many (vdW) materials that have a much lower 2D conductivity than graphite. A low 2D conductivity pushes resonances to low frequencies, much like large sample widths. Thus, one could imagine that small samples with even smaller 2D conductivities also recover a response that can be fit with the Drude model?

An example 2D conductivity of such a low-conductivity 2D material is given in Fig. 6.31a. The DC conductivity is 4.64 mS and thereby much smaller than in the previous case. In general, it should also be noted here that it is much more difficult to detect such low-conductivity materials, as the changes to the THz pulse are much smaller, reducing the signal-to-noise ratio. However, active impedance engineering using the graphite gate can enhance sensitivity towards low-conductivity 2D materials, as detailed in Sect. A.8 of the Appendix.

The cavity conductivity calculation is given in Fig. 6.31b, c. When choosing a typical scattering rate of 0.12 THz a broad peak appears (see red traces in Fig. 6.31b, c). When reducing the scattering rate it becomes clear that this broad peak consists of many different cavity modes that overlap in frequency as the linewidths are much larger than their separations in frequency. Let's now assume that we measured the red traces obtained for σ_{cavity} with $\Gamma = 0.12$ THz without bandwidth below 0.1 THz, which can be a low-frequency limit for on-chip THz circuitry. In that case, the obtained spectrum would appear as shown in Fig. 6.31d, resembling the Drude-model. If we fit the data to a Drude model, we obtain values for $\Gamma = 0.31$ THz and the DC conductivity $\sigma_0 = 2.07$ mS, However, these values are fundamentally different from the actual intrinsic parameters that served as an input for the analytical theory (see Fig. 6.31e). One therefore has to be very careful when detecting a sub-wavelength conductivity spectrum that looks like the Drude model, particularly if one doesn't have spectral resolution below 0.1 THz, as it could be the sum of many self-cavity modes. This effect can be disguised even more if the shape of the 2D material is random and deviates from a perfect rectangular shape, leading to inhomogeneous broadening of the individual cavity modes (see Sect. 6.5.1).

The map shown in Fig. 6.32 indicates when cavity effects have to be considered and in which limits a Drude-model response can be expected. This provides some guidance for future on-chip THz spectroscopy projects on various different materials. If both sample width and 2D conductivity of the sample of interest are known, it is simple to see if one operates in the cavity- or Drude-response-regime. Both regimes are separated by a curve,[4] that was calculated when assuming that a Drude response is still detectable if the conductivity amplitudes of the real parts of σ_{cavity} and σ_{2D} are less than 10 % different at a frequency of 10 GHz, as shown in the right panel of Fig. 6.32.

Interestingly, it seems that for typical vdW sample dimensions and 2D conductivities, self-cavity effects strongly influence the conductivity spectrum (see Fig. 6.32). In contrast, a Drude response could be measurable by increasing the sample's dimensions.

While the calculation in Fig. 6.32a focuses on a heterostructure embedded in a coplanar strip geometry with $W = 2W_1 + W_2$, $W_1 = W_2$, and $d_{\text{hBN}} = 25$ nm, additional calculations were performed for geometries with varying hBN thicknesses and for smaller coplanar strip gaps (see Fig. 6.33). Larger hBN thicknesses push the boundary between the two regimes further towards larger widths or lower conductivities, as an increased hBN thickness pushes the resonance frequencies of the cavity modes to higher frequencies.

Furthermore, the calculation for coplanar strips with small gaps reveal that cavity effects must be considered over a broader range of 2D conductivities, largely independent of the sample width (see Fig. 6.33b). This arises because, in configurations with very narrow coplanar strip gaps, the cavity modes of varying

[4] The curve was calculated for a scattering rate of $\Gamma = 0.1$ THz, however, the curve appeared to be very robust against various scattering rates.

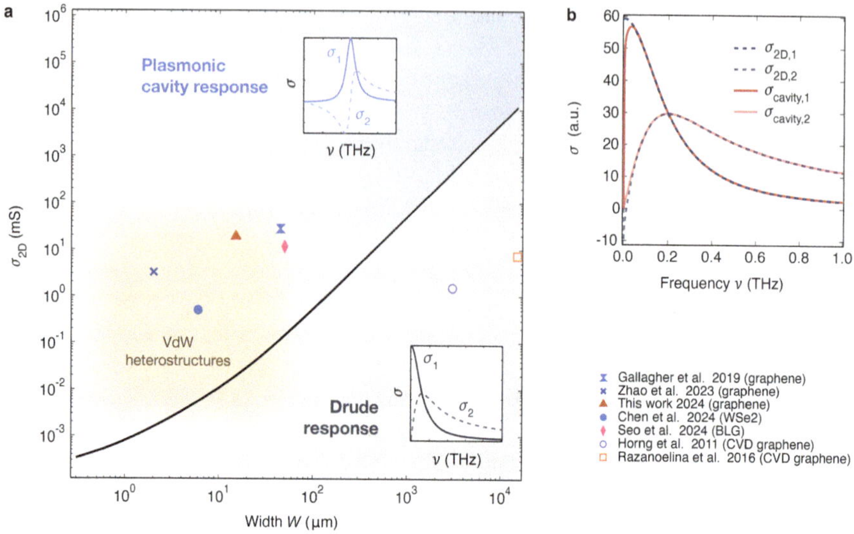

Fig. 6.32 Cavity response vs Drude response map. The black curve was calculated to obtain the maximum 2D conductivity for a given sample width that still allows to fit the obtained near-field conductivity obtained with on-chip THz spectroscopy with the Drude model. The curve marks the boundary between the cavity and Drude model regime at which the real parts of σ_{cavity} and σ_{2D} are only different by 10 % at a frequency of 10 GHz and σ_{cavity} and σ_{2D} match nicely for frequency in the typical spectral sensitivity range (~0.1 THz to 1 THz), as shown in the right panel. The data points provide a selection of literature values for built and measured 2D material samples: Solid symbols indicate measurements with on-chip THz spectroscopy [5–7, 21], empty symbols indicate far-field measurements on large-area CVD grown graphene [3, 4]. (Parts of the figure adapted from [1] under a CC BY 4.0 license)

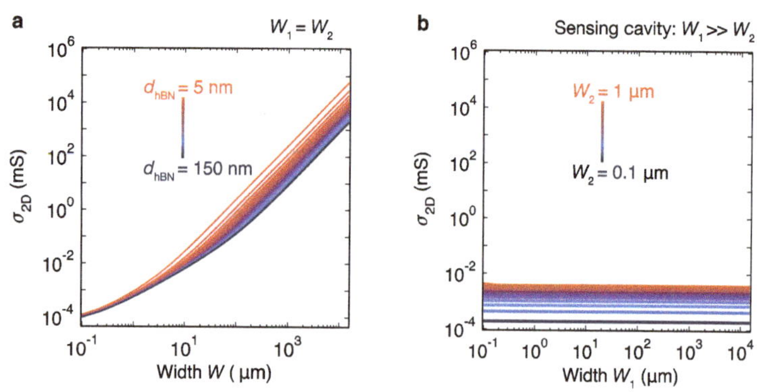

Fig. 6.33 Cavity response vs Drude response for (**a**) different hBN thicknesses for a coplanar strip geometry with $W = 2W_1 + W_2$, $W_1 = W_2$ and for (**b**) different coplanar strip gap widths, where then only W_1 is scaled, instead of the entire coplanar strip width. Cavity effects have to be considered for a σ_{2D} and W tuple lying above the calculated curves

momenta exhibit similar amplitudes, whereas for wider gaps, the amplitudes of higher-order cavity modes decrease with increasing node number (see Fig. 6.26). Consequently, a Drude-like response is more easily achieved in geometries where the gap size approaches the width of the metal traces, whereas cavity effects remain significant in more heavily screened 2D materials. This highlights the importance of finite-size effects in vdW heterostructures more generally and suggests that these effects could influence a wide array of devices, as 2D materials are often uniformly screened by electrostatic gates, similar to the uniform screening found in narrow-gapped coplanar strip configurations.

6.7 Discussion

This chapter presented the discovery that graphite flakes, commonly used as electrostatic gates in vdW heterostructures, naturally function as plasmonic self-cavities for THz light due to their subwavelength size. This finding is based on the experimentally measured cavity conductivity spectra of 19 graphite heterostructures, studied using on-chip THz spectroscopy. While graphite heterostructures partially embedded beneath the coplanar strip (Sect. 6.2.1) did not show resonances within the spectral sensitivity range, those spanning the entire coplanar strip width exhibited cavity conductivity spectra with one or multiple resonances. These resonances can be well modeled by the Drude-Lorentz model (Sect. 6.2.2), with quality factors $Q = \nu_0/\Gamma$ in the range of 0.5 to 1.5 (see Table A.3 in the Appendix). The observed plasmonic resonances thus exhibit a hybrid character, originating from both screened and unscreened regions of the graphite layer—an effect that, to my knowledge, has not been experimentally observed before.

An analytical theory was developed in close collaboration with Dr. Marios H. Michael (Sect. 6.4) to explain the experimental findings. The analytical theory reveals that the observed resonances stem from 2D plasmonic standing waves of varying momenta, and is consistent with both experimental data and finite element simulations, which model the full 3D electromagnetic field propagation within the cavity. The THz pulse excites a current in the graphite layer, that can be described by the Drude model, which in turn generates an electric field that induces a further current in the graphite (see Fig. 6.15). This current reflects at discontinuities in the dielectric environment, particularly at the sample edges, forming standing waves of current density that are predominantly confined beneath the metal traces (Sects. 6.3 and 6.4.5). The wavelengths and momenta of these confined plasmonic resonances are therefore determined by the dielectric environment and the heterostructure's geometry [19–24].

These findings emphasize that subwavelength-sized conductive 2D materials inherently function as self-cavities. However, the dielectric environment in which the 2D material is embedded significantly impacts the excited currents and, consequently, the cavity conductivity spectrum. While the relative dielectric permittivity of the substrate and cavity length have negligible effects on the cavity spectrum, all other dielectric (relative dielectric permittivity of the insulator between 2D material and coplanar strip (hBN) and 3D bulk plasma frequency of the 2D material) and geometric parameters (W_0, W_1, W_2, d_{hBN} and d_{gr}) are critical in determining the resonance frequencies and amplitudes of the cavity modes (Sect. 6.5). Particularly the overall width of the coplanar strips and the material under investigation, and the 2D conductivity defined by the 3D bulk plasma frequency and thickness of the material, determine the resonance frequencies of the cavity modes. Decreasing the width and increasing the thickness (or 3D bulk plasma frequency) has the effect of increasing the resonance frequencies of the cavity modes.[5]

Importantly, the analytical theory has to be multiplied by a "filling" factor (Sect. 6.4.7) to match both experimental and full 3D electromagnetic simulations. This filling factor, treated as a free parameter in the theory, depends on the specific geometry of the coplanar strip and describes how effectively the electric field of the THz pulse excites the currents in the graphite layer. As the filling factor remains unaffected by the bulk plasma frequency of the material, carrier-density-dependent changes (as in graphene, see Chap. 8) can be detected. Therefore, when extracting intrinsic parameters of a 2D material, such as its conductivity, it is essential to account for the dielectric environment and, in the case of on-chip THz spectroscopy, the precise geometry of the coplanar strip.

The analytical theory, including the "filling" factor, provides an important key advancement: a central challenge for THz spectral probes applied to 2D materials is establishing a direct, analytical connection between THz response and intrinsic conductivity. While time-domain, near-field THz probes offer a path to extracting linear response functions without Kramers-Kronig transformations, all such techniques face a fundamental issue: the measurement setup inevitably perturbs the sample due to the long-range evanescent decay of collective modes in the out-of-plane direction. This probe–sample interaction must be accounted for, no matter if the probe is a coplanar strip, a tip or an emitter, to access intrinsic properties. Such a theory for probe-sample interactions was missing in both prior on-chip THz spectroscopy studies and alternative approaches like THz-SNOM and THz near-field microscopy [25, 26]. This thesis however shows for the first time an analytical mapping of the intrinsic optical conductivity of a 2D material to it's self-cavity response in on-chip THz spectroscopy, including the effectivity of the probe excitation (see Sect. 6.4.7). This progress is expected to significantly enhance on-chip THz spectroscopy as a method, establishing it as a powerful optical probe capable of fully capturing the intrinsic conductivity of vdW heterostructures.

While the analytical theory has been verified in the case of probing graphite with on-chip THz spectroscopy, the theory is applicable to a plethora of other 2D materials. In case of a metallic 2D material, simply the 3D bulk plasma frequency has to be adjusted to simulate the corresponding cavity conductivity

[5] As the size of a sample decreases, cavity modes shift to higher frequencies, and in larger samples, these modes tend toward zero frequency. This phenomenon is analogous to musical instruments, where the pitch of a sound (and the frequency) increases as the size of the instrument decreases.

spectrum. For other materials, the 2D conductivity model can be customised, meaning that the assumed Drude model for metallic layers can be replaced by any other conductivity model (examples are given in Chap. 10). This makes the analytical theory a precious tool for analysing detected THz conductivity spectra of micropatterned vdW heterostructures, facilitating both understanding and design of self-cavity modes. For example, for a given sample with specific bulk plasma frequency and thickness, the widths of the coplanar strip are an effective tuning knob for designing the resonance frequencies of the self-cavity modes. If a low-conducting transition metal dichalcogenides shall be investigated, the analytical theory can predict how much the overall coplanar strip width should be reduced to distribute a defined number of cavity modes over the spectral sensitivity range. Importantly, significant coplanar strip width reductions can only be performed over a length of a few hundreds of micrometers and not for the entire measurement arms, as otherwise losses and attenuation of the excited THz pulses significantly reduce the measurement bandwidth (Sect. 2.2.3). Thus, the coplanar strip widths have to be smoothly narrowed down in the sample region, also for the reference measurement arm at the same position, to account for the customised transmission line design.

In the opposite limit of investigating materials with particularly high conductivities such as superconductors, the coplanar strip widths have to be designed particularly wide in order to push the resonances of the superconducting self-cavity down into the spectral sensitivity range, as predicted by the analytical theory and already achieved in Ref. [27].

The hybrid nature of the plasmonic cavity modes, which involve both screened and unscreened regions, results in cavity modes for graphite flakes (with typical widths ranging from 9 μ m to 30 μ m) that fall within the spectral sensitivity range. However, if the graphite flake is etched to be only partially embedded beneath the coplanar strip, for example only the W_1 or W_2 region, the overall effective cavity width is reduced, pushing the resonances to higher frequencies. Consequently, for the series of investigated graphite test devices that span only sections of the coplanar strips (Sect. 6.2.1), one would expect much higher resonance frequencies as compared to a graphite sample spanning the entire coplanar strip. The theory verifies that these partial coverage devices should have higher resonance frequencies (see Table A.1 in the Appendix for predicted resonance frequencies of these test devices). However, many resonances of the test devices where graphite (or graphene) was placed only beneath the metal traces, not in the coplanar strip gap, did not show THz signals that significantly differed from the reference side. This indicates that the THz pulse can not effectively excite currents in materials positioned outside the central coplanar strip region.

In contrast, if a sample is placed exclusively within the gap region of the coplanar strip without extending beneath the metal traces, plasmonic standing waves can still be efficiently excited. Test devices TD5 and TD6 are consisting of a too narrow graphite flake to exhibit resonances within the spectral sensitivity range (see Table A.1). However, a further test device, TD7 (see Fig. A.1), consisting of a PdAu-graphite stack (see Fig. 6.4), was investigated. The 5 nm thick PdAu layer is expected to screen the graphite layer, reducing the graphite's plasmonic resonance frequency within the spectral sensitivity range. At the same time, the thin PdAu layer can also host 2D plasmonic standing waves, that are screened by the graphite layer. Given that PdAu is more metallic than graphite, PdAu plasmons are expected to resonate at a higher frequency, with greater spectral weight, compared to those in graphite. While the 3D plasma frequency of thin PdAu is unknown, the observed spectrum suggests the $m = 0$ mode in PdAu resonates just outside the spectral sensitivity range (see Fig. A.1 in the Appendix). Additional evidence supporting the excitation of plasmonic cavity modes in conductive 2D materials placed exclusively within the coplanar strip gap is provided in Ref. [21]. However, since contacting samples confined to the coplanar strip gap can be challenging, which is necessary for carrier density modulation in 2D materials like graphene, it is generally preferable to have the 2D material fully cover the coplanar strip.

In the case of 2D materials that span and extend beyond the coplanar strip, the dielectric environment, and particularly the screening, becomes inhomogeneous. This variation in screening leads to significant effects on the nature of the 2D plasmonic modes. The relative widths of coplanar strip gap (W_2) compared to the metal trace width (W_1) governs the extent of hybridization between the screened and unscreened plasmons in different regions (Sect. 6.5.3). This happens because the widths W_1 and W_2 establish the wave vectors q_{screened} and $q_{\text{unscreened}}$, which determine the resonance frequencies of the screened and unscreened plasmons. By increasing W_2 while keeping W_1 constant, the resonance frequency of the unscreened plasmon can be continuously decreased toward that of the screened plasmon, facilitating resonant coupling between them. However, even when the plasmons are not resonant or have different momenta, hybridization occurs due to the scattering of momentum and energy at the boundaries, ensuring that both quantities are conserved across the boundaries. This effect can be most prominently seen in the plotted current distributions for a heterostructure with fixed W_1 and varying W_2, as the wavelength of the screened plasmon in the W_1 region is strongly dependent on the width of the W_2 region.

In the other extreme limit of making the coplanar strip gap very small ($W_2 \approx 100\,\text{nm}$ and $W_0 = 0$), the plasmonic cavity modes behave as purely screened 2D plasmons ($\nu \propto q$) and the series of cavity modes have wavelengths of $\lambda = \frac{W}{2}, \frac{3W}{2}, \frac{5W}{2}, \ldots$, suppressing hybridization of screened and unscreened plasmons.[6] In case of a heterostructure with a single self-cavity, as discussed in this chapter, the coupling of screened and unscreened plasmons in a heterostructure probed with on-chip THz spectroscopy is not so crucial. A detected THz response can be understood using the developed theory, and intrinsic parameters of the investigated 2D material can be extracted. However, in case of a heterostructure with two self-cavities, such as for a graphene heterostructure with a graphite gate, this coupling can have a significant effect, as will be seen later in Chap. 8.

[6] In such a geometry, the coplanar strips itself could further be used as topgates, such that by applying a voltage between 2D material and metal traces, the carrier density of the 2D material can be tuned.

The theory further highlights that vdW heterostructures of typical dimensions (<=100 μ m) and 2D conductivities σ_{2D} should generally exhibit discrete cavity modes, in contrast to the typical Drude response of metals (Sect. 6.6). Thereby, both coplanar strip width and 2D conductivity determine which regime applies. For example, if a low-conducting transition metal dichalcogenide would be measured in a 3 μ m–3 μ m–3 μ m coplanar strip geometry, the cavity modes would be pushed towards lower resonance frequencies, with cavity mode frequency spacing lower than the mode linewidths. The broad spectrum that results may look like a Drude response, but fitting the data with a Drude model would yield significantly different intrinsic material properties.

Additionally, the exact frequency position of the cavity modes (for modes $m \neq 0$) is dependent on the material's width extending the coplanar strip. In case of highly irregular shaped 2D materials, this dependence can result in strong inhomogeneous broadening (Sect. 6.5.1). This insight highlights the role of finite-size effects on the measured near-field conductivity, which must be considered when extracting material parameters from on-chip THz spectroscopy data.

To observe a Drude response in a small sample, an alternative method is to establish direct ohmic contact with the coplanar strip. This allows the detection of the Drude response regardless of the sample size. However, as noted in Sect. A.9 of the Appendix, this ohmic contact geometry is more suited to robust grown or evaporated thin films and is less applicable to delicate vdW materials.

I would like to emphasize that in both regimes, cavity response or Drude response regime, intrinsic material parameters can be extracted from the experimental data such as DC conductivity, Drude weight, scattering rate and mean free path. However, doing so requires understanding the regime that applies, and taking into account the geometric effects into the analysis. The developed analytical theory now provides the needed guidelines to achieve this. As these on-chip techniques are applied to more complex materials (superconductors, correlated insulators, strange metals, etc), the finite-size effects will continue to play a role in all measured conductivity spectra. Thus, this chapter provides the basis for a new framework for investigating the low energy dynamics of microstructured materials that are necessary for interpreting all present and future on-chip THz experiments on vdW heterostructures.

Having established that all vdW heterostructures inherently form self-cavities, the next chapter will explore why this discovery is particularly intriguing and what new opportunities it presents for further research and applications in the promising field of cavity engineering.

References

1. Kipp G, Bretscher HM, Schulte B, Herrmann D, Kusyak K, Day MW, Kesavan S, Matsuyama T, Li X, Langner SM, Hagelstein J, Sturm F, Potts AM, Eckhardt CJ, Huang Y, Watanabe K, Taniguchi T, Rubio A, Kennes DM, Sentef MA, Baudin E, Meier G, Michael MH, McIver

JW (2024) Cavity electrodynamics of van der Waals heterostructures. Nat Phys 21:1926–1933 https://doi.org/10.1038/s41567-025-03064-8
2. Michael MH, Kipp G, Potts AM, Day MW, Matsuyama T, Meier G, Bretscher HM, McIver JW (2025) Resolving self-cavity effects in two-dimensional quantum materials. arXiv preprint arXiv:2505.12799
3. Horng J, Chen CF, Geng B, Girit C, Zhang Y, Hao Z, Bechtel HA, Martin M, Zettl A, Crommie MF et al (2011) Drude conductivity of Dirac fermions in graphene. Phys Rev B—Condensed Matter Mater Phys 83:165113. https://doi.org/10.1103/PhysRevB.83.165113
4. Razanoelina M, Bagsican FR, Kawayama I, Zhang X, Ma L, Murakami H, Vajtai R, Ajayan PM, Kono J, Tonouchi M (2016) Probing low-density carriers in a single atomic layer using terahertz parallel-plate waveguides. Opt Express 24:3885–3893. https://doi.org/10.1364/OE.24.003885
5. Gallagher P, Yang C-S, Lyu T, Tian F, Kou R, Zhang H, Watanabe K, Taniguchi T, Wang F (2019) Quantum-critical conductivity of the Dirac fluid in graphene. Science 364:158–162. https://doi.org/10.1126/science.aat8687
6. Chen S-D, Feng Q, Zhao W, Qi R, Zhang Z, Abeysinghe D, Uzundal C, Xie J, Taniguchi T, Watanabe K, Wang F (2024) Direct measurement of terahertz conductivity in a gated monolayer semiconductor. arXiv preprint arXiv:2409.17633
7. Seo J, Lu Z, Park S, Yang J, Xia F, Ye S, Yao Y, Han T, Shi L, Watanabe K et al (2024) On-chip terahertz spectroscopy for dual-gated van der Waals Heterostructures at cryogenic temperatures. Nano Lett. https://doi.org/10.1021/acs.nanolett.4c04137
8. Herrmann D (2023) Plasmon resonances of graphite heterostructures probed by on-chip THz spectroscopy. Master's thesis. University of Hamburg
9. Yoon H, Yeung KYM, Kim P, Ham D (2014) Plasmonics with two-dimensional conductors. Philos Trans A Math Phys Eng Sci 372:20130104. https://doi.org/10.1098/rsta.2013.0104
10. Kesavan S (2024) Engineering THz plasmonic cavity modes in microstructured van der Waals heterostructures. Master's thesis. University of Hamburg
11. Systemes D (2010) CST simulation software. https://space.mit.edu/RADIO/CST_online/mergedProjects/3D/special_struct/special_struct_layer_parameters_general.htm
12. Scheuch M (2012) Strong modulation of THz optical properties of layered materials by thermal and ultrafast heating. PhD thesis. Freie Universität Berlin
13. Dressel M, Grüner G (2002) Electrodynamics of solids: Optical properties of electrons in matter. Cambridge University Press
14. Yoon H, Yeung KY, Kim P, Ham D (2014) Plasmonics with two-dimensional conductors. Philos Trans Roy Soc A: Math Phys Eng Sci 372:20130104. https://doi.org/10.1098/rsta.2013.0104
15. Economou EN (1969) Surface plasmons in thin films. Phys Rev 182:539–554. https://doi.org/10.1103/PhysRev.182.539
16. Deuflhard P (2011) Newton methods for nonlinear problems. Affine invariance and adaptive algorithms. Springer.
17. Svintsov DA, Alymov GV (2023) Refraction laws for two-dimensional plasmons. Phys Rev B 108:L121410. https://doi.org/10.1103/PhysRevB.108.L121410
18. Giancoli DC (2010) Physik. ger. 3., erweiterte Auflage. ph - Physik. Pearson Studium
19. Graef H, Mele D, Rosticher M, Banszerus L, Stampfer C, Taniguchi T, Watanabe K, Bocquillon E, Fève G, Berroir J et al (2018) Ultra-long wavelength Dirac plasmons in graphene capacitors. J Phys: Mater 1:01LT02. https://doi.org/10.1088/2515-7639/aadd8c
20. Yoshioka K, Bernard G, Wakamura T, Hashisaka M, Sasaki K-I, Sasaki S, Watanabe K, Taniguchi T, Kumada N (2024) On-chip transfer of ultrashort graphene plasmon wave packets using terahertz electronics. Nat Electron 7:537–544. https://doi.org/10.1038/s41928-024-01197-x
21. Zhao W, Wang S, Chen S, Zhang Z, Watanabe K, Taniguchi T, Zettl A, Wang F (2023) Observation of hydrodynamic plasmons and energy waves in graphene. Nature 614:688–693. https://doi.org/10.1038/s41586-022-05619-8

22. Fei Z, Goldflam MD, Wu J-S, Dai S, Wagner M, McLeod A, Liu M, Post K, Zhu S, Janssen G et al (2015) Edge and surface plasmons in graphene nanoribbons. Nano Lett 15:8271–8276. https://doi.org/10.1021/acs.nanolett.5b03834
23. Kim S, Jang MS, Brar VW, Mauser KW, Kim L, Atwater HA (2018) Electronically tunable perfect absorption in graphene. Nano Lett 18:971–979. https://doi.org/10.1021/acs.nanolett.7b04393
24. Castilla S, Agarwal H, Vangelidis I, Bludov YV, Iranzo DA, Grabulosa A, Ceccanti M, Vasilevskiy MI, Kumar RK, Janzen E et al (2024) Electrical spectroscopy of polaritonic nanoresonators. Nat Commun 15:8635. https://doi.org/10.1038/s41467-024-52838-w
25. Handa T, Huang C-Y, Li Y, Olsen N, Chica DG, Xu DD, Sturm F, McIver JW, Roy X, Zhu X (2024) A 2D van der Waals material for terahertz emission with giant optical rectification. arXiv preprint arXiv:2411.09421
26. Helmrich F, Khanonkin I, Kroner M, Scalari G, Faist J, Imamoglu A, Nova T (2024) Ultra-strong Terahertz Coupling in a van der Waals Heterostructure. arXiv preprint arXiv:2408.00189
27. Potts AM, Nayak AK, Nagel M, Kaj K, Stamenic B, John DD, Averitt RD, Young AF (2023) On-chip time-domain terahertz spectroscopy of superconducting films below the diffraction limit. Nano Lett 23:3835–3841. https://doi.org/10.1021/acs.nanolett.3c00412

Chapter 7
Cavity Engineering of Quantum Materials

Abstract Building on the previous chapter, which revealed the "self-cavity" behavior of van der Waals materials probed by on-chip THz spectroscopy, this chapter introduces the principles of cavity engineering. It defines and distinguishes the four known light-matter coupling regimes (weak, strong, ultrastrong, and deep strong) and emphasizes the unique suitability of 2D materials for controlling macroscopic material properties via cavity engineering. Finally, it explores how these concepts can be applied to self-cavities, for instance by achieving ultrastrong coupling between two self-cavities, such as a graphene layer and its underlying graphite gate.

A key goal of condensed matter physics is to explore and control extraordinary material properties, which could enable advances in future technology. Chapter 1 showed that vdW heterostructures are highly tunable through in situ control of carrier density, the application of magnetic fields or pressure, and by twisting different vdW layers. While these tuning mechanisms already offer vast opportunities for discovering and controlling novel physics in vdW heterostructures, introducing new axes of control is always advantageous, as additional tuning knobs might offer access to even more or enhanced phenomena.

One such new axis of control is delivered by strong light-matter interactions. While light-matter interactions are typically small, as defined by the fine structure constant $\alpha \approx 1/137$ [2], they can be increased by either using very strong light pulses, thus many photons, or by confining the electromagnetic field into confined spaces, "cavities" [3]. When a material is placed inside a cavity, even few or zero photons (quantum fluctuations) can achieve strong light-matter interaction strengths, as a single photon oscillating between the cavity's boundaries can repeatedly interact with the material. The present chapter delves into the promising realm of cavity engineering: the design of cavities to tailor light-matter interactions and potentially manipulate the thermodynamic ground state (see Fig. 7.1) [4–14].

G. Kipp, *Self-Cavity Electrodynamics of Van der Waals Heterostructures*, Springer Theses, https://doi.org/10.1007/978-3-032-31979-1_7

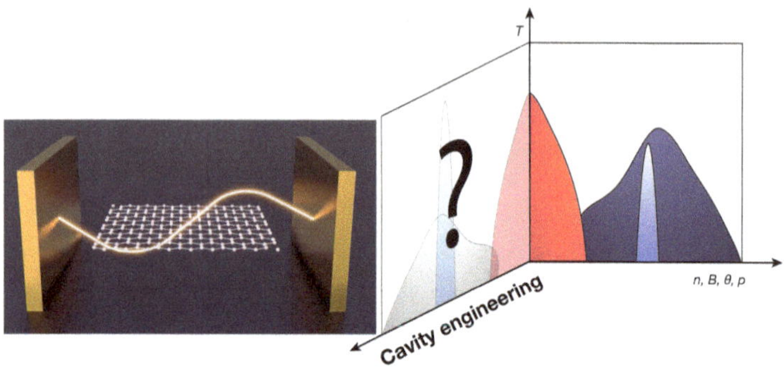

Fig. 7.1 Cavity engineering. A cavity, such as an optical cavity formed by two parallel mirrors, greatly enhances the likelihood of a photon interacting with a sample placed inside, as the photon reflects multiple times between the mirrors. This enables interactions between light and matter to become significantly stronger, potentially enabling the manipulation of the thermodynamic ground state. This raises the question: how can the rich phenomena found in 2D quantum materials be further enhanced or tailored through cavity engineering?

The chapter is structured in the following way: Sect. 7.1 introduces the concept of confining electromagnetic fields by imposing strict boundaries, such as by mirrors of an optical cavity. Section 7.2 discusses quantum fluctuations and how they define or alter material properties. Section 7.3 explores the Purcell effect, demonstrating that matter can be influenced by the confined space of a cavity, even in the absence of light, with only quantum fluctuations interacting with the material inside. Section 7.4 describes the four different coupling regimes as defined in literature based on the comparison of the coupling strength g to either the losses of both cavity and matter, or by comparing g to the frequency ν_0, at which cavity and matter couple resonantly. Section 7.5 evaluates the suitability of 2D materials for cavity engineering and in Sect. 7.6, the principles of cavity engineering are scrutinised for the case of self-cavities in vdW heterostructures.

7.1 Electromagnetic Fields in Cavities

In an infinitely large space, the wave vector k of an electromagnetic field can take any continuous value. However, when boundaries are imposed on the field, such as confining light within a cavity of length l, the wave vector k becomes restricted to discrete values. In this scenario, k must satisfy periodic boundary conditions, meaning, for example, that the field at one side of the cavity must align with the field at the opposite side [15, 16]:

$$k = \frac{2\pi}{l}n, \tag{7.1}$$

where n are integer values. These discrete values for the wave vector k are called *modes* of the electromagnetic field, which show up as resonances in the frequency-domain.

Intriguingly, the excitation of discrete cavity modes requires not many photons. Even in vacuum, where classical electromagnetic fields are absent, *quantum fluctuations* can be discretized and modified by imposing strict spatial boundaries [16], as discussed in the following section.

7.2 Quantum Fluctuations

It has been known since the development of quantum mechanics in the early 20th century that vacuum is not actually empty space, but rather full of fluctuating electromagnetic fields with a zero-point energy of:

$$E = \sum_k \frac{1}{2}\hbar\omega_k. \tag{7.2}$$

These quantum fluctuations are said to perturb or stimulate atoms and are responsible for several fundamental physical processes including [17, 18]:

- **Casimir force:** Casimir showed in 1948 that two parallel, uncharged perfect electrical conductor plates experience an attractive force due to quantum fluctuations [19].
- **VdW forces:** One type of vdW force, the "London-vdW force" is attributed to quantum fluctuations as it arises due to spatial fluctuations of electrons and nuclei that cause temporary dipole moments, even in neutral atoms without permanent dipoles. The fluctuating dipole in one atom then polarizes nearby atoms, inducing dipole moments in them too. This leads to an attractive force between the atoms, or, in case of vdW materials or heterostructures, to a weak force between individual one-atomic thick layers [18, 20].
- **The Lamb shift:** The Lamb shift describes the energy difference between the $2S_{1/2}$ and $2P_{1/2}$ levels of the hydrogen atom, which should be degenerate as described by the original Dirac theory [21]. The splitting arises due to the interaction between the electron and the quantized electromagnetic field due to quantum fluctuations [18].
- **Spontaneous emission:** Spontaneous emission is the process by which an excited atom releases energy by emitting a photon and transitioning to a lower energy state, without any external stimulation. This phenomenon arises from quantum fluctuations, which effectively act as the trigger for the emission, and it plays a fundamental role in enabling laser operation [18].

These examples above highlight that quantum fluctuations can be detected via their interaction with matter. Intriguingly, these interactions can be altered inside a cavity, as Edward Mills Purcell discovered in the early 1940s.

7.3 Purcell Effect

Edward Mills Purcell discovered that the presence of a cavity increases the spontaneous emission rate of a quantum system (in vacuum) by altering the vacuum's electromagnetic fluctuations [22, 23]. According to Fermi's golden rule, the transition rate of an atom coupled to vacuum fluctuations (or to their modified counterpart within a cavity) is directly proportional to the density of accessible final photon states. To quantify the enhancement in spontaneous emission, Purcell therefore compared the density of photon states in free space as given by (n is the refractive index of the cavity material) [24]

$$\rho_{\text{free space}} = \frac{8\pi n^3 \nu^2}{c^3}, \tag{7.3}$$

to the density of photon states in a cavity as given by

$$\rho_{\text{cavity}} = \frac{1}{V\Delta\nu} = \frac{Q}{V\nu}. \tag{7.4}$$

V is the volume of the cavity and $Q = \nu/\Delta\nu$ is the quality factor of the cavity [24]. The Purcell factor, that defines how much the spontaneous emission of an atom is increased due to the presence of a cavity is therefore [23]:

$$f_{\text{Purcell}} = \frac{\rho_{\text{cavity}}}{\rho_{\text{free space}}} = \frac{1}{8\pi}\frac{c^3}{\nu^3 n^3}\frac{Q}{V}, \tag{7.5}$$

where V is the mode volume of the cavity, ν is the frequency of the quantum fluctuation, $Q = \nu/\Delta\nu$ is the quality factor of the cavity and n is the refractive index of the cavity material. The Purcell factor can be increased by:

- operating at low frequencies ν,
- reducing the mode volume V,
- increasing the quality factor Q,

whereby the operation at low frequencies has a more dominant effect than high Q-factors for achieving high Purcell factors.

The Purcell effect shows that no external driving forces are necessary; the quantum fluctuations alone, in cavities, can be used to manipulate matter.

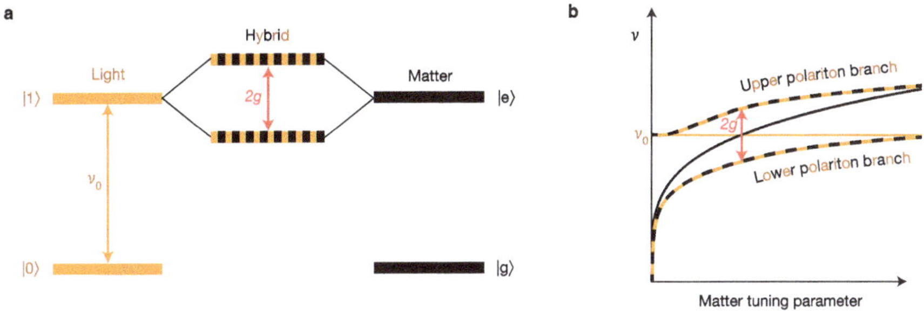

Fig. 7.2 Formation of light-matter hybrids. (**a**) When cavity light and a matter excitation resonate and couple, they form light-matter hybrid states. The coupling strength is characterized by the mode splitting, $2g$, at resonance, ν_0: the larger the splitting, the greater the coupling strength, g. (**b**) Another characteristic of the formation of hybrid modes is the formation of avoided crossings. If the matter excitation (dark blue) is tuned via some tuning parameter into resonance with the cavity light mode (yellow), the two hybrid modes (yellow-dark blue dashed curves) repel each other in frequency, thereby forming upper and lower polariton branches of mixed character of the original uncoupled light and matter modes

7.4 Coupling Regimes

The Purcell effect becomes evident even with weak light-matter coupling, but the potential to engineer new thermodynamic ground states emerges when the coupling strength increases [13, 25–27]. In stronger coupling regimes, confined light in a cavity and matter excitations form hybrid light-matter modes that combine properties of both [28]. When cavity light and matter excitation are in resonance, these hybrid modes exhibit a characteristic mode splitting, known as "Rabi splitting," where the modes shift apart by an amount defined as twice the coupling strength g (see Fig. 7.2a).

Another hallmark of enhanced coupling is the formation of avoided crossings, or "anticrossings". Hybrid modes not only form precisely at resonance but also within a frequency range around it. As a matter mode, tunable by an external parameter (such as carrier density), approaches resonance with a cavity light mode, an avoided crossing forms, resulting in the formation of upper and lower polariton branches (see Fig. 7.2b).

Thus, high coupling strengths g lead to more altered frequency scales as defined by larger splittings and larger avoided crossings of the hybrid modes. However, strong is not just strong: The coupling strength g can be evaluated in different ways, and the field generally defines four different coupling strength regimes in relation to the resonance frequency of light and matter modes ν_0, or decay rates of atomic γ and cavity κ excitations (see Fig. 7.3a–d):

- Weak coupling regime ($g \ll \gamma, \kappa, \nu_0$)
- Strong coupling regime ($\gamma, \kappa \ll g \ll \nu_0$)

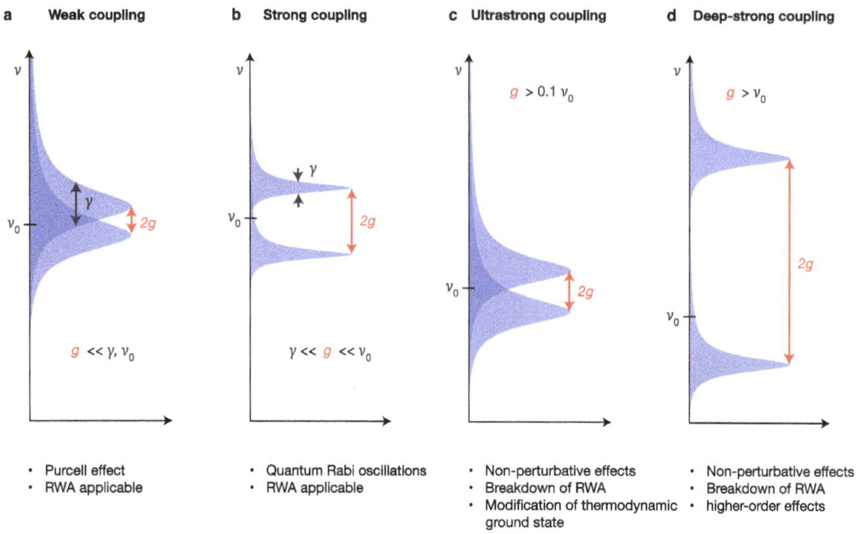

Fig. 7.3 Coupling regimes for enhanced light-matter interactions. As shown in Fig. 7.2, the coupling strength is defined by the mode splitting $2g$ of the two hybrid modes, and dependent on how large g is compared to the decay rates of the matter mode (γ) or cavity mode (κ) or compared to the resonant frequency ν_0, four different coupling regimes are defined, because different physical characteristics exist in each of the regimes. For simplicity, $\gamma = \kappa$ here. (**a**) In the weak coupling regime, the coupling strength g is much smaller than the linewidths γ and the resonance frequency of the uncoupled modes ν_0. In this regime, the rotating wave approximation (RWA) is applicable and the Purcell effect can be observed. (**b**) In the strong coupling regime, the coupling strength g becomes larger than γ, but is still significantly smaller than the energy scale of the system. The RWA is still a valid approximation and a typical characteristic of this regime are quantum Rabi oscillations, that show that information can be exchanged coherently between light and matter modes. (**c**) The ultrastrong coupling regime is defined as $g > 0.1\nu_0$, independent of γ. (**d**) In the deep-strong coupling regime, $g > \nu_0$, also independent of the linewidth. The RWA breaks down in the ultrastrong and deep-strong coupling regimes, and non-perturbative effects are observable, which can also lead to asymmetric avoided crossings [29]. In order to engineer the thermodynamic ground state of a material, the cavity system has to operate at least in the ultrastrong coupling regime, or in the deep strong coupling regime. (See also Ref. [26])

- Ultrastrong coupling regime ($g > 0.1\nu_0$)
- Deep strong coupling regime ($g > \nu_0$).

These different definitions of the regimes are made because different physical characteristics exist, as discussed in the following.

7.4.1 Weak Coupling Regime

In the weak coupling regime ($g \ll \gamma, \kappa, \nu_0$), the coupling strength g is much smaller than the resonance frequencies and decay rates of the system (see Fig. 7.3c). This means that if light-matter hybrids form that are split in frequency by a small amount ($2g$), the overall resonance just looks broadened as the linewidths are significantly larger than the frequency splitting.

The rotating wave approximation (RWA)[1] is fully applicable in this regime and the Hamiltonian describing the system is the Jaynes-Cummings Hamiltonian [30]. A typical observable in the weak-coupling regime is the Purcell effect (see Sect. 7.3).

7.4.2 Strong Coupling Regime

In the strong coupling regime ($\gamma, \kappa \ll g \ll \nu_0$), an atomic system can absorb and re-emit a photon multiple times before it leaks from the cavity [31, 32], allowing coherent exchange of information. Critically, this can only occur when the scattering rates are very small compared to the coupling strength g [33].

In the strong coupling regime, the RWA and the Jaynes-Cummings model still hold true[2] and a typical characteristic is the Rabi splitting in the frequency domain or the Rabi oscillation[3] in the time domain [34]. The strong coupling regime is the prerequisite to study coherent quantum dynamics between light and matter [25, 26], such as processes between a single atom and a single photon [35].

7.4.3 Ultrastrong Coupling Regime

While the coupling is considered weak or strong depending on how large g is compared to the losses γ and κ of the system, ultrastrong coupling is not, as intuitively expected, a regime with even stronger coupling than in the strong coupling regime [26]. The ultrastrong coupling regime is independent of dissipation, and instead of comparing g to the losses, g is compared to the resonant frequency ν_0 of the modes (see Fig. 7.3) [25]. Thus, a system can be in the ultrastrong coupling regime, without being in the strong coupling regime, when dissipation is high [26].

[1] The RWA is a simplified theoretical description of light-matter interactions in quantum mechanics, where rapidly oscillating terms in the interaction Hamiltonian are neglected. This approximation assumes that only energy-conserving processes, such as photon absorption and emission, contribute dominantly to the system dynamics, allowing for easier analysis of phenomena [16].

[2] If the system is not simultaneously in the ultrastrong coupling regime, too (see below).

[3] Rabi oscillations describe the phenomenon of a cyclic energy transfer between an atom and an electromagnetic field. An atom alternates between emitting and reabsorbing photons, whereby it changes symmetrically between its ground state and its excited state at the "Rabi" frequency. Rabi oscillations occur at high frequencies and small mode volumes, when both the coupling strength and the number of involved photons is high. However, the visibility of these oscillations in the time domain is largely defined by the scattering rate of a material or by the quality factor Q of a cavity. A low quality factor leads to rapid damping of the oscillations, while a high Q-value preserves the coherence of the system, making the Rabi oscillations more pronounced and long-lived [16].

The ultrastrong coupling regime is generally defined for $g > 0.1\nu_0$, or for the normalized coupling strength being

$$\boxed{\eta = \frac{g}{\nu_0} > 0.1}. \tag{7.6}$$

In this regime, the rotating wave approximation breaks down as the counter-rotating terms cannot be neglected in the case of the coupling strength g becoming comparable or larger than the energy difference of light and matter modes. This means that the mixing of the states becomes significant and high frequency oscillating terms become important in the interaction Hamiltonian, that defines the thermodynamic ground state. As a result, simulations and experiment deviate from expected predictions using the Jaynes-Cummings model [36]. This leads to the possibility to observing qualitatively new physics that are not dependent on the dissipation rates of the cavity or the matter involved.

While the threshold of 0.1 serves as a practical reference point rather than a strict boundary, cavity induced changes to a material become increasingly likely as the normalized coupling strength η increases [26]. A series of fascinating phenomena are proposed for this regime as highlighted by many theoretical publications. For example, cavities operating in the ultrastrong coupling regime are predicted to

- **Enhance superconductivity** [6, 7, 37–40],
- **Manipulate ferroelectricity** [8, 41, 42],
- **Control topology** [12].

Pioneering experiments have demonstrated an enhancement of ferromagnetism [43], a change in the critical temperature of a metal to insulator transition [44] and the modification of integer [45] and fractional [46] quantum Hall states in an empty cavity, all of which were achieved with relatively low cavity quality factors of order ~ 5. This highlights that the strong coupling regime (low dissipation) is not a prerequisite for cavity engineering of the thermodynamic ground state, but the establishment of the cavity system in the ultrastrong coupling regime is desirable to perform cavity engineering of materials.

7.4.4 Deep Strong Coupling Regime

If the normalized coupling strength

$$\eta = \frac{g}{\nu_0} > 1, \tag{7.7}$$

the coupling of the system is located in the so called "deep strong coupling regime". Here, the RWA breaks down similarly as for the case of ultrastrong coupling and higher-order perturbative processes can not only become observable but may even dominate [26].

The most striking phenomenon in the deep strong coupling regime is the decoupling of light and matter, leading to the surprising reversal of the Purcell effect (see Sect. 7.3). Instead of an increased spontaneous emission rate with stronger coupling, the emission rate diminishes as the coupling strength grows, with no energy exchange occurring between light and matter for sufficiently large couplings. Although this effect has been theoretically predicted for many years [47–49], experimental confirmation has only recently emerged as more experimental platforms have begun successfully reaching the deep strong coupling regime [50].

7.5 Suitability of 2D Materials for Cavity Engineering

Having explored the four distinct coupling regimes, revealing that entering of the ultrastrong coupling regime ($g/\nu_0 > 0.1$) is desirable to fundamentally alter material properties, the question arises: what potential do these principles hold for 2D materials and how well-suited are they for performing cavity engineering?

As discussed in Chap. 1, vdW heterostructures can host strongly correlated phases, where adding or removing only a few electrons from a 2D material can change the macroscopic properties of the sample for example from a superconducting phase to a correlated insulating phase (see Fig. 7.4). This high degree of competition between microscopic interactions means that small perturbations are enough to cause macroscopic changes [51], suggesting that perturbations induced by a cavity could be sufficient to tip the balance between two competing phases [13, 48, 52–56].

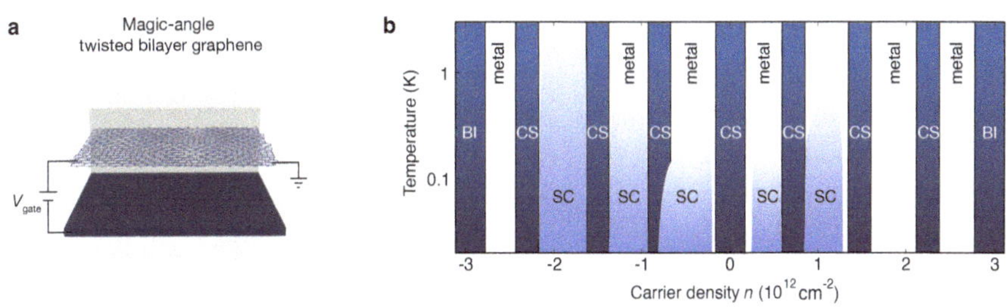

Fig. 7.4 Strongly correlated phases in twisted bilayer graphene. (**a**) Two graphene sheets twisted by 1.1 degrees ("magic angle") are encapsulated in insulating hBN and placed onto a metallic gate. By applying a voltage V_{gate} between the gate and the magic-angle twisted bilayer graphene, the charge carrier density of the twisted graphene can be tuned. (**b**) Depending on the charge carrier density and temperature, a plethora of different macroscopic properties such as band insulating (BI), correlated insulating (CS), superconducting (SC) or metallic phases can be observed [57]. Thereby, only a few more or fewer electrons can tip the phase from for example a correlated insulator to a superconducting phase

Secondly, light-matter hybrids like plasmon, magnon, and phonon polaritons, that are already signatures of enhanced light-matter interaction [58], can form effectively in vdW materials due to their strong in-plane coupling and weak interlayer interactions. These collective excitations naturally have small mode volumes V, due to the two-dimensionality of vdW heterostructures and their sub-wavelength size for THz light that is resonant to the characteristic energy scale of the emergent phenomena in these materials. This opens the pathway for large Purcell factors ($f_{\text{Purcell}} \propto 1/V$, see Eq. 7.5).

Furthermore, both the Eq. 7.5 for the Purcell factor and the definition of the ultra-strong coupling regime ($\eta = g/\nu_0 > 0.1$) suggest that cavity engineering becomes increasingly possible if the cavity-matter system operates at low frequencies. As illustrated in Fig. 1.4 in the introduction, many emergent phenomena in vdW heterostructures manifest at low, THz frequencies. Intuitively, one might therefore expect that the low energy scale of these phenomena would naturally lend itself to achieving large normalized coupling strengths, η. However, the coupling strength, g, is itself frequency-dependent, and this dependence is heavily influenced by the specific material coupling to the cavity mode. Therefore, no definitive conclusions about increased normalized coupling strengths can be drawn without simulating the frequency dependence of g for the material in question.

However, given their sensitivity to small perturbations and strong mode confinement to small volumes, vdW materials are ideally suited for cavity engineering.

7.6 Discussion

This chapter highlights the promising frontier of cavity engineering. Typically, cavity engineering has been performed by embedding a material within a resonant structure (usually a cavity composed of two planar, conductive mirrors), enabling modulation of the material's characteristic energy scales or even its macroscopic properties through interaction with a cavity mode. However, the discovery presented in Chap. 6 offers a transformative perspective: many conductive 2D quantum materials, as components of vdW heterostructures, inherently form self-cavities for THz light, eliminating the need for external mirrors altogether.

This realization challenges the conventional distinction between cavity and material. In micropatterned vdW heterostructures, the cavity and the quantum material form an inseparable entity, preventing the typical tuning or matching of material and cavity as independent components. This leads to a compelling question: how can one meaningfully perform cavity engineering when the cavity is not an external structure but an emergent feature of the material itself?

Beyond the general finding that 2D materials form THz self-cavities, Chap. 6 further demonstrated that the dielectric environment plays a fundamental role in shaping the current density profile of these plasmonic modes, thereby defining the effective conductivity landscape of the self-cavity (see Fig. 6.15e, f). Much like

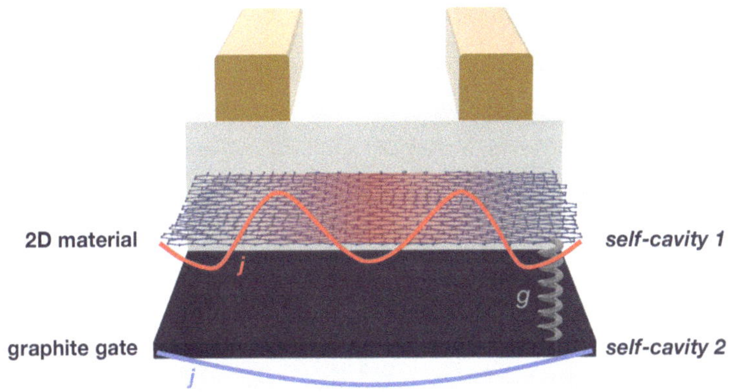

Fig. 7.5 Heterostructure comprising a 2D material and a graphite gate. Both the 2D material and the graphite gate act as self-cavities for THz light, opening the possibility of coupling between the two layers, illustrated schematically by the connecting spring. (Figure adapted from [1] under a CC BY 4.0 license)

the effect of a conventional cavity, the dielectric surroundings and subwavelength dimensions of a 2D material intricately tailor its response.

This pronounced sensitivity to the dielectric environment opens new avenues for cavity engineering. In many vdW heterostructures, graphite flakes are employed as electrostatic gates to tune quantum phases in adjacent 2D materials, which are often separated by only tens of nanometers. Given that both the gate and the gated material can independently support self-cavity modes in the THz range (see Sect. 6.6), it is natural to ask whether the standing current density waves in each layer can interact. Specifically, could a plasmonic self-cavity formed in a graphite gate couple ultrastrongly to the self-cavity of a gated 2D material (e.g. graphene), thereby altering or controlling its electrodynamic properties (see Fig. 7.5)?

To investigate this compelling possibility, the following chapter explores heterostructures composed of graphene and a graphite gate using on-chip THz spectroscopy.

References

1. Kipp G, Bretscher HM, Schulte B, Herrmann D, Kusyak K, Day MW, Kesavan S, Matsuyama T, Li X, Langner SM, Hagelstein J, Sturm F, Potts AM, Eckhardt CJ, Huang Y, Watanabe K, Taniguchi T, Rubio A, Kennes DM, Sentef MA, Baudin E, Meier G, Michael MH, McIver JW (2025) Cavity electrodynamics of van der Waals heterostructures. Nat Phys 21:1926–1933. https://doi.org/10.1038/s41567-025-03064-8
2. Kinoshita T (1996) The fine structure constant. Rep Progr Phys 59:1459. https://doi.org/10.1088/0034-4885/59/11/003
3. Claassen M (2025) Designing matter with photons and many electrons. YouTube

4. Schlawin F, Kennes DM, Sentef MA (2022) Cavity quantum materials. Appl Phys Rev 9. https://doi.org/10.1063/5.0083825
5. Hübener H, De Giovannini U, Schäfer C, Andberger J, Ruggenthaler M, Faist J, Rubio A (2021) Engineering quantum materials with chiral optical cavities. Nat Mater 20:438–442. https://doi.org/10.1038/s41563-020-00801-7
6. Eckhardt CJ, Chattopadhyay S, Kennes DM, Demler EA, Sentef MA, Michael MH (2023) Theory of resonantly enhanced photo-induced superconductivity. Nat Commun 15:2300. https://doi.org/10.1038/s41467-024-46632-x
7. Sentef MA, Ruggenthaler M, Rubio A (2018) Cavity quantum-electrodynamical polaritonically enhanced electron-phonon coupling and its influence on superconductivity. Sci Adv 4:eaau6969. https://doi.org/10.1126/sciadv.aau6969
8. Curtis JB, Michael MH, Demler E (2023) Local fluctuations in cavity control of ferroelectricity. Phys Rev Res 5:043118 (2023). https://doi.org/10.1103/PhysRevResearch.5.043118.
9. Masuki K, Ashida Y (2024) Cavity moiré materials: controlling magnetic frustration with quantum light-matter interaction. Phys Rev B 109:195173. https://doi.org/10.1103/physrevb.109.195173
10. Bloch J, Cavalleri A, Galitski V, Hafezi M, Rubio A (2022) Strongly correlated electron–photon systems. Nature 606:41–48. https://doi.org/10.1038/s41586-022-04726-w
11. Boström EV, Sriram A, Claassen M, Rubio A (2023) Controlling the magnetic state of the proximate quantum spin liquid α-RuCl3 with an optical cavity. NPJ Comput Mater 9:202. https://doi.org/10.1038/s41524-023-01158-6
12. Wang X, Ronca E, Sentef MA (2019) Cavity quantum electrodynamical Chern insulator: towards light-induced quantized anomalous Hall effect in graphene. Phys Rev B 99:235156. https://doi.org/10.1103/PhysRevB.99.235156
13. Riolo R, Tomadin A, Mazza G, Asgari R, MacDonald AH, Polini M (2024) Tuning Fermi liquids with sub-wavelength cavities. arXiv preprint arXiv:2403.20067
14. Andberger J, Graziotto L, Sacchi L, Beck M, Scalari G, Faist J (2024) Terahertz chiral subwavelength cavities breaking time-reversal symmetry via ultrastrong light-matter interaction. Phys Rev B 109:L161302. https://doi.org/10.1103/PhysRevB.109.L161302
15. Quantum Optics—02 Lecture 16 Review Maxwell equations Vector Potentials. Educational Documentaries, YouTube (2020)
16. Loudon R (2010) The quantum theory of light, vol 3. Oxford Science Publications
17. Scully MO, Zubairy MS (1997) Quantum optics. Cambridge University Press
18. Milonni PW (1994) The quantum vacuum. Milonni PW (ed) Academic Press, San Diego. https://doi.org/10.1016/B978-0-08-057149-2.50005-9
19. Casimir HBG (1948) On the attraction between two perfectly conducting plates. Kon Ned Akad Wetensch Proc 51:793–795.
20. Casimir HBG, Polder D (1948) The influence of retardation on the London-van der Waals forces. Phys Rev 73:360–372. https://doi.org/10.1103/PhysRev.73.360
21. Lamb WE, Retherford RC (1947) Fine structure of the hydrogen atom by a microwave method. Phys Rev 72:241–243. https://doi.org/10.1103/PhysRev.72.241
22. Purcell EM, Torrey HC, Pound RV (1946) Resonance absorption by nuclear magnetic moments in a solid. Phys Rev 69:37–38. https://doi.org/10.1103/PhysRev.69.37
23. Purcell EM (1995) Spontaneous emission probabilities at radio frequencies. In: Burstein E, Weisbuch C (eds) Confined electrons and photons: new physics and applications. Springer, Boston, pp 839–839. https://doi.org/10.1007/978-1-4615-1963-8_40
24. Kleppner D (1981) Inhibited spontaneous emission. Phys Rev Lett 47:233–236. https://doi.org/10.1103/PhysRevLett.47.233
25. Forn-Díaz P, Lamata L, Rico E, Kono J, Solano E (2019) Ultrastrong coupling regimes of light-matter interaction. Rev Mod Phys 91:025005. https://doi.org/10.1103/RevModPhys.91.025005
26. Frisk Kockum A, Miranowicz A, De Liberato S, Savasta S, Nori F (2019) Ultrastrong coupling between light and matter. Nat Rev Phys 1:19–40. https://doi.org/10.1038/s42254-018-0006-2

27. Bayer A, Pozimski M, Schambeck S, Schuh D, Huber R, Bougeard D, Lange C (2017) Terahertz light–matter interaction beyond unity coupling strength. Nano Lett 17:6340–6344. https://doi.org/10.1021/acs.nanolett.7b03103
28. Garcia-Vidal FJ, Ciuti C, Ebbesen TW (2021) Manipulating matter by strong coupling to vacuum fields. Science 373:eabd0336. https://doi.org/10.1126/science.abd0336
29. Zhang Q, Lou M, Li X, Reno JL, Pan W, Watson JD, Manfra MJ, Kono J (2016) Collective non-perturbative coupling of 2D electrons with high-quality-factor terahertz cavity photons. Nat Phys 12:1005–1011. https://doi.org/10.1038/nphys3850
30. Jaynes E, Cummings F (1963) Comparison of quantum and semiclassical radiation theories with application to the beam maser. Proc IEEE 51:89–109. https://doi.org/10.1109/PROC.1963.1664
31. Wallraff A, Schuster DI, Blais A, Frunzio L, Huang RS, Majer J, Kumar S, Girvin SM, Schoelkopf RJ (2004) Strong coupling of a single photon to a superconducting qubit using circuit quantum electrodynamics. Nature 431:162–167. https://doi.org/10.1038/nature02851
32. Schuster DI, Houck AA, Schreier JA, Wallraff A, Gambetta JM, Blais A, Frunzio L, Majer J, Johnson B, Devoret MH, Girvin SM, Schoelkopf RJ (2007) Resolving photon number states in a superconducting circuit. Nature 445:515–518. https://doi.org/10.1038/nature05461
33. Greuter L, Starosielec S, Kuhlmann AV, Warburton RJ (2015) Towards high-cooperativity strong coupling of a quantum dot in a tunable microcavity. Phys Rev B 92:045302. https://doi.org/10.1103/PhysRevB.92.045302
34. Niu Y, Gao L, Xu H, Wei H (2023) Strong coupling of multiple plasmon modes and excitons with excitation light controlled active tuning. Nanophotonics 12:735–742. https://doi.org/10.1515/nanoph-2022-0701
35. Haroche S (2013) Nobel Lecture: Controlling photons in a box and exploring the quantum to classical boundary. Rev Mod Phys 85:1083–1102. https://doi.org/10.1103/RevModPhys-85-1083
36. Li X, Bamba M, Zhang Q, Fallahi S, Gardner GC, Gao W, Lou M, Yoshioka K, Manfra MJ, Kono J (2018) Vacuum Bloch–Siegert shift in Landau polaritons with ultrahigh cooperativity. Nat Photon 12:324–329. https://doi.org/10.1038/s41566-018-0153-0
37. Curtis JB, Raines ZM, Allocca AA, Hafezi M, Galitski VM (2019) Cavity quantum Eliashberg enhancement of superconductivity. Phys Rev Lett 122:167002. https://doi.org/10.1103/PhysRevLett.122.167002
38. Allocca AA, Raines ZM, Curtis JB, Galitski VM (2019) Cavity superconductorpolaritons. Phys Rev B 99:020504. https://doi.org/10.1103/PhysRevB.99.020504
39. Chakraborty A, Piazza F (2021) Long-range photon fluctuations enhance photon-mediated electron pairing and superconductivity. Phys Rev Lett 127:177002. https://doi.org/10.1103/PhysRevLett.127.177002
40. Kozin VK, Thingstad E, Loss D, Klinovaja J (2025) Cavity-enhanced superconductivity via band engineering. Phys Rev B 111. https://doi.org/10.1103/physrevb.111.035410
41. Ashida Y, İmamoğlu A, Faist J, Jaksch D, Cavalleri A, Demler E (2020) Quantum electrodynamic control of matter: cavity-enhanced ferroelectric phase transition. Phys Rev X 10:041027. https://doi.org/10.1103/PhysRevX.10.041027
42. Latini S, Shin D, Sato SA, Schäfer C, De Giovannini U, Hübener H, Rubio A (2021) The ferroelectric photo ground state of SrTiO3: cavity materials engineering. Proc Natl Acad Sci 118:e2105618118. https://doi.org/10.1073/pnas.2105618118
43. Thomas A, Devaux E, Nagarajan K, Rogez G, Seidel M, Richard F, Genet C, Drillon M, Ebbesen TW (2021) Large enhancement of ferromagnetism under a collective strong coupling of YBCO nanoparticles. Nano Lett 21:4365–4370. https://doi.org/10.1021/acs.nanolett.1c00973
44. Jarc G, Mathengattil SY, Montanaro A, Giusti F, Rigoni EM, Sergo R, Fassioli F, Winnerl S, Dal Zilio S, Mihailovic D et al (2023) Cavity-mediated thermal control of metalto-insulator transition in 1T-TaS2. Nature 622: 487–492. https://doi.org/10.1038/s41586-023-06596-2
45. Appugliese F, Enkner J, Paravicini-Bagliani GL, Beck M, Reichl C, Wegscheider W, Scalari G, Ciuti C, Faist J (2022) Breakdown of topological protection by cavity vacuum fields in the integer quantum Hall effect. Science 375:1030–1034. https://doi.org/10.1126/science.abl5818

46. Enkner J, Graziotto L, Boriçi D, Appugliese F, Reichl C, Scalari G, Regnault N, Wegscheider W, Ciuti C, Faist J (2024) Enhanced fractional quantum Hall gaps in a two-dimensional electron gas coupled to a hovering split-ring resonator. arXiv preprint arXiv:2405.18362
47. De Liberato S (2014) Light-matter decoupling in the deep strong coupling regime: the breakdown of the Purcell effect. Phys Rev Lett 112:016401. https://doi.org/10.1103/PhysRevLett.112.016401
48. Ashida Y, Imamoğlu A, Demler E (2021) Cavity quantum electrodynamics at arbitrary light-matter coupling strengths. Phys Rev Lett 126:153603. https://doi.org/10.1103/PhysRevLett.126.153603
49. Mercurio A, Macrì V, Gustin C, Hughes S, Savasta S, Nori F (2022) Regimes of cavity QED under incoherent excitation: from weak to deep strong coupling. Phys Rev Res 4:023048. https://doi.org/10.1103/PhysRevResearch.4.023048
50. Mueller NS, Okamura Y, Vieira BG, Juergensen S, Lange H, Barros EB, Schulz F, Reich S (2020) Deep strong light–matter coupling in plasmonic nanoparticle crystals. Nature 583:780–784. https://doi.org/10.1038/s41586-020-2508-1
51. Di Battista G, Fong KC, Diez-Carlon A, Watanabe K, Taniguchi T, Efetov DK (2024) Ultra-low carrier density superconducting bolometers with single photon sensitivity based on magic-angle twisted bilayer graphene. arXiv preprint arXiv:2403.02049
52. Herzig Sheinfux H, Orsini L, Jung M, Torre I, Ceccanti M, Marconi S, Maniyara R, Barcons Ruiz D, Hötger A, Bertini R et al (2024) High-quality nanocavities through multimodal confinement of hyperbolic polaritons in hexagonal boron nitride. Nat Mater 1–7. https://doi.org/10.1038/s41563-023-01785-w
53. Dirnberger F, Quan J, Bushati R, Diederich GM, Florian M, Klein J, Mosina K, Sofer Z, Xu X, Kamra A et al (2023) Magneto-optics in a van der Waals magnet tuned by self-hybridized polaritons. Nature 620:533–537. https://doi.org/10.1038/s41586-023-06275-2
54. Lin Z, Xiao C, Nguyen D-P, Arwas G, Ciuti C, Yao W (2023) Remote gate control of topological transitions in moiré superlattices via cavity vacuum fields. Proc Natl Acad Sci 120. https://doi.org/10.1073/pnas.2306584120
55. Helmrich F, Khanonkin I, Kroner M, Scalari G, Faist J, Imamoglu A, Nova T (2024) Ultra-strong Terahertz Coupling in a van der Waals Heterostructure. arXiv preprint arXiv:2408.00189
56. Vitalone RA, S Jessen B, Jing R, Rizzo DJ, Xu S, Hsieh V, Cothrine M, Mandrus DG, Wehmeier L, Carr GL et al (2024) Charge transfer plasmonics in bespoke graphene/$\alpha - RuCl_3$ cavities. ACS Nano. https://doi.org/10.1021/acsnano.4c08441
57. Lu X, Stepanov P, Yang W, Xie M, Aamir MA, Das I, Urgell C, Watanabe K, Taniguchi T, Zhang G, Bachtold A, MacDonald AH, Efetov DK (2019) Superconductors, orbital magnets and correlated states in magic-angle bilayer graphene. Nature 574:653–657. https://doi.org/10.1038/s41586-019-1695-0
58. Yu H, Peng Y, Yang Y, Li Z-Y (2019) Plasmon-enhanced light–matter interactions and applications. NPJ Comput Mater 5:45. https://doi.org/10.1038/s41524-019-0184-1

Chapter 8
Self-cavity Coupling Between Graphene and Its Graphite Gate

Abstract This chapter investigates gate-tunable heterostructures composed of two self-cavities: a graphene layer and a graphite gate. A first device is designed so that the graphite cavity resonance lies outside the spectral range, isolating the response of the graphene cavity. Surprisingly, the observed graphene plasmonic modes exhibit a carrier-density dependence that is significantly suppressed compared to that of uncoupled graphene plasmons, prompting an extension of the analytical theory to describe the near-field "cavity conductivity" of coupled self-cavities. The extended theory identifies two distinct coupling mechanisms that mediate hybridization between the self-cavities. Leveraging this insight, a second gate-tunable graphene heterostructure is designed to spectrally resolve ultrastrong coupling between graphene and graphite plasmonic modes. Furthermore, the theory establishes a framework for designing devices that either probe unperturbed "cavity sensing" phenomena or enable intentionally tailored "cavity control" in gate-tunable quantum materials.

Building on the discovery of self-cavity effects in vdW heterostructures at THz frequencies (Chap. 6) and the promising possibilities of cavity engineering (Chap. 7), this chapter explores how plasmonic modes in a graphene self-cavity can be modulated by the presence of a graphite gate. Given their finite dimensions, both graphene and graphite naturally form self-cavities for THz light, making their interaction in typical tens to hundred nanometer thick heterostructures highly probable. As the plasmonic resonances of graphite gates typically fall in the GHz–THz range that corresponds to the same μ eV-meV energy scale as the phenomena they electrically control (see Chap. 1), it could even be possible that graphite self-cavity modes are generally relevant for shaping the low-energy physics of vdW heterostructures.

G. Kipp, *Self-Cavity Electrodynamics of Van der Waals Heterostructures*, Springer Theses, https://doi.org/10.1007/978-3-032-31979-1_8

This chapter is structured as follows: To resolve self-cavity modes of gate-tunable hBN-encapsulated graphene, a first heterostructure (Device GG1) is designed with a thick hBN layer between graphene and graphite such that the lowest frequency self-cavity mode of the graphite layer ($m = 0$ mode) is placed outside the spectral sensitivity range. This should enable the exclusive characterization of the graphene layer's self-cavity modes (Sect. 8.1). However, the detected spectrum of Device GG1 featured unexpectedly distorted carrier-density dependencies of the three detected graphene plasmonic modes. To explain this experimental finding, the analytical theory for heterostructures consisting of a single self-cavity (Sect. 6.4) was extended to capture the conductivity spectrum and current density distributions of a heterostructure containing two self-cavities (Sect. 8.2).

The extended analytical theory shows that hybridized symmetric and antisymmetric modes form due to currents that oscillate in-phase and out-of-phase in the two self-cavities (Sect. 8.2.1). As symmetric and antisymmetric modes are additionally dependent on the screening by a metal in close proximity, similar as typical 2D plasmons, the theory provides in total four dispersion relations that thereby increase the complexity of the boundary equations (Sect. 8.2.2). The method for calculating the cavity conductivity of the heterostructure comprising two self-cavities is detailed in Sect. 8.2.3, and the good agreement of the extended theoretical model with full 3D electromagnetic simulations is demonstrated in Sect. 8.2.4. The theory provides a full understanding of the distorted carrier-density dependencies of the three detected graphene plasmonic modes in Device GG1 (Sect. 8.2.5).

Due to the multimodal coupling nature and two different types of cavity-coupling (current-current hybridization and coupling between screened and unscreened modes) contributing to the overall hybridization, the tunable coupling mechanisms are discussed in detail in Sect. 8.3.

Based on the understanding of the two coupling mechanisms, the coupling between a graphene self-cavity and a graphite self-cavity is maximized in a next device (Device GG2) to reach the ultrastrong coupling regime (Sect. 8.4). Moreover, design principles to construct heterostructures which either minimise or maximise cavity-coupling strengths to either sense or control the THz electrodynamics of collective modes in vdW materials are discussed in Sect. 8.6.

8.1 Spectrally Resolving Gate-Tunable Graphene Plasmonic Modes

To spectrally resolve gate-tunable plasmonic self-cavity modes in graphene, a first device, Device GG1, is designed with the graphite cavity resonance frequency centered around $\sim$1.13 THz, intentionally positioned outside the experimental spectral sensitivity range. This allows for the clear spectral characterization of individual graphene plasmonic modes. The high resonance frequency of the graphite

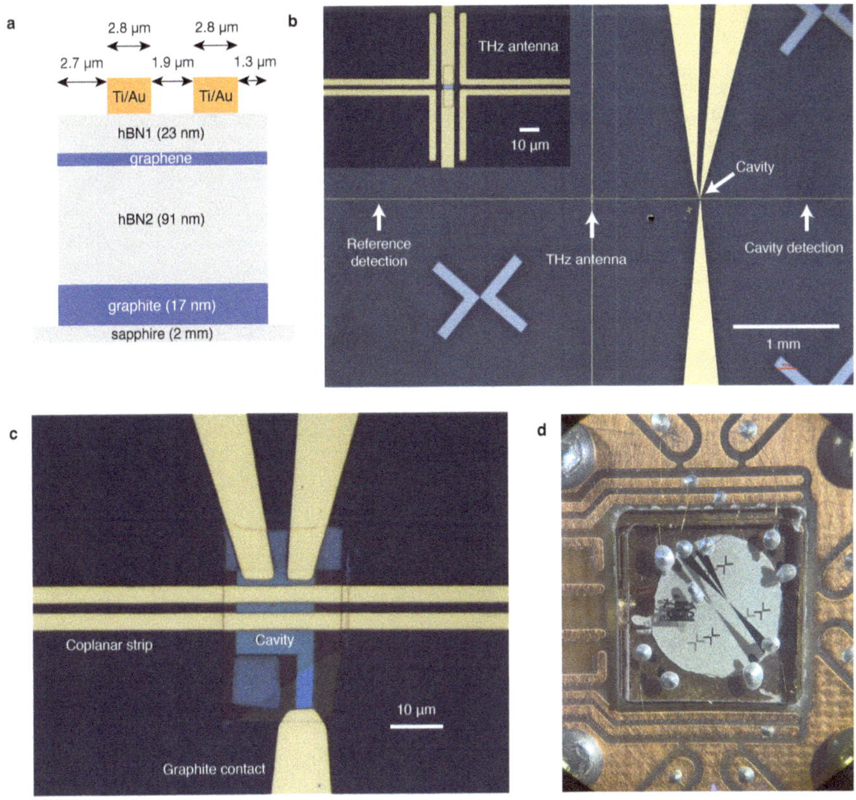

Fig. 8.1 Heterostructure geometry and optical micrographs of Device GG1. (**a**) Cross-section with geometric dimensions of the heterostructure. (**b**) Overview of the circuitry with a heterostructure consisting of both graphene and graphite self-cavities embedded in the coplanar strip on the right. Top left inset shows a magnified picture of the THz antenna. (**c**) Optical micrograph of the gate-tunable vdW heterotructure with edge contacts to the graphite (bottom contact) and to the graphene (other two contacts). The close proximity of the contacts to the gold traces were not observed to play any role in the observed conductivities. (**d**) Device mounted on a sample holder, with wire connections from coplanar strip arms, THz antenna contacts and graphene and graphite contacts to the contacts on the sample holder. (Parts of the figure adapted from [1] under a CC BY 4.0 license)

$m = 0$ mode was achieved by making both the hBN layer between the graphene and graphite layer (d_{hBN}) and the graphite layer itself (d_{gr}) thick (see Fig. 8.1a).[1]

Device GG1 features a precision-cut graphite gate with one edge contact to the gate and two contacts to the rectangular shaped graphene flake (see Fig. 8.1c). However, only one of the two placed graphene contacts was actually contacting the

[1] Device GG1 was built together with Sivasruthi Kesavan [3] and Sara Maria Langner [4] as part of their Master's and Bachelor's theses, using the device fabrication protocol presented in Chap. 3.

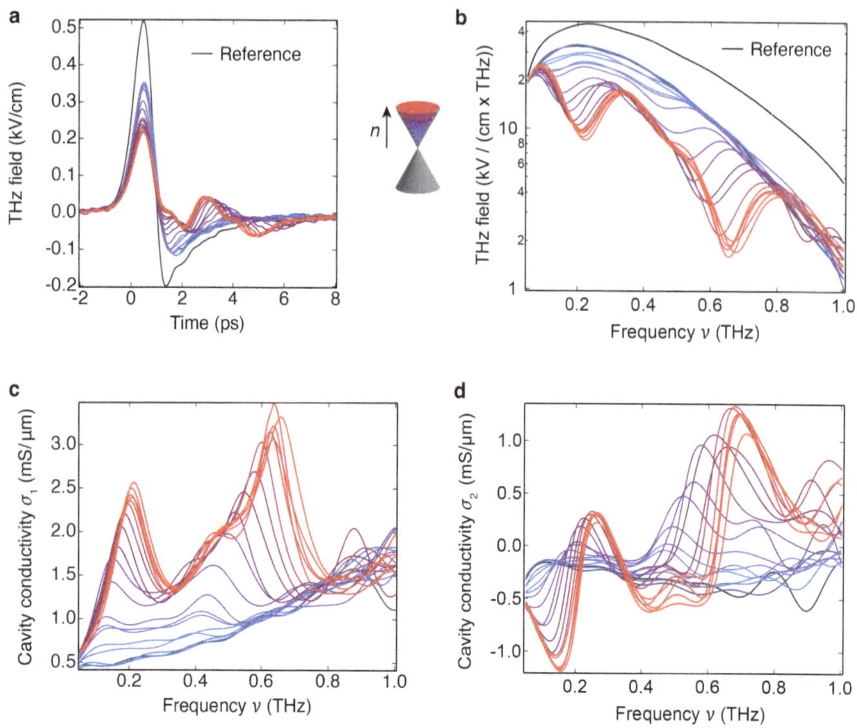

Fig. 8.2 On-chip THz spectroscopy data for Device GG1. (**a**) Experimentally measured THz transients of the reference (black trace) and increasing graphene carrier densities (colored traces). While the dark blue trace was taken at graphene's charge neutrality point, the red trace was measured at $n = 3.5 \times 10^{12}$ cm−2. The electric field strengths were calculated from the measured voltages using Equation 4.3. (**b**) Real part of the Fourier transforms for the reference (black) and for different graphene carrier densities. (**c**) Real and (**d**) imaginary parts of the cavity conductivity obtained from the Fourier transforms shown in (**b**). (Figure adapted from [1] under a CC BY 4.0 license)

graphene, preventing DC transport measurements to be performed on that device, but allowing to perform on-chip THz measurements as described in the following section.

8.1.1 On-chip THz Measurements

On-chip THz measurements were prepared with the methods described in Sect. 4.2 and data was taken at 20 K. As shown in Fig. 8.2a, the measured data show that the transmission decreases as the sample reflectivity increases with increasing carrier density. Furthermore, the time-domain traces on the reference side and on the sample side at the charge neutrality point of graphene do not differ as much as the traces for the bare graphite heterostructure discussed in Fig. 6.2, already indicating that the graphite self-cavity mode, that was designed to be outside the spectral sensitivity range, has a minor contribution to the THz pulse transmission. Time-domain oscillations appear with increasing graphene carrier density, indicating a non-trivial frequency-dependence of the optoelectronic response.

8.1.2 Cavity Conductivity of Graphene Plasmonic Modes

The cavity conductivity of the vdW heterostructure comprised of both graphene and graphite was calculated for each measured back gate voltage by Fourier transforming the time-domain signals to obtain the transmission coefficients (see Fig. 8.2b), which then serve as inputs for the conductivity extraction analysis procedure discussed in Sect. 5.3. Full 3D electromagnetic simulations were performed to extract the exact ratio of a field transmitted through the sample as compared to an empty circuit, similar to the device discussed in Sect. 6.1, to obtain high precision in the conductivity magnitudes. Ambiguities in the ratio of the empty coplanar strip to the sample side only leads to a systematic uncertainty of $\sim$10 % on the amplitude of the data and does not affect the resonance frequency positions. As only one DC contact to the graphene was functional in Device GG1, the charge neutrality point was found by using the conductivity where the spectral weight was minimized.

The graphene carrier density could then be calculated from the applied back gate voltages using the relation $n = \frac{\epsilon_0 \epsilon_{\mathrm{hBN}} (V_{\mathrm{g}} - V_{\mathrm{cnp}})}{d_{\mathrm{hBN2}} e}$. At the lowest carrier density, close to charge neutrality, the cavity conductivity data grows slowly with increasing frequency. This is consistent with the expectation that charge neutral graphene is largely transparent to THz fields, such that only the rising edge of the graphite self-cavity resonance is captured, which peaks outside the experimentally accessible spectral range at around 1.13 THz, as the analytical theory predicts.

As the graphene becomes more conductive, three resonances appear on top of this background, which grow in amplitude and blueshift as the carrier density increases. This can be explained as the formation of confined plasmonic standing waves (self-cavity modes) in the graphene [5–11]. Higher frequency resonances originate from higher order plasmonic self-cavity modes. The labelling $s = 0, 1, 2$ corresponds to the number of nodes in the current density in the graphene layer beneath each coplanar strip, similar to the labeling $m = 0, 1, 2$ of the graphite self-cavity modes.

The cavity conductivity spectra in Fig. 8.2c, d were fit with a function comprised by a sum of four Drude-Lorentz models to extract the linewidth, amplitude and resonance frequency of each mode ($s = 0, 1, 1', 2$) as a function of carrier density. The additional $s = 1'$ mode captured a peak that can be seen as a shoulder in Figs. 8.2c and 8.3a, which occurs due to a slight asymmetry of the heterostructure beneath the coplanar strip (see Fig. 8.1c). The sloped line (see Fig. 8.3a, g, h) captures the conductivity of plasmons, such as the graphite plasmon and higher order graphene plasmonic modes, whose resonance frequencies are centered outside

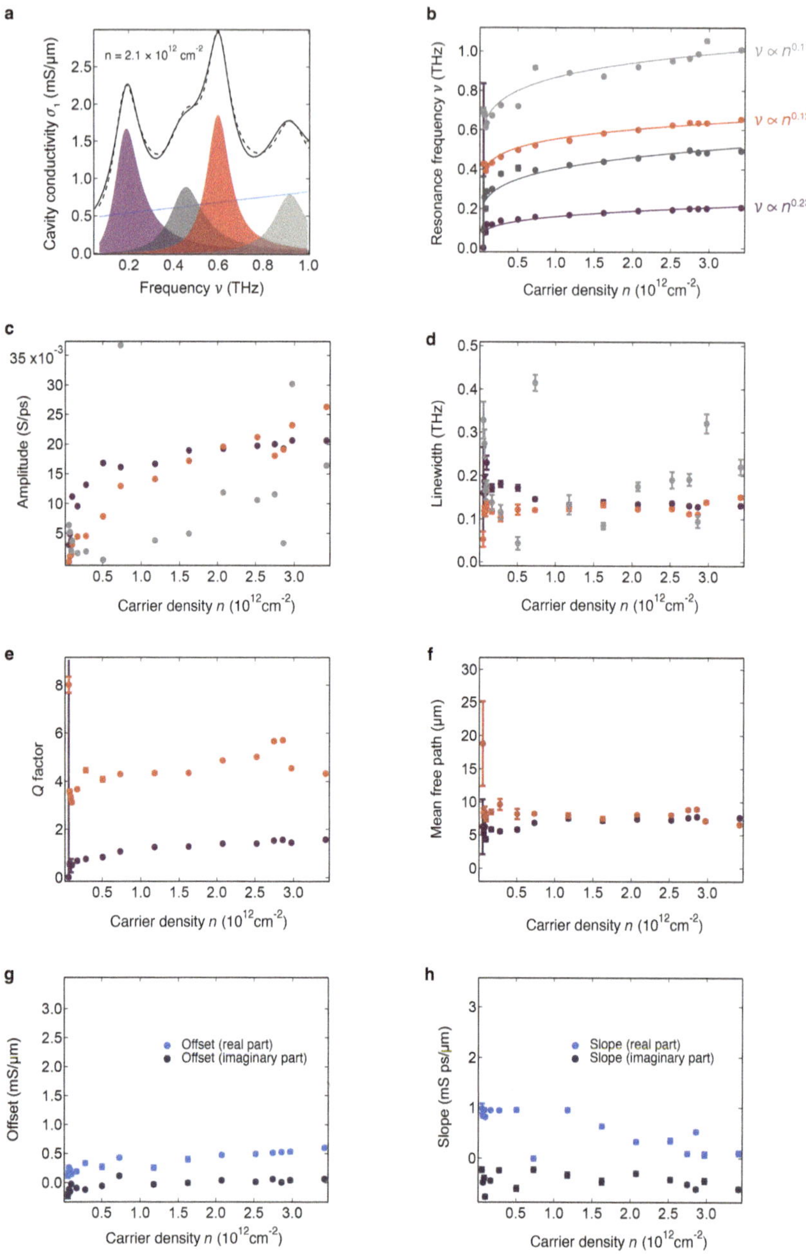

Fig. 8.3 Fit parameters from Device GG1. (**a**) Fit result at a graphene carrier density of $n = 2.1 \times 10^{12}$ cm^{-2}. Real and imaginary parts of the cavity conductivity were fit to a sum of four Drude-Lorentz models and an offset with a slope. The dark grey mode ($s = 1'$ occurs due to device asymmetry (graphene which extends further on one side than the other). (**b**) Resonance frequencies, (**c**) amplitudes, (**d**) linewidths, (**e**) quality factors, (**f**) mean free paths, (**g**) offsets and (**h**) slopes as a function of graphene carrier density. (Figure adapted from [1] under a CC BY 4.0 license)

of the experimental sensitivity range. The low-frequency tail of these modes contributes to a finite, frequency-dependent offset, and thus the real part of the offset and slope were constrained to be positive.

The fits of the multiple self-cavity modes then enable the extraction of various different parameters. For example, the quality factor Q of the self-cavity modes can be extracted by taking the ratio of the resonance frequency, ν, to the linewidth, Γ. Exemplary quality factors for the $s = 0, 2$ modes are shown in Fig. 8.3e.

The mean free path of the graphene carriers can also be calculated from the fit parameters. Assuming $l_{\mathrm{mfp}} = \frac{v_F}{\Gamma}$ with a Fermi velocity of graphene of $v_F = 1 \times 10^6$ m/s [12], an electronic mean free path of $\sim(9.1 \pm 0.2)$ μm was obtained using a representative linewidth value for the $s = 2$ mode at $n = 2.85 \times 10^{12}$ cm−2. As the mean free path matches roughly the width of the device (the coplanar strip width is 7.5 μm, the heterostructure width is $W = 11.5$ μm; see Fig. 8.3), this means that the graphene self-cavity exhibited ballistic transport.[2]

The corresponding quality factor of this mean free path is $Q \sim 5.7 \pm 0.1$, defined as $Q \sim \nu_{s=2}/\Gamma_{s=2}$, where $\nu_{s=2}$ is the resonance frequency and $\Gamma_{s=2}$ is the linewidth. Since the quality factor, like the mean free path, is fundamentally constrained by the vdW heterostructure dimensions, the measured graphene self-cavity is close to the maximal quality achievable for this geometry and frequency range.

Moreover, the fit parameters can further be used to track the resonance frequencies of each self-cavity mode as a function of graphene carrier density. Such an approach can deliver the electronic band dispersion of the sample, if the system is uncoupled and in the absence of electronic correlations [13]. From the fit parameters one finds that the $s = 0$ mode scales as $\nu \propto n^{0.23 \pm 0.03}$, in good agreement with the expectation of a Dirac band structure, where 2D plasmons scale as $\nu \propto n^{0.25}$ (as derived in Sect. 6.4.2). However, the higher order plasmon modes, $s = 1, 2$ scale as $\nu \propto n^{0.12 \pm 0.01}$ and $\nu \propto n^{0.11 \pm 0.01}$ respectively, deviating significantly from the Dirac band structure expectation (see Fig. 8.3b).

Comparable deviating power laws were also observed in an additional device with similar geometry, shown in Appendix A.3 (Fig. A.3).

The anomalously suppressed power-law scalings of the graphene self-cavity plasmon modes are intriguing as they hint at possible coupling to the graphite cavity resonance near 1.13 THz: the closer a graphene self-cavity mode approaches this frequency, the more it deviates from the expected $\nu \propto n^{0.25}$ scaling.

To investigate this hypothesis and pinpoint the origin of the suppressed power laws, the analytical theory was extended to capture the THz electrodynamics of a heterostructure hosting not just one, but two plasmonic self-cavities: A 2D material and an electrostatic graphite gate, as discussed in the following section.

[2] We will see later in Sects. 8.2.1 and 8.4.6 that the graphene self-cavity modes are indeed in the ballistic transport regime, assuming a plasmonic standing wave from sample edge to sample edge, however the modes are slightly broadened due to the hybridization with the oscillating currents in the graphite layer, that has a larger scattering rate than graphene.

8.2 Theoretical Description of Heterostructures with Two Self-cavities

The analytical theory that captures the THz electrodynamics of integrated graphite heterostructures (see Sect. 6.4) was extended to capture the cavity conductivity of a heterostructure that additionally hosts a 2D material, such as graphene. A typical cross-section of such a cavity is shown in Fig. 8.11a. The model assumes that the heterostructure is perfectly symmetric across a central mirror plane and, as in the case of a single layer introduced above, divides the cavity up into five regions (0 to IV), that are either screened by the coplanar strip or unscreened.

When the 2D material such as graphene is metallic, 2D plasmons can be excited, in addition to the plasmons that are excited in the graphite layer. If the graphene were infinitely far away from the graphite, the resonances of these graphene plasmon modes would be determined by the geometry and screening environment of the self-cavity in a similar manner to the method described for the bare graphite self-cavities above.

As the graphene and graphite are brought closer, the charge densities in each layer interact through the Coulomb force and hybrid, symmetric and antisymmetric modes (see Fig. 8.4) are formed [13, 14]. Such plasmon–plasmon interactions were already observed in different former experiments, for example by [15], where individual graphene ribbon modes were split into two hybridized modes, with increasing splitting at decreasing ribbon separations.

As the here discussed heterostructures feature both graphene and graphite plasmons, that are only spatially separated by a few tens of nanometers, it is very likely that hybridized plasmon modes are observed. This chapter shows how currents can be expanded in terms of hybrid, symmetric and antisymmetric modes that form in the graphene and graphite layers (Sect. 8.2.1) and how current densities and potentials are matched in each layer across the inhomogeneous dielectric environment provided by the coplanar strip (Sect. 8.2.2). Section 8.2.3 explains how

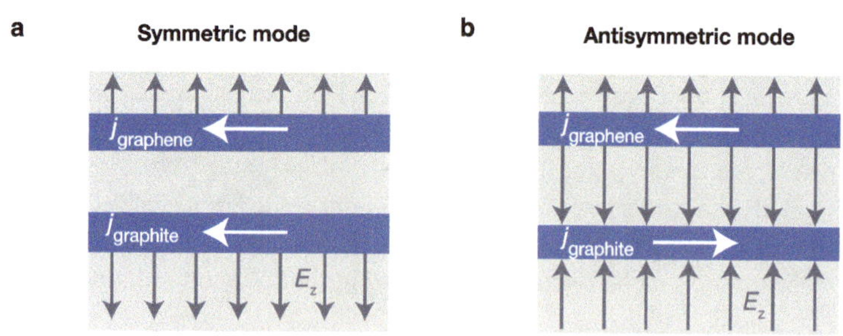

Fig. 8.4 Symmetric and antisymmetric modes. (**a**) Currents in graphene and graphite layers oscillate in-phase (in the same direction) in the case of a symmetric mode and (**b**) out-of-phase (in opposite direction) in the case of an antisymmetric mode. (Figure adapted from [1] under a CC BY 4.0 license)

the cavity conductivity of a heterostructure consisting of two self-cavities can be extracted and Sect. 8.2.5 shows how the extended analytical theory matches well with full 3D electromagnetic simulations. Finally, Device GG1 is calculated in Sect. 8.2.5, revealing the origin of the experimentally detected suppressed power law scalings of the graphene self-cavity modes.

8.2.1 *Symmetric and Antisymmetric Modes*

This section derives the dispersion relations of screened and unscreened symmetric and antisymmetric modes of a graphene—graphite heterostructure from Maxwell's equations, following a similar approach as for the derivations of the simpler screened and unscreened 2D plasmons shown in Sect. 6.4.2.

A visualization of symmetric and anti-symmetric modes is given in Fig. 8.4. Symmetric modes are "unscreened" modes, where the currents in the two layers oscillate in-phase with no electric field between the two hybridizing conductive layers, such that the heterostructure could be treated as just a thicker 2D material. In contrast, the antisymmetric mode is a screened mode, where currents oscillate out-of-phase, and there is a strong electric field between the layers.

8.2.1.1 Screened Symmetric and Antisymmetric Modes

First, the dispersion relations of symmetric and antisymmetric modes of a heterostructure consisting of a graphene layer that is screened by a gold transmission line in a distance d_{hBN1} and that is separated from a graphite layer by a distance of d_{hBN2} are derived (see Fig. 8.5).

The electric and magnetic fields at the boundaries of the metal interface (1), at the position of the graphene (2) and at the graphite (3) can be defined as:

$$E_{\mathrm{x},1}(z,x,t) = t_1 e^{-qz+iqx-i\omega t} + r_1 e^{qz+iqx-i\omega t} \tag{8.1}$$

$$\begin{aligned} B_{\mathrm{y},1}(z,x,t) &= \frac{\epsilon_1}{c^2}\frac{\frac{\partial}{\partial t}e^{-qz+iqx-i\omega t}}{-\frac{\partial}{\partial z}e^{-qz+iqx-i\omega t}} t_1 e^{-qz+iqx-i\omega t} \\ &+ \frac{\epsilon_1}{c^2}\frac{\frac{\partial}{\partial t}e^{qz+iqx-i\omega t}}{-\frac{\partial}{\partial z}e^{qz+iqx-i\omega t}} r_1 e^{qz+iqx-i\omega t} \end{aligned} \tag{8.2}$$

$$E_{\mathrm{x},2}(z,x,t) = t_2 e^{-qz+iqx-i\omega t} + r_2 e^{qz+iqx-i\omega t} \tag{8.3}$$

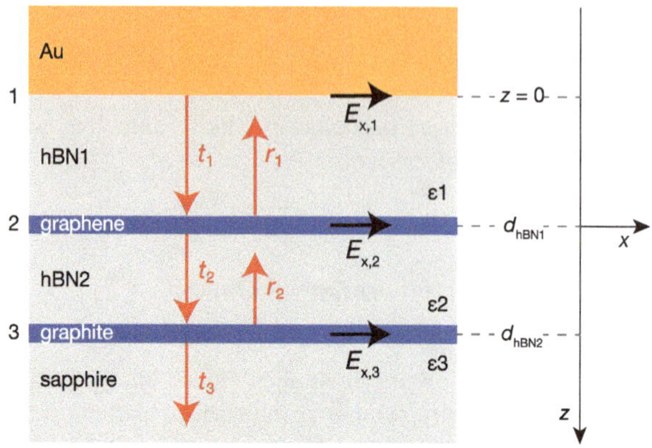

Fig. 8.5 Screened graphene and graphite. A graphene and graphite heterostructure on a sapphire substrate is covered by a gold layer. By considering the reflection and transmission of an electromagnetic wave in the z-direction through the stack, the dispersion relations for symmetric and antisymmetric modes forming in this screened heterostructure can be found. (Figure adapted from [2] under a CC BY 4.0 license)

$$B_{y,2}(z,x,t) = \frac{\epsilon_2}{c^2}\frac{\frac{\partial}{\partial t}e^{-qz+iqx-i\omega t}}{-\frac{\partial}{\partial z}e^{-qz+iqx-i\omega t}}t_2 e^{-qz+iqx-i\omega t} + \frac{\epsilon_2}{c^2}\frac{\frac{\partial}{\partial t}e^{qz+iqx-i\omega t}}{-\frac{\partial}{\partial z}e^{qz+iqx-i\omega t}}r_2 e^{qz+iqx-i\omega t} \tag{8.4}$$

$$E_{x,3}(z,x,t) = t_3 e^{-qz+iqx-i\omega t} \tag{8.5}$$

$$B_{y,3}(z,x,t) = \frac{\epsilon_3}{c^2}\frac{\frac{\partial}{\partial t}e^{-qz+iqx-i\omega t}}{-\frac{\partial}{\partial z}e^{-qz+iqx-i\omega t}}t_3 e^{-qz+iqx-i\omega t} \tag{8.6}$$

Now, five boundary conditions can be defined from the expressions above. First of all, t and x are set to zero as the spatial and temporal components should match across the interface, so they cancel out. The only thing that is different is z.

At the metal interface between hBN1 and the gold metal trace, z is zero:

$$E_{x,1}(0,0,0) \equiv \boxed{0 = t_1 + r_1}. \tag{8.7}$$

For the second equation, the electric fields in x-direction at the position of the graphene can be matched (the electric field is conserved across the interface):

$$E_{x,1}(d_{hBN1},0,0) = E_{x,2}(0,0,0), \tag{8.8}$$

which leads to the following equation:

$$\boxed{r_1 e^{d_{\mathrm{hBN1}} q} + t_1 e^{-d_{\mathrm{hBN1}} q} \equiv r_2 + t_2}. \tag{8.9}$$

The difference in magnetic field is given by the current in 2D and the current is given by $j_{\mathrm{2D}} = \sigma_{\mathrm{2D}}(\omega) E$. Assuming that the 2D graphene conductivity has a Drude form

$$\sigma_{\mathrm{2D}}(\omega) = \frac{e^2 E_f}{\hbar^2 \pi \epsilon_0} \frac{1}{-i\omega + \Gamma}, \tag{8.10}$$

the difference between the magnetic fields (and the graphene current) is given by:

$$B_{\mathrm{y,2}}(0,0,0) - B_{\mathrm{y,1}}(d_{\mathrm{hBN1}},0,0) \equiv -\frac{e^2 E_f}{\hbar^2 \pi \epsilon_0} \frac{E_{\mathrm{x,1}}(d_{\mathrm{hBN1}},0,0)}{(-i\omega + \Gamma_1)c^2}, \tag{8.11}$$

with Γ_1 being the decay rate of the graphene plasmon. The equation above can be reformulated to the following term:

$$\boxed{\frac{i\omega}{c^2 q}\left(-e^{d_{\mathrm{hBN1}} q} \epsilon_1 r_1 + \epsilon_2 r_2 + e^{-d_{\mathrm{hBN1}} q} \epsilon_1 t_1 - \epsilon_2 t_2\right) \equiv -\frac{e^2 E_f}{\hbar^2 \pi \epsilon_0} \frac{r_1 e^{d_{\mathrm{hBN1}} q} + t_1 e^{-d_{\mathrm{hBN1}} q}}{c^2(-i\omega + \Gamma_1)}}. \tag{8.12}$$

At the position of the graphite, the electric fields can be matched again:

$$E_{\mathrm{x,2}}(d_{\mathrm{hBN2}},0,0) \equiv E_{\mathrm{x,3}}(0,0,0), \tag{8.13}$$

leading to this equation:

$$\boxed{r_2 e^{d_{\mathrm{hBN2}} q} + t_2 e^{-d_{\mathrm{hBN2}} q} \equiv t_3}. \tag{8.14}$$

Next, the differences of the magnetic fields around the graphite layer can also be rewritten in order to express this difference as a function of the 2D current in the graphite:

$$B_{\mathrm{y,3}}(0,0,0) - B_{\mathrm{y,2}}(d_{\mathrm{hBN2}},0,0) \equiv -\frac{\omega_{\mathrm{3D}}^2 d_{\mathrm{gr}} E_{\mathrm{x,1}}(d_{\mathrm{hBN1}},0,0)}{(-i\omega + \Gamma_1)c^2}. \tag{8.15}$$

This equation can be reformulated to:

$$\boxed{\frac{i\omega}{c^2 q}\left(e^{d_{\mathrm{hBN2}} q} \epsilon_2 r_2 + e^{-d_{\mathrm{hBN2}} q} \epsilon_2 t_2 - \epsilon_3 t_3\right) \equiv -\frac{d_{\mathrm{gr}} t_3 \omega_{\mathrm{3D}}^2}{(-i\omega + \Gamma_2)c^2}}, \tag{8.16}$$

with Γ_2 being the decay rate of the graphite plasmon.

By solving the five boxed equations above, expanding the numerator of the rather complex solution for small hBN thicknesses ($d_{\mathrm{hBN1}} \rightarrow 0$, $d_{\mathrm{hBN2}} \rightarrow 0$), treating the case of no dissipation ($\Gamma_1 = 0$, $\Gamma_2 = 0$) and only selecting the dominant terms, two solutions for the plasmon wave vector q are found:

$$q_{1,\mathrm{sc}} = \sqrt{\frac{\left(\epsilon\omega^2(d_{\mathrm{hBN1}}(A_{\mathrm{gr}} + \omega_{3\mathrm{D}}^2 d_{\mathrm{gr}}) + d_{\mathrm{hBN2}}\omega_{3\mathrm{D}} d_{\mathrm{gr}}) - \epsilon\omega^2\sqrt{-4A_{\mathrm{gr}} d_{\mathrm{gr}}\omega_{3\mathrm{D}} d_{\mathrm{hBN1}} d_{\mathrm{hBN2}} + (-A_{\mathrm{gr}}(d_{\mathrm{hBN1}} + d_{\mathrm{hBN2}}))^2}\right)}{2A_{\mathrm{gr}} d_{\mathrm{gr}} d_{\mathrm{hBN1}} d_{\mathrm{hBN2}}\omega_{3\mathrm{D}}^2}}, \tag{8.17}$$

$$q_{2,\mathrm{sc}} = \sqrt{\frac{\left(\epsilon\omega^2(d_{\mathrm{hBN1}}(A_{\mathrm{gr}} + \omega_{3\mathrm{D}}^2 d_{\mathrm{gr}}) + d_{\mathrm{hBN2}}\omega_{3\mathrm{D}} d_{\mathrm{gr}}) + \epsilon\omega^2\sqrt{-4A_{\mathrm{gr}} d_{\mathrm{gr}}\omega_{3\mathrm{D}} d_{\mathrm{hBN1}} d_{\mathrm{hBN2}} + (-A_{\mathrm{gr}}(d_{\mathrm{hBN1}} + d_{\mathrm{hBN2}}))^2}\right)}{2A_{\mathrm{gr}} d_{\mathrm{gr}} d_{\mathrm{hBN1}} d_{\mathrm{hBN2}}\omega_{3\mathrm{D}}^2}}, \tag{8.18}$$

where A_{gr} is:

$$A_{\mathrm{gr}} = \frac{e^2 v_F \sqrt{n}}{\epsilon_0 \hbar \sqrt{\pi}}, \tag{8.19}$$

with the charge of an electron e, the Fermi velocity of graphene v_F, the graphene carrier density n, the vacuum permittivity ϵ_0 and the reduced Planck constant $\hbar$.

To achieve maximum precision for the plasmon dispersion relations of screened symmetric and antisymmetric modes, Newton's method is used again, as described in Sect. 6.4.3 for heterostructures hosting only a single self-cavity.

Equations 8.17 and 8.18 give the plasmon wave vectors as a function of frequency as well as material parameters and dimensions, which are required in the boundary equations discussed in Sect. 8.2.2 below. However, typically, dispersion relations are defined such that the plasmon wave vector is the input and the output is some energy or frequency ($E(q)$ or $\omega(q)$). Thus, it would be helpful to solve the five boxed equations in terms of ω or $\nu = \omega/(2\pi)$ instead of q (or transforming equations 8.17 and 8.18 in terms of ω). The two solutions are:

$$\nu_{1,\mathrm{sc}}(q_{1,\mathrm{sc}}) = \frac{\sqrt{\frac{2q_{1,\mathrm{sc}}^2 A_{\mathrm{gr}} d_{\mathrm{gr}} d_{\mathrm{hBN1}} d_{\mathrm{hBN2}}\omega_{3\mathrm{D}}^2}{\left(\epsilon(d_{\mathrm{hBN1}}(A_{\mathrm{gr}} + \omega_{3\mathrm{D}}^2 d_{\mathrm{gr}}) + d_{\mathrm{hBN2}}\omega_{3\mathrm{D}} d_{\mathrm{gr}}) - \epsilon\sqrt{-4A_{\mathrm{gr}} d_{\mathrm{gr}}\omega_{3\mathrm{D}} d_{\mathrm{hBN1}} d_{\mathrm{hBN2}} + (-A_{\mathrm{gr}}(d_{\mathrm{hBN1}} + d_{\mathrm{hBN2}}))^2}\right)}}}{2\pi}, \tag{8.20}$$

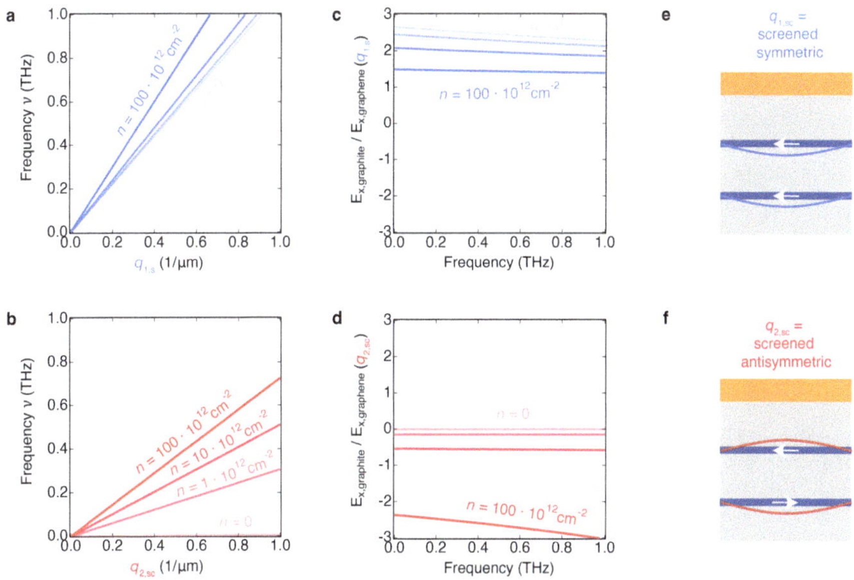

Fig. 8.6 Identifying the mode character of $\nu_{1,\mathrm{sc}}$ and $\nu_{2,\mathrm{sc}}$ given in equations 8.20 and 8.21. (**a**) $\nu_{1,\mathrm{sc}}(q_{1,\mathrm{sc}})$ and (**b**) $\nu_{2,\mathrm{sc}}(q_{2,\mathrm{sc}})$ for four different graphene carrier densities. For $n = 0$, $\nu_{2,\mathrm{sc}}(q_{2,\mathrm{sc}})$ vanishes in contrast to $\nu_{1,\mathrm{sc}}(q_{1,\mathrm{sc}})$, indicating that $\nu_{2,\mathrm{sc}}(q_{2,\mathrm{sc}})$ has graphene-like character. (**c**) Fraction of the electric field in x-direction in the graphite layer and the electric field in x-direction in the graphene layer for $q_{1,\mathrm{sc}}$ and (**d**) for $q_{2,\mathrm{sc}}$. The fractions for $q_{1,\mathrm{sc}}$ only give positive results indicating that the electric fields and thus the currents flow in-phase (in the same direction) in graphene and graphite layer as typical for (**e**) symmetric modes. In contrast, the fractions for $q_{2,\mathrm{sc}}$ (**d**) only give negative results indicating that the electric fields and thus the currents flow out-of-phase (in the opposite direction) in graphene and graphite layer as typical for (**f**) antisymmetric modes

$$\nu_{2,\mathrm{sc}}(q_{2,\mathrm{sc}}) = \frac{\sqrt{\dfrac{2q_{2,\mathrm{sc}}^2 A_{\mathrm{gr}} d_{\mathrm{gr}} d_{\mathrm{hBN1}} d_{\mathrm{hBN2}} \omega_{\mathrm{3D}}^2}{\left(\begin{array}{l}\epsilon(d_{\mathrm{hBN1}}(A_{\mathrm{gr}} + \omega_{\mathrm{3D}}^2 d_{\mathrm{gr}}) + d_{\mathrm{hBN2}}\omega_{\mathrm{3D}} d_{\mathrm{gr}}) \\ \quad + \epsilon\sqrt{-4A_{\mathrm{gr}} d_{\mathrm{gr}} \omega_{\mathrm{3D}} d_{\mathrm{hBN1}} d_{\mathrm{hBN2}} + (-A_{\mathrm{gr}}(d_{\mathrm{hBN1}} + d_{\mathrm{hBN2}}))^2}\end{array}\right)}}}{2\pi}. \tag{8.21}$$

The dispersions $\nu_{1,\mathrm{sc}}(q_{1,\mathrm{sc}})$ and $\nu_{2,\mathrm{sc}}(q_{2,\mathrm{sc}})$ are plotted for four different graphene carrier densities in Fig. 8.6a, b. As shown in the figure, the resonance frequencies are increasing with increasing carrier densities at a fixed q and overall, the solutions for $q_{1,\mathrm{sc}}$ are energetically higher than for $q_{2,\mathrm{sc}}$. Please note that the dispersion relations are drawn as continuous lines, corresponding to a continuum of 2D plasmon modes that could be sustained in a heterostructure of infinite width. However, due to the finite size of the self-cavities, only discrete modes with specific momenta can be excited (such as modes comparable to quarter, three-quarter, and five-quarter wavelength-like modes as shown in Sect. 6.3).

It would be helpful to know if the solutions $q_{1,\mathrm{sc}}(\nu_{1,\mathrm{sc}})$ and $q_{2,\mathrm{sc}}(\nu_{2,\mathrm{sc}})$ correspond to the excitation of symmetric modes, where the current in the graphene and graphite layers oscillate in-phase, or antisymmetric modes, where the current oscillates out-of-phase. This can be found out by comparing the in-plane electric fields in each of the two layers for each of the two solutions $q_{1,\mathrm{sc}}$ and $q_{2,\mathrm{sc}}$. The ratio of in-plane electric fields should be positive for symmetric modes (as currents oscillate in-phase) and negative for antisymmetric modes (as currents oscillate out-of-phase).

Solving the boundary equations and assuming no scattering (Γ=0) reveals that the in-plane electric field in the graphite layer is given by

$$E_{\mathrm{x,graphite}}(q,\omega) = \frac{2(\epsilon_1\omega^2\cosh(d_{\mathrm{hBN1}}q) - A_{\mathrm{gr}}q\sinh(d_{\mathrm{hBN1}}q))}{(\epsilon_3\omega^2 - d_{\mathrm{gr}}q\omega_{\mathrm{3D}}^2)\cosh(d_{\mathrm{hBN2}}q) + \epsilon_2\omega^2\sinh(d_{\mathrm{hBN2}}q)}, \tag{8.22}$$

and the in-plane electric field in the graphene layer is given by

$$E_{\mathrm{x,graphene}}(q,\omega) = \frac{\begin{array}{c}2(\epsilon_1\omega^2\cosh(d_{\mathrm{hBN1}}q) - A_{\mathrm{gr}}q\sinh(d_{\mathrm{hBN1}}q))(\epsilon_2\omega^2\cosh(d_{\mathrm{hBN2}}q)\\ + (\epsilon_3\omega^2 - d_{\mathrm{gr}}q\omega_{\mathrm{3D}}^2)\sinh(d_{\mathrm{hBN2}}q))\end{array}}{\epsilon_2\omega^2((\epsilon_3\omega^2 - d_{\mathrm{gr}}q\omega_{\mathrm{3D}}^2)\cosh(d_{\mathrm{hBN2}}q) + \epsilon_2\omega^2\sinh(d_{\mathrm{hBN2}}q))}. \tag{8.23}$$

Next, the following ratio can be formed

$$\frac{E_{\mathrm{x,sc,graphite}}}{E_{\mathrm{x,sc,graphene}}}(q) = \frac{\epsilon_2\omega^2}{\epsilon_2\omega^2\cosh(d_{\mathrm{hBN2}}q) + (\epsilon_3\omega^2 - d_{\mathrm{gr}}q\omega_{\mathrm{3D}}^2)\sinh(d_{\mathrm{hBN2}}q)}, \tag{8.24}$$

where $q_{1,\mathrm{sc}}$ and $q_{2,\mathrm{sc}}$ can be inserted for q. $E_{\mathrm{x,graphite}}/E_{\mathrm{x,graphene}}(q,\omega)$ is shown in Fig. 8.6 for a typical sample with $d_{\mathrm{gr}} = 8$ nm, $d_{\mathrm{hBN1}} = 15$ nm and $d_{\mathrm{hBN2}} = 25$ nm, $\nu_{\mathrm{3D}} = 118$ THz and assuming $\Gamma_1 = 0$ and $\Gamma_2 = 0$ for a selection of different carrier density as a function of $\nu = \omega/2\pi$.

As shown in Fig. 8.6, the ratio $\frac{E_{\mathrm{x,sc,graphite}}}{E_{\mathrm{x,sc,graphene}}}$ is always positive for $q_{1,\mathrm{sc}}$, meaning that the electric fields (and thereby the currents) in the graphene and in the graphite layer are pointing in the same direction, and are always negative (or zero for $n = 0$) for $q_{2,\mathrm{sc}}$, and that the electric fields (and thereby the currents) are pointing in opposite directions for this solution. Therefore, the solution for $q_{1,\mathrm{sc}}$ corresponds to the excitation of a symmetric mode and the solution for $q_{2,\mathrm{sc}}$ corresponds to the excitation of an antisymmetric mode. This also matches the finding shown in Fig. 8.6a, that the symmetric mode has larger resonance frequencies than its antisymmetric counterpart, which is due to the larger Coulomb interactions for symmetric, in-phase modes.

Section 8.1 talked about graphene and graphite self-cavity modes. The question arises, which of the symmetric and antisymmetric modes, that both contain material parameters of the graphene and the graphite layer, behaves more like a graphene or more like a graphite plasmon. This can be answered by investigating the transformation of equations $\nu_{1,\mathrm{sc}}(q_{1,\mathrm{sc}})$ and $\nu_{2,\mathrm{sc}}(q_{2,\mathrm{sc}})$ in the limit of very small graphene carrier densities:

$$\lim_{n\to 0} \nu_{1,\mathrm{sc}} = \sqrt{\frac{q_{1,\mathrm{sc}}^2 \nu_{3D}^2 d_{\mathrm{gr}}(d_{\mathrm{hBN1}} + d_{\mathrm{hBN2}})}{\epsilon_{\mathrm{hBN}}}}, \tag{8.25}$$

and

$$\lim_{n\to 0} \nu_{2,\mathrm{sc}} = \sqrt{\frac{q_{2,\mathrm{sc}}^2 A_{\mathrm{gr}}}{\epsilon_{\mathrm{hBN}} \left(\frac{1}{d_{\mathrm{hBN1}}} + \frac{1}{d_{\mathrm{hBN2}}}\right)}}, \tag{8.26}$$

with $A_{\mathrm{gr}} = \frac{e^2 v_F \sqrt{n}}{\epsilon_0 \hbar \sqrt{\pi}}$, where $\sqrt{n}$ is the carrier density of graphene, e is the electron charge and v_F the Fermi velocity in graphene. When comparing Eqs. 8.25 and 8.26 with Eqs. 6.47 and 6.44, one finds that in the limit of small graphene carrier densities, the solution for the screened symmetric mode transforms into the dispersion for a graphite plasmon that is screened by the metal in distance $d_{\mathrm{hBN1}} + d_{\mathrm{hBN2}}$ and the solution for the antisymmetric mode transforms into the dispersion relation for a double screened graphene plasmon (screened from the metal above and the graphite below, see Sect. A.4 in the Appendix). Furthermore, the electric field ratio shown in Fig. 8.6 is maximum for the symmetric mode and zero for the antisymmetric mode at zero graphene carrier density. Thus, the dispersion of the symmetric mode is predominantly determined by the graphite ("graphite-like"), whereas the antisymmetric mode is largely "graphene-like".

To understand better, in which limits the symmetric mode indeed behaves mostly as a "graphite-like" mode and the antisymmetric mode as a "graphene-like" mode, the resonance frequencies of the screened symmetric and antisymmetric modes are plotted as a function of graphene carrier densities for different heterostructures with varied graphite thicknesses $d_{\mathrm{gr}} = 3, 8, 30$ nm ($d_{\mathrm{hBN1}} = 15$ nm and $d_{\mathrm{hBN2}} = 25$ nm, $\nu_{3D} = 118$ THz) in Fig. 8.7.

As shown in Fig. 8.7, at small carrier densities, the symmetric and antisymmetric modes match the unhybridized graphite and graphene modes, respectively, for all graphite thicknesses. However, for larger carrier densities, the discrepancy between antisymmetric and symmetric and unhybridized graphene and screened graphite modes increases. The effect of hybridization is not so pronounced in the case of a thick graphite flake (see Fig. 8.7c), but the thinner the graphite, the more pronounced the difference between the hybrid modes and the uncoupled graphene and graphite modes. This is due to the reduction in 2D conductivity for thinner graphite, and has two consequences:

1. A homogeneously screened heterostructure with a thick graphite flake as discussed in Sect. 8.1 hosts a symmetric mode that is weakly dependent on the graphene carrier density n, and an antisymmetric mode, that behaves like an uncoupled graphene plasmonic mode. Thus, calling the symmetric mode the

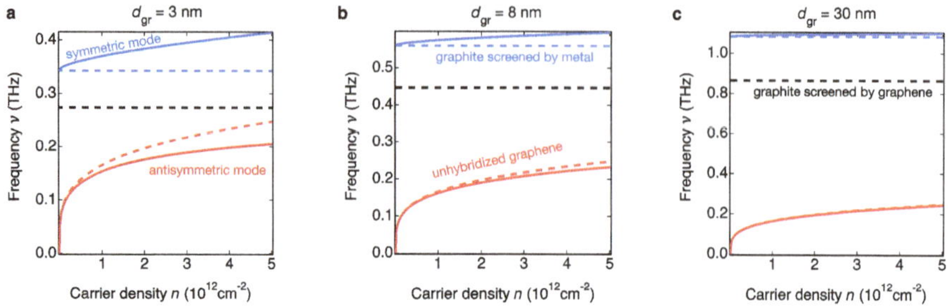

Fig. 8.7 Resonance frequencies of screened symmetric and antisymmetric modes for different graphite thicknesses. (**a**) For a thin graphite layer of 3 nm thickness, symmetric (solid blue line) and antisymmetric (solid red line) modes differ from a pure graphite mode that would exist in the absence of graphene (dashed blue line). Additionally, the power law of the antisymmetric mode appears suppressed when compared to a pure graphene mode, that is just screened by the graphite layer and that scales with $\nu \propto n^{0.25}$ (red dashed line). (**b, c**) With increasing graphite layer thickness, the difference between (anti)symmetric modes and pure graphene/graphite plasmons is decreasing

"graphite mode" and the antisymmetric mode the "graphene mode" is still a valid simplification.

2. If the conductivity of the graphite layer is comparable to or lower than that of the graphene layer, the hybridized symmetric and antisymmetric plasmon modes can deviate significantly from their unhybridized counterparts, potentially even approaching the deep-strong coupling regime (see Sect. 8.3.1).

While hybridization between screened graphene and graphite self-cavity modes can, in principle, be suppressed by using a sufficiently thick graphite layer, Device GG1 displayed a suppression of the graphene self-cavity resonances despite a graphite thickness of 17 nm. This indicates that the suppression in this case stems from the inhomogeneous dielectric environment instead. Specifically, the heterostructure is only partially and unevenly screened by the coplanar strip, leaving other regions unscreened, similar to the bare graphite heterostructure case discussed in Sect. 6.4. To account for this, let's next derive the dispersion relations for the unscreened symmetric and antisymmetric modes in the following section.

8.2.1.2 Unscreened Symmetric and Antisymmetric Modes

As a next step, the boundary equations for a graphene layer that is not screened by a gold transmission line but that is separated from a graphite layer by a distance d_{hBN2} have to be derived (see Fig. 8.8). Solving these boundary equations will give the plasmon wave vectors q of the hybridizing graphene and graphite layers for the unscreened case.

The boundary equations can be derived very similarly to the screened case as done in the previous chapter. The difference in the unscreened case is that there is

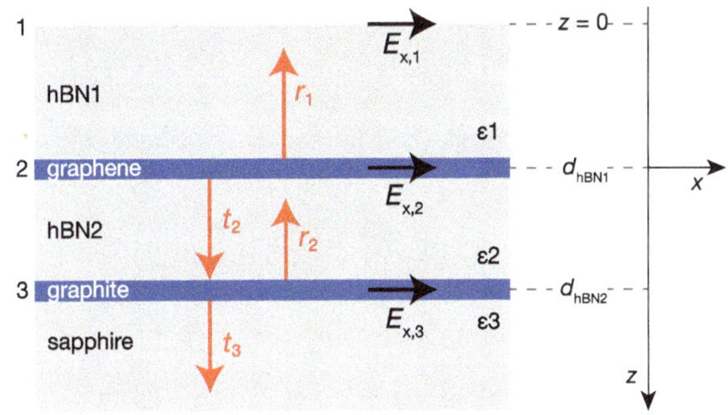

Fig. 8.8 Unscreened graphene and graphite. A graphene and graphite heterostructure on a sapphire substrate that is not covered by a gold layer. By considering the reflection and transmission of an electromagnetic wave in the z-direction through the stack, the dispersion relations for symmetric and antisymmetric modes forming in this unscreened heterostructure can be found

no reflection r_1 from the metal interface of the gold transmission lines and therefore, the result will be independent of the thickness of the top hBN layer (d_{hBN1}).

The electric and magnetic fields for the unscreened case above the graphene (1), at the position of the graphene (2) and at the position of the graphite (3) are given in the following:

$$E_{\text{x},1}(z, x, t) = t_1 e^{-qz+iqx-i\omega t} \tag{8.27}$$

$$B_{\text{y},1}(z, x, t) = \frac{\epsilon_1}{c^2} \frac{\frac{\partial}{\partial t} e^{-qz+iqx-i\omega t}}{-\frac{\partial}{\partial z} e^{-qz+iqx-i\omega t}} t_1 e^{-qz+iqx-i\omega t} \tag{8.28}$$

$$E_{\text{x},2}(z, x, t) = t_2 e^{-qz+iqx-i\omega t} + r_2 e^{qz+iqx-i\omega t} \tag{8.29}$$

$$\begin{aligned} B_{\text{y},2}(z, x, t) = &\frac{\epsilon_2}{c^2} \frac{\frac{\partial}{\partial t} e^{-qz+iqx-i\omega t}}{-\frac{\partial}{\partial z} e^{-qz+iqx-i\omega t}} t_2 e^{-qz+iqx-i\omega t} \\ &+ \frac{\epsilon_2}{c^2} \frac{\frac{\partial}{\partial t} e^{qz+iqx-i\omega t}}{-\frac{\partial}{\partial z} e^{qz+iqx-i\omega t}} r_2 e^{qz+iqx-i\omega t} \end{aligned} \tag{8.30}$$

$$E_{\text{x},3}(z, x, t) = t_3 e^{-qz+iqx-i\omega t} \tag{8.31}$$

$$B_{\text{y},3}(z, x, t) = \frac{\epsilon_3}{c^2} \frac{\frac{\partial}{\partial t} e^{-qz+iqx-i\omega t}}{-\frac{\partial}{\partial z} e^{-qz+iqx-i\omega t}} t_3 e^{-qz+iqx-i\omega t} \tag{8.32}$$

From here, four boundary conditions can be derived from the expressions above. t and x are again set to zero as the spatial and temporal components should match across the interface, so they cancel out in the boundary conditions. The only thing that is different is z. (t and x are only kept in the expressions above as derivatives have to be taken first.)

First of all, the in-plane electric field above the graphene and at the graphene interface should be equal:

$$E_{\mathrm{x},1}(0,0,0) = E_{\mathrm{x},2}(0,0,0), \tag{8.33}$$

which then gives the following equation:

$$\boxed{t_1 = r_2 + t_2}. \tag{8.34}$$

Next, the difference between the magnetic fields is given by:

$$B_{\mathrm{y},2}(0,0,0) - B_{\mathrm{y},1}(0,0,0) \equiv -A_{\mathrm{gr}} \frac{E_{\mathrm{x},1}(0,0,0)}{(-i\omega + \Gamma_1)c^2}, \tag{8.35}$$

with Γ_1 being the decay rate of the graphene. The equation above can be reformulated to:

$$\boxed{\frac{i\omega}{c^2 q}(\epsilon_2 r_2 + \epsilon_1 t_1 - \epsilon_2 t_2) \equiv -A_{\mathrm{gr}} \frac{t_1}{c^2(-i\omega + \Gamma_1)}}. \tag{8.36}$$

At the position of the graphite, the electric fields can be matched again:

$$E_{\mathrm{x},2}(d_{\mathrm{hBN2}},0,0) = E_{\mathrm{x},3}(0,0,0), \tag{8.37}$$

giving this as a third equation:

$$\boxed{r_2 e^{d_{\mathrm{hBN2}} q} + t_2 e^{-d_{\mathrm{hBN2}} q} \equiv t_3}. \tag{8.38}$$

We can now also write the differences of the magnetic fields around the graphite in order to express this difference as a function of the 2D current in the graphite:

$$B_{\mathrm{y},3}(0,0,0) - B_{\mathrm{y},2}(d_{\mathrm{hBN2}},0,0) \equiv -\frac{\omega_{3\mathrm{D}}^2 d_{\mathrm{gr}} E_{\mathrm{x},1}(d_{\mathrm{hBN1}},0,0)}{(-i\omega + \Gamma_1)c^2}. \tag{8.39}$$

This equation can be reformulated to:

$$\boxed{\frac{i\omega}{c^2 q}\left(e^{d_{\mathrm{hBN2}} q}\epsilon_2 r_2 + e^{-d_{\mathrm{hBN2}} q}\epsilon_2 t_2 - \epsilon_3 t_3\right) \equiv -\frac{d_{\mathrm{gr}} t_3 \omega_{3\mathrm{D}}^2}{(-i\omega + \Gamma_2)c^2}}, \tag{8.40}$$

with Γ_2 being the decay rate of the graphite.

By solving the four boxed equations above, series expanding the rather complex solution for a small bottom hBN thickness ($d_{\mathrm{hBN2}} \rightarrow 0$), treating the case of no dissipation ($\Gamma_1 = 0, \Gamma_2 = 0$) and only selecting the most dominant terms, two solutions for the plasmon wave vector q can be found:

$$\boxed{q_{1,\mathrm{unscreened}} = \frac{\omega^2(\epsilon_1 + \epsilon_2)}{\omega_{3\mathrm{D}}^2 d_{\mathrm{gr}} + A_{\mathrm{gr}}}}, \tag{8.41}$$

$$\boxed{q_{2,\mathrm{unscreened}} = \sqrt{\frac{\epsilon_2\omega^2(A_{\mathrm{gr}} + d_{\mathrm{gr}}\omega_{3\mathrm{D}}^2)}{A_{\mathrm{gr}} d_{\mathrm{gr}} \omega_{3\mathrm{D}}^2 d_{\mathrm{hBN2}}}}}. \tag{8.42}$$

These two equations can be transformed into

$$\nu_{1,\mathrm{unscreened}}(q_{1,\mathrm{unscreened}}) = \sqrt{\frac{q_{1,\mathrm{unscreened}}(\nu_{3\mathrm{D}}^2 d_{\mathrm{gr}} + A_{\mathrm{gr}}/(4\pi^2))}{\epsilon_1 + \epsilon_2}}, \tag{8.43}$$

$$\nu_{2,\mathrm{unscreened}}(q_{2,\mathrm{unscreened}}) = q_{2,\mathrm{unscreened}} \sqrt{\frac{A_{\mathrm{gr}} d_{\mathrm{gr}} 4\pi^2 \nu_{3\mathrm{D}}^2 d_{\mathrm{hBN2}}}{2\pi\epsilon_2(A_{\mathrm{gr}} + d_{\mathrm{gr}} 4\pi^2 \nu_{3\mathrm{D}}^2)}}. \tag{8.44}$$

To also achieve maximum precision for the plasmon dispersion relations of unscreened symmetric and antisymmetric modes, Newton's method is applied, as described in Sect. 6.4.3 for heterostructures hosting only a single self-cavity.

Equations 8.43 and 8.44 can be plotted for typical values as a function of q, as shown in Fig. 8.9a. While the first solutions is $\nu \propto \sqrt{q}$ and yields resonance frequency solutions that are far exceeding the spectral sensitivity range (>1 THz) for typical q-values, the second solution is linear in q and shows resonances within the spectral sensitivity range.

Due to its square root dispersion relation and its high energy solutions, one can expect that $\nu_{1,\mathrm{unscreened}}(q_{1,\mathrm{unscreened}})$ is the solution for the symmetric mode, as it behaves as an unscreened mode with no electric field between the two hybridizing conductive layers, and $\nu_{2,\mathrm{unscreened}}(q_{2,\mathrm{unscreened}})$ is the solution for the lower-energy antisymmetric mode.

A verification of the mode character of the two solutions is given by plotting the fraction of the in-plane electric field in the graphite layer and the in-plane electric field in the graphene layer, as done in case for the screened heterostructure above: If the fraction gives positive results for a given dispersion relation, the character of the mode is symmetric, if the results are negative, the mode is an antisymmetric mode.

In the case of unscreened graphene and graphite layers, the electric field in x-direction in the graphite divided by the electric field in x-direction in the graphene is identical to the electric field fraction in the screened case as the fraction is independent of the screening:

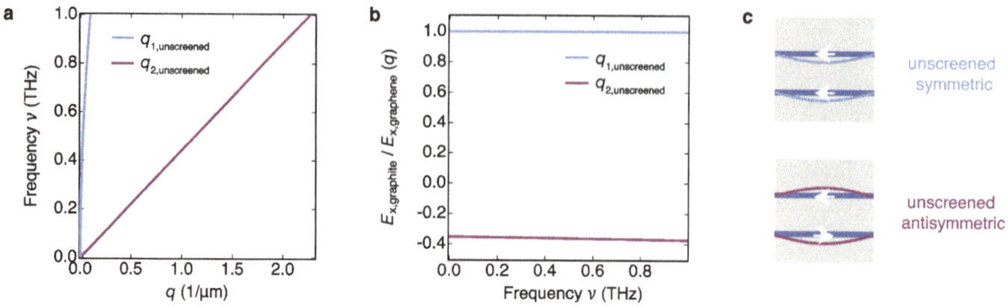

Fig. 8.9 Dispersion relations for unscreened symmetric and antisymmetric modes in a heterostructure consisting of two self-cavities. (**a**) Resonance frequencies as a function of q for the two dispersion relation solutions given by Eqs. 8.43 and 8.44. (**b**) Fraction of the electric field in x-direction in the graphite layer and the electric field in x-direction in the graphene layer for $q_{1,\text{unscreened}}$ and $q_{2,\text{unscreened}}$. As the electric field fraction as a function of $q_{1,\text{unscreened}}$ yields only positive values, and the fraction of $q_{c,\text{unscreened}}$ shows only negative values, the first solution corresponds to the solution for the symmetric mode, where currents oscillate in-phase in the two layers, and the second solution corresponds to the antisymmetric mode, where the currents oscillate out-of-phase, as sketched in (**c**)

$$\frac{E_{\text{x,unscreened,graphite}}}{E_{\text{x,unscreened,graphene}}}$$

$$= \frac{\epsilon_2\omega^2}{\epsilon_2\omega^2\cosh\left(d_{\text{hBN2}}q_{un}\right) + (\epsilon_3\omega^2 - d_{\text{gr}}q_{un}\omega_{\text{3D}}^2)\sinh(d_{\text{hBN2}}q_{un})}, \tag{8.45}$$

where ϵ_2 is the dielectric constant of hBN, ϵ_3 is the dielectric constant of sapphire, d_{hBN2} is the thickness of the hBN between the graphene and the graphite, d_{gr} is the thickness of the graphite and ω_{3D} is the 3D plasma frequency of graphite.

Figure 8.9b shows plots of the electric field in x-direction in the graphite divided by the electric field in x-direction in the graphene (Eq. 8.45) for the two solutions for q. The electric field fraction only yields positive values for $q_{1,\text{unscreened}}$, verifying that this is the solution for the symmetric mode where the electric fields (and thereby the current) in the graphene and in the graphite layer are pointing in the same direction. The electric field fraction for $q_{2,\text{unscreened}}$ exhibits only negative values, verifying that this is the solution for an antisymmetric mode where the electric fields (and thereby the currents) in the graphene and in the graphite layer are pointing in opposite directions.

Interestingly, one can find two limiting cases for the antisymmetric mode for very small graphene dopings ($A_{\text{gr}} \to 0$) and for very large graphene dopings ($A_{\text{gr}} \to \infty$) when investigating the dispersion relation as given by Eq. 8.44: For small graphene dopings, the $d_{\text{gr}}\omega_{\text{3D}}^2$ becomes dominant and A_{gr} is like a small perturbation such that the $d_{\text{gr}}\omega_{\text{3D}}^2$ cancel out and the dispersion relation is as in the case of a graphene plasmon that is screened by the graphite. For unrealistically large graphene dopings,

$A_{gr} \gg d_{gr}\omega_{3D}^2$ and the A_{gr} terms cancel out such that the antisymmetric mode gets graphite-plasmon character.

In summary, both in the case of screened graphene and graphite layers and in the case of unscreened graphene and graphite layers two modes form: a symmetric mode and an antisymmetric mode. The different dispersion relations for screened and unscreened modes are compared in the following section.

8.2.1.3 Comparing Screened and Unscreened Dispersion Relations

Before applying the derived dispersion relations of screened and unscreened, symmetric and antisymmetric modes to establish the boundary equations for a heterostructure consisting of two self-cavities, it is helpful to compare them first in some limiting cases to better understand their characteristics and typical energy scales.

Figure 8.10 presents the four calculated dispersion relations (shown in blue, bright blue, purple, and magenta) for three representative graphene carrier densities. To assess the extent of hybridization, two additional reference curves are included: the dispersion of an isolated graphite 2D plasmon screened by the coplanar strip metal in the absence of graphene (black dashed line), and that of a "double-screened" graphene 2D plasmon screened by both the coplanar strip and an idealized metallic gate (grey dashed line, see Sect. A.4 in the Appendix for the derivation of such an "double-screened" 2D plasmon). These comparisons highlight how the symmetric and antisymmetric hybrid modes deviate from their uncoupled counterparts.

At a graphene carrier density of $n = 0.95 \times 10^{12}$ cm^{-2}, the screened graphite and double-screened graphene plasmon dispersions do not coincide with the hybridized modes (Fig. 8.10a), confirming their hybrid character. Notably, the screened antisymmetric mode exhibits the lowest resonance frequency at a given q, while the unscreened antisymmetric mode lies only slightly higher. In contrast, the unscreened symmetric mode, which is unique in its $\nu_0 \propto \sqrt{q}$ scaling, lies well above the screened symmetric mode, typically beyond the accessible spectral range. This distinction becomes crucial for understanding the full coupling mechanism between two integrated self-cavities later in Sect. 8.3.

At low graphene carrier densities n, the screened symmetric mode (blue) lies on top of the dispersion for the screened 2D graphite plasmon (black dashed line, see Fig. 8.10b), with a dispersion relation given by Eq. 8.25. Also the dispersion of the screened antisymmetric mode (red) lies on top of the dispersion for a double-screened graphene plasmon (grey dashed line, see Fig. 8.10b) that is screened from the metal strip at a distance of d_{hBN1} and perfectly screened by the graphite at a distance d_{hBN2}. This highlights that in the limit of low graphene carrier densities, the symmetric modes still behave mostly as unhybridized graphite plasmons and antisymmetric modes as unhybridized graphene plasmons.

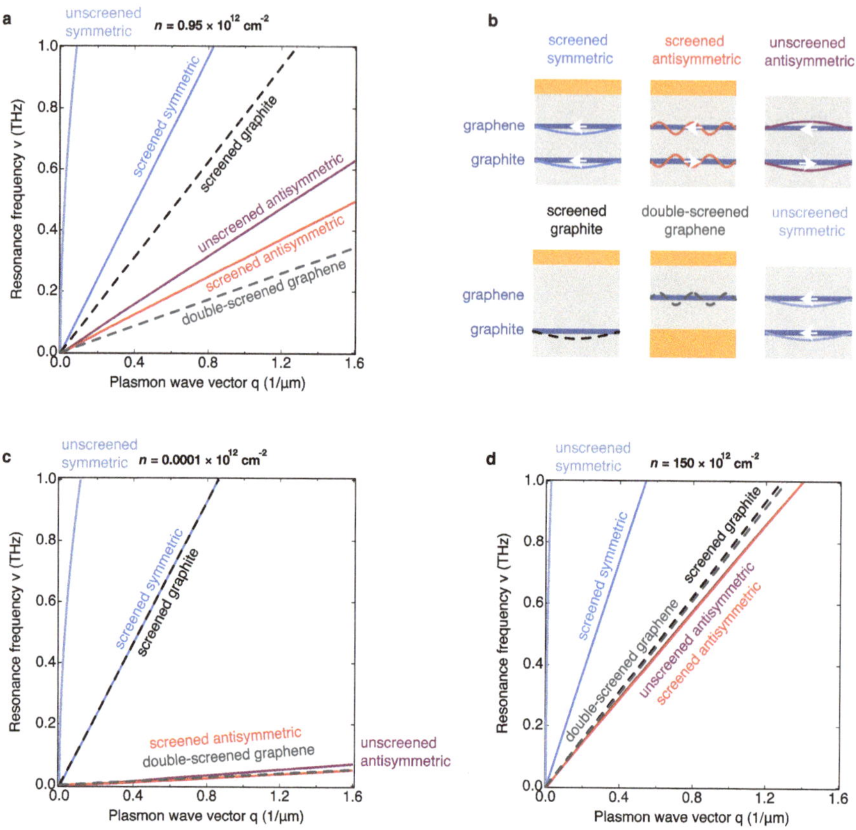

Fig. 8.10 Comparison of dispersion relations. (**a**) Dispersion relations for screened and unscreened, symmetric and antisymmetric modes, and screened graphite and double-screened graphene modes (see Section A.4 in the Appendix) for a typical graphene-graphite heterostructure with $d_{\text{gr}} = 9$ nm, $d_{\text{hBN1}} = d_{\text{hBN2}} = 9$ nm and $\nu_{3D} = 118$ THz at a graphene carrier density of $n = 0.95 \times 10^{12}$ cm^{-2}. (**b**) Legend for other panels with current density and screening sketches for the different modes. (**c**) Dispersion relations for a small graphene carrier density of $n = 0.0001 \times 10^{12}$ cm^{-2}. In this limit, the screened symmetric mode acts like a screened graphite plasmon and the screened antisymmetric mode acts like a double-screened graphene plasmon, screened both by the metal strip above and the graphite underneath (see (**b**)). (**d**) Dispersion relations in the limit of a very large graphene carrier density of $n = 150 \times 10^{12}$ cm^{-2}, where screened and unscreened antisymmetric modes lie on top of each other. (Figure adapted from [1] under a CC BY 4.0 license)

The analytical theory is further validated in the high carrier-density limit: at very large graphene carrier densities, the screened and unscreened antisymmetric modes should converge, as the highly conductive graphene effectively acts as a perfect screen for the graphite mode, rendering the presence of the distant coplanar strip negligible. As shown in Fig. 8.10d, both modes coincide at $n = 150 \times 10^{12}$ cm^{-2}, confirming the theory's consistency.

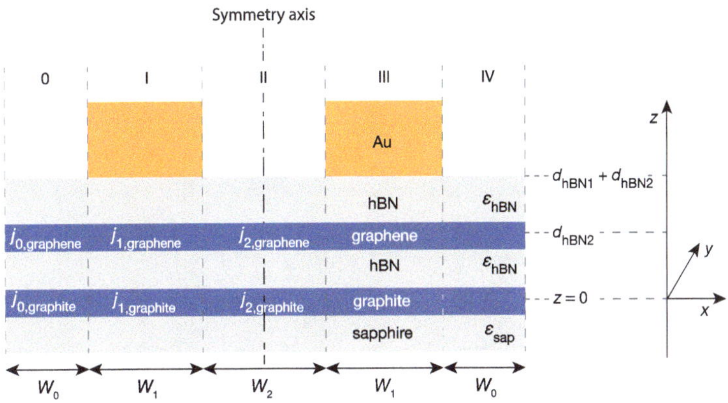

Fig. 8.11 Cross-section of a graphene-graphite heterostructure embedded beneath coplanar strips. The heterostructure can be divided into five regions (0 to IV), that are either screened by the coplanar strip (I and III) or unscreened (0, II and IV). (Figure adapted from [1] under a CC BY 4.0 license)

It's important to note that the dispersion relations are plotted as continuous lines, representing a continuum of 2D plasmon modes for an infinitely wide heterostructure. In reality, the finite-sized self-cavities support only discrete modes with specific momenta, such as quarter and three-quarter wavelength modes (see Sect. 6.3).

8.2.2 Boundary Equations

In this section, current densities and potentials are matched in both conductive layers of a graphene-graphite heterostructure (see Fig. 8.11), similar to the case of a heterostructure consisting of a single self-cavity (see Sect. 6.4.4), considering the different dispersions of screened and unscreened, symmetric and antisymmetric modes (see Sect. 8.2.1) to derive the cavity conductivity of the entire integrated graphene-graphite heterostructure as a function of frequency.

Dispersion relations and electric field fractions were given for the dissipationless case ($\Gamma_1 = 0$ and $\Gamma_2 = 0$) in the previous sections. However, the full, complex solutions that include dissipation are needed for computing the cavity conductivity as a function of frequency, that are found by solving the equations given in Sect. 8.2.1 with $\Gamma_1 \neq 0$ and $\Gamma_2 \neq 0$.

To capture the response of the entire cavity, with broken translational symmetry due to the finite size and different momenta $q_{1,sym}$ (screened symmetric), $q_{1,antisym}$ (screened antisymmetric), $q_{2,sym}$ (unscreened symmetric) and $q_{2,antisym}$ (unscreened antisymmetric), five different regions have to be considered. This is done by expressing the currents j and potentials V in the graphene and graphite layers in terms of transmission t and reflection r coefficients and in terms of the in-plane electric field of the graphene in each region.

Matching these expressions at the boundaries in each layer, while considering the symmetry of the heterostructure around the middle of the coplanar strip, leads to twelve boundary equations in total (for exact expressions of j and V see Sect. A.6 in the Appendix):

$$\boxed{\begin{aligned}
(1) &= j_{1,\text{graphene}}(W_1, t) = j_{2,\text{graphene}}(-W_2/2, t)\\
(2) &= V_{1,\text{graphene}}(W_1, t) = V_{2,\text{graphene}}(-W_2/2, t)\\
(3) &= j_{1,\text{graphite}}(W_1, t) = j_{2,\text{graphite}}(-W_2/2, t)\\
(4) &= V_{1,\text{graphite}}(W_1, t) = V_{2,\text{graphite}}(-W_2/2, t)\\
(5) &= r_{2,\text{as}} = t_{2,\text{as}}\\
(6) &= r_{2,\text{sym}} = t_{2,\text{sym}}\\
(7) &= j_{1,\text{graphene}}(0, t) = j_{0,\text{graphene}}(W_0, t)\\
(8) &= V_{1,\text{graphene}}(0, t) = V_{0,\text{graphene}}(W_0, t)\\
(9) &= j_{1,\text{graphite}}(0, t) = j_{0,\text{graphite}}(W_0, t)\\
(10) &= V_{1,\text{graphite}}(0, t) = V_{0,\text{graphite}}(W_0, t)\\
(11) &= j_{0,\text{graphene}}(0, t) = 0\\
(12) &= j_{0,\text{graphite}}(0, t) = 0
\end{aligned}} \tag{8.46}$$

The first four boundary equations 8.1–8.4 match currents and potentials at the boundary between region I and II, accounting for the external electric field E_{ext} that excites the plasmons in the graphene and graphite layer in the coplanar strip gap (region II). The symmetry around the middle of the coplanar strip leads to boundary equations 8.5 and 8.6. Boundary equations 8.7–8.10 match currents and potentials at the boundary between region 0 and I. Boundary equations 8.11 and 8.12 set the current at the graphene and graphite edges to zero.

8.2.3 Computing the Cavity Conductivity

The twelve boundary equations can be solved in terms of the reflection coefficients $r_{2,\text{as}}$ and $r_{2,\text{sym}}$ in the coplanar strip gap. The total current density, which is proportional to the cavity conductivity that is detected with on-chip THz spectroscopy from a graphene—graphite heterostructure, can again be obtained by an RPA (random phase approximation) conductivity approach (see Sect. 6.4.5 for the bare cavity case). In contrast to the bare graphite cavity solution discussed in Sect. 6.4.5, symmetric and antisymmetric modes have to be considered in a graphene-graphite heterostructure. The total current in graphene (or other 2D material) amounts to:

$$\begin{aligned} j_{\mathrm{tot,graphene}} &= \sigma_{\mathrm{graphene}}(\omega) E_{\mathrm{ext}} + j_{\mathrm{ind}} \\ &= \sigma_{\mathrm{graphene}}(\omega) E_{\mathrm{ext}} + \alpha\omega E_{\mathrm{ext}} \\ &= \left(\sigma_{\mathrm{graphene}}(\omega) + \alpha(\omega)\right) E_{\mathrm{ext}} \\ &= \sigma_{\mathrm{RPA,graphene}} E_{\mathrm{ext}} \end{aligned} \tag{8.47}$$

where $\sigma_{\mathrm{RPA,graphene}}$ is the RPA conductivity of the graphene layer and $\alpha(\omega)$ is the response function of transmitted and reflected waves of symmetric and antisymmetric modes in the coplanar strip gap region of the graphene layer:

$$\begin{aligned} \alpha(\omega) &= \alpha_{\mathrm{as}}(\omega) + \alpha_{\mathrm{sym}}(\omega) \\ &= \sigma_{\mathrm{graphene}}(\omega) \left(t_{2,\mathrm{as}} e^{i q_{\mathrm{un,as}} x} + r_{2,\mathrm{as}} e^{-i q_{\mathrm{un,as}} x} \right. \\ &\quad \left. + t_{2,\mathrm{sym}} e^{i q_{\mathrm{un,sym}} x} + r_{2,\mathrm{sym}} e^{-i q_{\mathrm{un,sym}} x} \right). \end{aligned} \tag{8.48}$$

For $r_2 = t_2$ (see boundary equations (5) and (6)) and in the limit of small q, this term simplifies to:

$$\begin{aligned} \alpha(\omega) &= \sigma_{\mathrm{graphene}}(\omega) \left(2 r_{2,\mathrm{as}} \cos\left(q_{\mathrm{un,as}} x\right) + 2 r_{2,\mathrm{sym}} \cos\left(q_{\mathrm{un,sym}} x\right)\right) \\ &= 2\sigma_{\mathrm{graphene}}(\omega) \left(r_{2,\mathrm{sym}} + r_{2,\mathrm{as}}\right), \end{aligned} \tag{8.49}$$

such that the total current in the graphene (or other 2D material) layer is:

$$j_{\mathrm{tot,graphene}} = \sigma_{\mathrm{graphene}}(\omega) \left(1 + 2\left(r_{2,\mathrm{sym}} + r_{2,\mathrm{as}}\right)\right) E_{\mathrm{ext}}. \tag{8.50}$$

The total current in the graphite (or other electrostatic gate material) can be calculated in the same way as for graphene as follows:

$$j_{\mathrm{tot,graphite}} = \sigma_{\mathrm{graphite}}(\omega) \left(1 + 2 \frac{E_{\mathrm{x,unscreened,graphite}}}{E_{\mathrm{x,unscreened,graphene}}} \left(r_{2,\mathrm{sym}} + r_{2,\mathrm{as}}\right)\right) E_{\mathrm{ext}}. \tag{8.51}$$

where $E_{\mathrm{x,unscreened,graphite}} / E_{\mathrm{x,unscreened,graphene}}$ is Eq. 8.45, giving the electric field in the graphite layer in terms of the electric field in the graphene layer.

Thus, the total current density of the graphene—graphite heterostructure is

$$\begin{aligned} j_{\mathrm{tot}} &= j_{\mathrm{tot,graphene}} + j_{\mathrm{tot,graphite}} \\ &= \left(\sigma_{\mathrm{graphene}}(\omega) \left(1 + 2\left(r_{2,\mathrm{sym}} + r_{2,\mathrm{as}}\right)\right)\right. \\ &\quad \left. + \sigma_{\mathrm{graphite}}(\omega) \left(1 + 2 \frac{E_{\mathrm{x,unscreened,graphite}}}{E_{\mathrm{x,unscreened,graphene}}} \left(r_{2,\mathrm{sym}} + r_{2,\mathrm{as}}\right)\right)\right) E_{\mathrm{ext}} \end{aligned} \tag{8.52}$$

The external electric field E_{ext} can be set to 1 in the theory, as it just serves as a multiplicative factor, which then allows to extract the cavity conductivity as a function of frequency as follows:

$$\boxed{\begin{aligned}\sigma(\omega) &= \sigma_{\mathrm{graphene}}(\omega)\left(1+2\left(r_{2,\mathrm{sym}}+r_{2,\mathrm{as}}\right)\right) \\ &\quad + \sigma_{\mathrm{graphite}}(\omega)\left(1+2\frac{E_{\mathrm{x,unscreened,graphite}}}{E_{\mathrm{x,unscreened,graphene}}}\left(r_{2,\mathrm{sym}}+r_{2,\mathrm{as}}\right)\right)\end{aligned}}, \tag{8.53}$$

where the conductivities of graphene and graphite layers are given by the Drude model (see Sect. 5.2.1):

$$\begin{aligned}\sigma_{\mathrm{graphene}}(\omega) &= \frac{A_{\mathrm{graphene}}}{-i\omega+\Gamma_1} \\ \sigma_{\mathrm{graphite}}(\omega) &= \frac{\omega_{\mathrm{3D}}^2 d_{\mathrm{gr}}}{-i\omega+\Gamma_2}\end{aligned} \tag{8.54}$$

with $A_{\mathrm{graphene}} = \frac{e^2 v_F \sqrt{n}}{\epsilon_0 \hbar \sqrt{\pi}}$, where e is the electron charge, v_F is the Fermi velocity in graphene, n is the graphene carrier density, ω_{3D} is the 3D bulk plasma frequency of graphite, d_{gr} is the thickness of the graphite layer, Γ_1 is the scattering rate of graphene and Γ_2 is the scattering rate of graphite.

Thus, the overall cavity conductivity of a heterostructure consisting of two conductive layers is composed of the individual contributions of the two layers, including their hybridization. While experimentally, only the overall conductivity can be detected, the analytical theory allows to disentangle the overall conductivity into contributions from each of the individual layers.

The filling factor C_{filling}, as discussed in Sect. 6.4.7 also has to be considered for the cavity conductivity of a heterostructure consisting of two self-cavities. If the two 2D materials are only separated by a thin hBN layer ($d_{\mathrm{hBN2}} < 25\,\mathrm{nm}$), assuming that the filling factor is equal for both materials is a good approximation (see Fig. 6.17). However, if the two 2D materials are separated by a thick hBN layer ($d_{\mathrm{hBN2}} \gg 20\,\mathrm{nm}$), the filling factor varies (see Fig. 6.17). This has the effect that the individual contributions of graphene and graphite layer to the overall cavity conductivity have to be multiplied by individual filling factors (see Eq. 8.53). A complete theoretical model for the filling factor is still under development.

8.2.4 Comparison to Full 3D Electromagnetic Simulations

In this section, the results obtained with the extended analytical theory for heterostructures with two conductive layers is compared to full 3D electromagnetic simulations, to validate the theory. As the full 3D electromagnetic simulations were optimized for graphite heterostructures and showed good agreement with the measurements discussed in Chap. 6, a dual-graphite heterostructure was simulated here, instead of one combining graphene and graphite layers. As shown in Fig. 8.12, the results from this dual-graphite heterostructure agree well with the analytical theory. The analytical theory further explains the mode characteristics of the six resonances observed below 1 THz. The two strongest resonances correspond to symmetric ("bottom-graphite-like") modes, while the weaker ones arise from antisymmetric ("upper-graphite-like") modes. The symmetric modes are labeled as $m = 0, 1, 2, \ldots$, similar to the case of the bare graphite heterostructure discussed in Chap. 6 and the antisymmetric modes are labeled as $s = 0, 1, 2, \ldots$.

Having demonstrated that the extended analytical theory can describe the cavity response of heterostructures consisting of two conductive layers well, let's next simulate Device GG1.

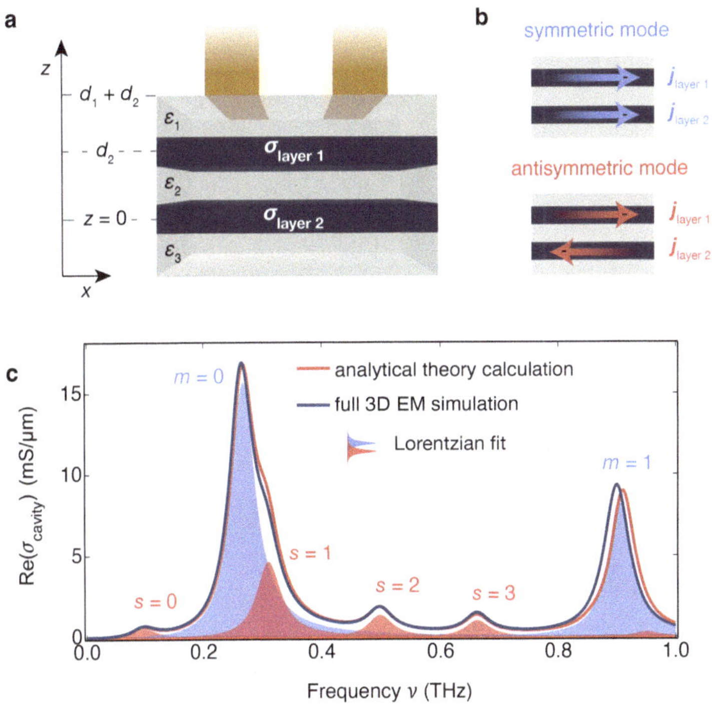

Fig. 8.12 Comparison of analytical theory with full 3D electromagnetic simulations. (**a**) Analytical theory calculations (red curve) and full 3D electromagnetic simulations (black) of a dual-conductive-layer heterostructure with the parameters: $W_0 = 1\,\mu\text{m}$, $W_1 = 3.5\,\mu\text{m}$, $W_2 = 2.5\,\mu\text{m}$, $d_{\text{layer 2}} = d_{\text{layer 1}} = 4\,\text{nm}$, $d_{\text{d1}} = 20\,\text{nm}$, $d_{\text{h2}} = 20\,\text{nm}$, $l = 7\,\mu\text{m}$, $\nu_{3\text{D}} = 118\,\text{THz}$, $\tau^{-1} = 0.05\,\text{THz}$ and $C_{\text{filling}} = 372\,\text{mm}^{-1}$. The result of the analytical theory was fit to a sum of Drude-Lorentz models, which are shaded in blue or magenta. (**b**) The blue shaded Drude-Lorentz models correspond to the excitation of symmetric modes, while the red shaded Drude-Lorentz models are from the excitation of antisymmetric modes. (Figure adapted from [2] under a CC BY 4.0 license)

8.2.5 Theoretical Calculation of the Experimental Data

In this section, analytical theory calculations are performed for the device geometry of Device GG1 (see Fig. 8.1) to gain a better understanding of the experimentally measured suppressed power-law scalings of the graphene plasmonic modes discussed in Sect. 8.1. The heterostructure of GG1 was placed slightly asymmetrically underneath the coplanar strip to avoid covering a bubble with the coplanar strips, such that $W_0 = 2.7\,\mu\text{m}$ on one side and $W_0 = 1.3\,\mu\text{m}$ on the other side (see Fig. 8.1). As the analytical theory calculations assumes a symmetric heterostructure, the averaged width of $2\,\mu\text{m}$ was used for the calculations. A scattering rate $\tau^{-1} = 0.14\,\text{THz}$ was chosen both for the graphite and for the graphene layer to capture comparable linewidths to the experimental data.

Comparing amplitudes of the $s = 1$ mode in the experimental data measured at the largest graphene doping to the analytical theory for the cavity geometry parameters, very good agreement is found when choosing a filling factor of $180\,\text{mm}^{-1}$. This is consistent with the filling factor of the graphene layer, as only the graphene modes are detectable within the spectral sensitivity range. Given that the graphite layer is separated from the graphene by a 91 nm thick hBN layer, the filling factor for the graphite layer is expected to be lower (see Fig. 6.17, and also Sects. 6.4.7 and 8.2.3). To determine the filling factor of the graphite layer, full 3D electromagnetic simulations of the heterostructure, without the graphene layer (or with graphene at the charge neutrality point), are compared to calculations of the analytical theory. These calculations extend beyond the experimental spectral sensitivity range to capture the graphite $m = 0$ mode. Through this method, a filling factor of $161\,\text{mm}^{-1}$ is determined for the graphite layer. However, since the deviation of the individual filling factors is small and the graphite (symmetric) mode falls outside the spectral sensitivity range, the adjustment of the graphite layer's filling factor is omitted in the analytical theory calculations for Device GG1.

Figure 8.13 plots the results of the real part of the calculated cavity conductivity up to 1.5 THz, using similar linewidths as extracted in experiment. Three modes appear below 1 THz with excellent qualitative agreement to the experimental data. Examining the calculated conductivity outside of the experimental range shows that when the graphene $s = 3, 4$ modes cross with the graphite self-cavity mode (dashed line, Fig. 8.13b), they hybridise and form avoided crossings, splitting into upper and lower polariton branches. The calculations reveal that at high carrier densities, the observed $s = 1, 2$ modes also form lower branches of avoided crossings occurring at carrier densities outside of the experimentally accessible range.

To evidence the role of plasmonic mode hybridization, a similar graphene cavity was calculated, but where the graphite layer was replaced by an infinitely thick perfect metal layer (see Fig. 8.14a). In this geometry, the graphene layer is still screened by a metal located at the same distance away as in the cavity with the graphite layer. However, it does not hybridize with back gate modes, as the plasmonic resonances would be so far away in frequency that the effect of hybridization is negligible. Such a calculation reveals a plasmonic spectrum of the

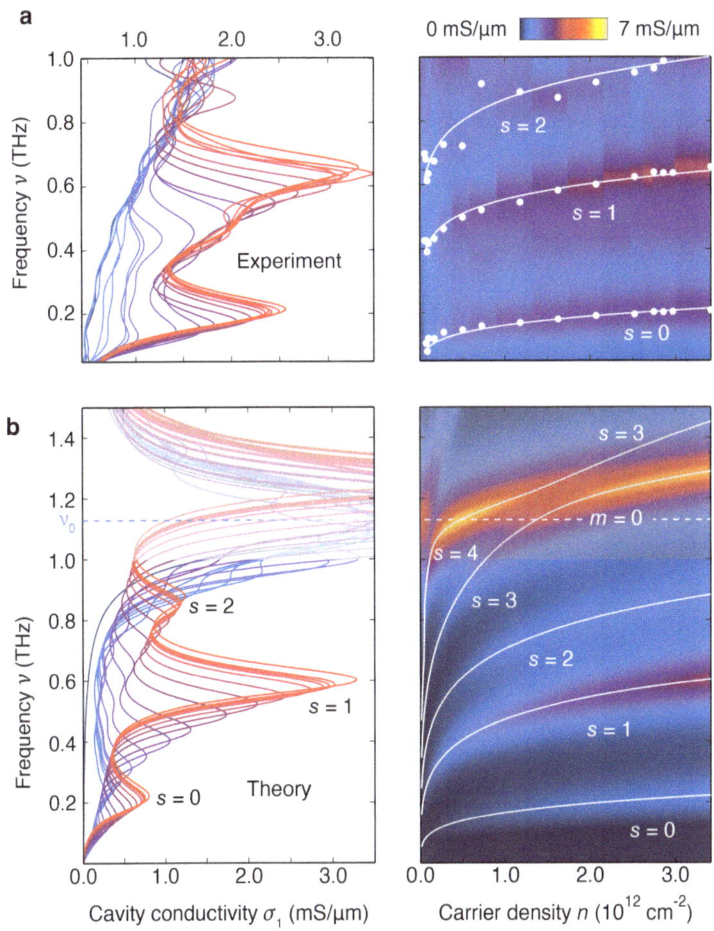

Fig. 8.13 Comparison of experimental data of Device GG1 to analytical theory calculations. (**a**) Experimentally obtained real part of the cavity conductivity spectra. The cavity conductivity of the bare graphite self-cavity (dark blue trace) increases slowly in conductivity with increasing frequency, and is expected to peak at ~1.13 THz. Three resonances appear on top of this background that increase in amplitude and shift to higher frequencies with increasing graphene carrier density (from blue to purple to red). The resonance frequencies as a function of graphene carrier density of these three resonances are shown in the right panel. (**b**) Analytical theory capturing the cavity conductivity for the same device shown in Fig. 8.22, identifying the three experimentally detected resonances as the $s = 0, 1, 2$ graphene-like (antisymmetric) self-cavity modes. The extended frequency range of the theory captures the avoided crossings between the $s = 3, 4$ modes to the graphite self-cavity mode (dashed line). (Figure adapted from [1] under a CC BY 4.0 license)

cavity with undistorted graphene modes $s = 0, 1, 2, 3, 4$ (see Fig. 8.14b). When the analytical theory calculations for thin and infinitely thick graphite gates are compared, the effect of reduced resonance frequencies for the $s = 1, 2, 3$ modes

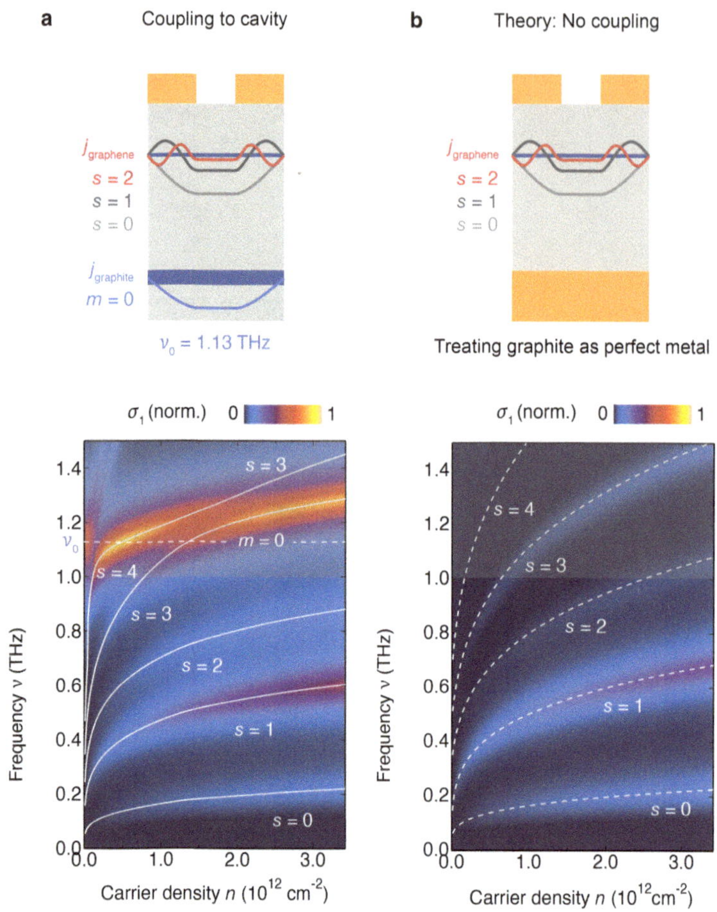

Fig. 8.14 Analytical theory calculations for Device GG1: No coupling versus coupling. (**a**) Cavity conductivity spectrum as shown in Fig. 8.13. Through the coupling with the cavity $m = 0$ mode, the graphene $s = 0, 1, 2, ..$ modes are significantly suppressed in resonance frequency when compared to the uncoupled scenario shown in (**b**): Graphite of the heterostructure is treated as a perfect, thick metal forming a cavity with no coupling. All graphene modes ($s = 0, 1, 2, 3, 4$) are undistorted and follow the typical $\nu \propto n^{0.25}$ power laws. The cavity conductivity spectra in (**a**) and (**b**) are normalized and comparable between the two panels. (Figure adapted from [1] under a CC BY 4.0 license)

through hybridization with the $m = 0$ mode becomes visible. The experimentally detected, suppressed power-law behaviours of the graphene $s = 1, 2$ modes are thus a consequence of the hybridization with the plasmonic graphite $m = 0$ mode.

As the avoided crossings that appear in GG1 occur outside the experimental spectral sensitivity range, only the analytical theory can be used to calculate the coupling strength between the graphene and graphite modes. As the avoided crossing between the $s = 3$ and the $m = 0$ mode appears most prominently in the calculation shown in Figs. 8.13b and 8.14a, the normalized coupling strength η can be calculated at the carrier density at which the hybridized modes are equally split by $2g$ around the uncoupled resonance frequency of the graphite mode, ν_0, as:

$$\eta = \frac{g}{\nu_0} = \frac{0.064\,\text{THz}}{1.13\,\text{THz}} = 0.057. \tag{8.55}$$

Since this value remains below 0.1, it indicates that the coupling in Device GG1 has not yet reached the ultrastrong coupling regime (see Sect. 7.4.3). However, by designing the heterostructure such that the graphite self-cavity mode is positioned at a lower resonance frequency, within the center of the experimental spectral sensitivity range, ultrastrong coupling can be achieved. However, before experimentally exploring such a heterostructure in Sect. 8.4, let's first discuss the two coupling mechanisms that enabled the near ultrastrong coupling of self-cavity modes of different momenta in this heterostructure.

8.3 Coupling Mechanisms of Self-cavities

This section explores the tunable, multimodal coupling mechanisms of self-cavities that can lead to ultrastrong coupling of plasmonic modes of different momenta. As discussed in Sect. 6.4.9, plasmonic self-cavity modes can be modeled as driven, damped harmonic oscillators, making it intuitive to represent plasmonic modes as simple pendula. To therefore better illustrate the concept of plasmon coupling, Fig. 8.15 employs a classical pendulum analogy: two coupled pendula that either oscillate in-phase or out-of-phase (Fig. 8.15a).

A 2D material placed within a homogeneous dielectric environment with no nearby metal exhibits a plasmonic self-cavity mode of purely unscreened character, akin to a single oscillating pendulum (Fig. 8.15b). However, when the 2D material is situated in a micropatterned dielectric environment, such as that provided by the metal traces of a coplanar strip, unscreened and screened plasmons (represented by yellow and dark blue pendula) can couple at the boundaries between screened and unscreened regions. This interaction leads to the formation of a hybrid self-cavity mode (depicted by a yellow-dark blue dashed current density) with both screened and unscreened characteristics (Fig. 8.15c). Since unscreened plasmons have a much higher resonance frequency than screened ones (see Sect. 6.4.2), the self-cavity modes of typical graphite heterostructures still predominantly exhibit screened plasmon behavior, as the resonance condition of screened and unscreened modes occurs outside the spectral sensitivity range (see Sect. 6.5.3).

In a heterostructure composed of two self-cavities, the dynamics become more intricate. Initially, each layer can be viewed as an independent pendulum. However,

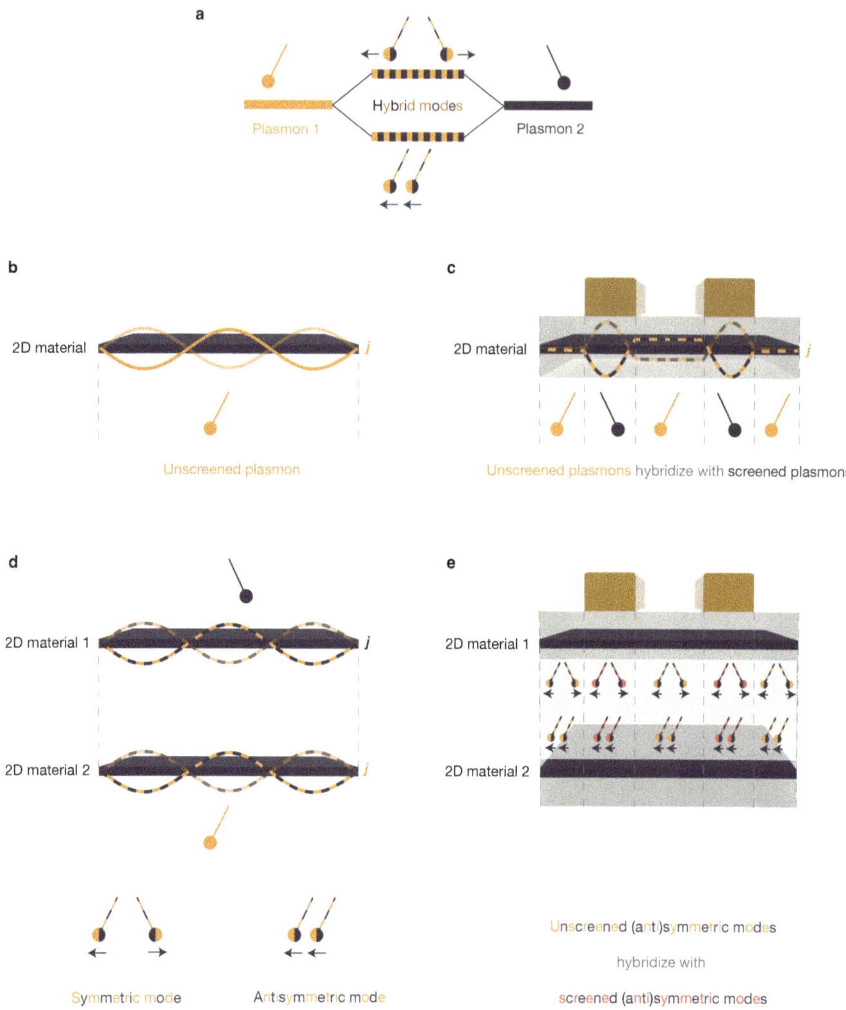

Fig. 8.15 Pendulum analogy of coupling plasmons. (**a**) A plasmon can be described as a damped harmonic oscillator, similar to a pendulum. Two pendula (plasmonic modes) can couple to either oscillate in- or out-of-phase. (**b**) A 2D material in a homogeneous dielectric environment with no metal in close proximity forms a self-cavity mode of purely unscreened character, that can be described by a single pendulum. (**c**) In contrast, a 2D material placed in a micropatterned dielectric environment excites both unscreened (yellow pendula) and screened (dark blue pendula) plasmonic modes, that hybridize at the dielectric boundaries provided by the metal strip and that lead to a hybrid self-cavity mode with both screened and unscreened plasmon character (denoted as the yellow-dark blue dashed standing wave). (**d**) If two 2D materials (two pendula) are placed in close proximity, they can couple to form symmetric and antisymmetric modes. (**e**) If two 2D materials placed in close proximity are integrated into a micropatterned dielectric environment, screened and unscreened symmetric and antisymmetric modes can couple at the dielectric boundaries provides by the metal edges

their proximity and comparable conductivities enable interaction, leading to the formation of symmetric and antisymmetric modes (Fig. 8.15d). While this "current–current hybridization" was previously
explored in Sect. 8.2.1, the key principles for either enhancing or suppressing this type of coupling are summarized in the following Sect. 8.3.1.

But moreover, in a micropatterned heterostructure, these symmetric and antisymmetric modes from the screened and unscreened regions can also couple at the dielectric boundaries (Fig. 8.15e). This second type of coupling of two self-cavities is discussed in detail in Sect. 8.3.2.

Lastly, the multimodal coupling between multiple graphene self-cavity modes and the $m = 0$ graphite self-cavity mode is discussed in Sect. 8.3.3, highlighting how multiple avoided crossings occur simultaneously as the graphene carrier density is varied.

8.3.1 *Current-Current Hybridization*

The first coupling type that occurs in heterostructures consisting of two conductive layers (plasmonic self-cavities) is the current-current hybridization that leads to the formation of symmetric and antisymmetric modes (see Sect. 8.2.1). As already discussed in Sect. 8.2.1.1 and shown in Fig. 8.7, the amount of current-current hybridization is strongly dependent on how comparable the 2D conductivities of the two conductive layers is, when they are only separated by a few tens of nanometer thick hBN layer. If the conductivity of the graphite layer is comparable or even lower than that of the graphene layer, the hybridization can become so strong that it is about to reach the "deep strong coupling" regime (see Fig. 8.16), where $\eta > 1$ (Sect. 7.4.4).

In the limit of separating the two conductive layers by a large distance, far exceeding the thickness of typical vdW heterostructures, the current-current hybridization of two layers with comparable conductivities would also vanish. However, if one wants to suppress this type of hybridization in realistic gate-tunable vdW heterostructures, the way forward is to make the graphite layer very thick (such that $\sigma_{\mathrm{graphite}} \gg \sigma_{\mathrm{2D\ material}}$) and the hBN layer between the 2D material and graphite layer as thick as possible. This approach is also necessary to minimize hybridization induced linewidth broadening, that will be discussed later in Sect. 8.4.6.

While the current-current hybridization explains the blue-shifting of the resonance frequency of the graphite mode with increasing graphene carrier density and the suppressed power laws of the graphene plasmons (see Fig. 8.7), it doesn't explain the formation of avoided crossings between modes of different momenta, such as the $s = 2$ and $m = 0$ mode as detected for Device GG2. The mechanism behind this type of coupling is explained in the following Sect. 8.3.2.

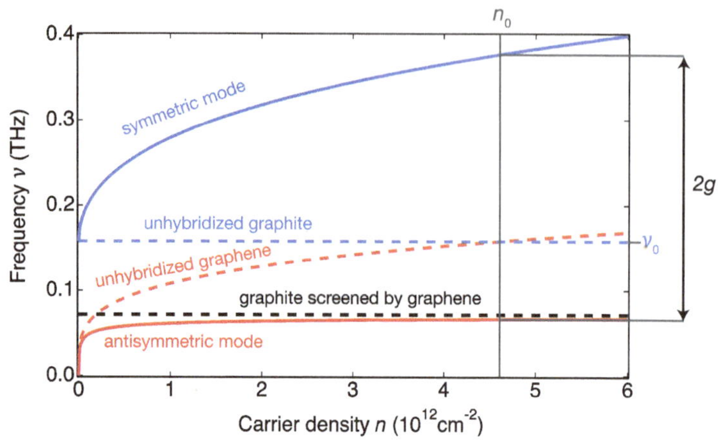

Fig. 8.16 Enhanced current-current hybridization almost entering the deep-strong coupling regime. Screened uncoupled graphene and graphite modes (dashed curves) are plotted in comparison to hybridized symmetric and antisymmetric modes. The calculations were performed for a heterostructure with the following parameters: $q = 2\pi/4W$, $W = 3$ μm, $d_{gr} = 1$ nm, $d_{hBN1} = 20$ nm and $d_{hBN2} = 5$ nm. The normalized coupling strength of this heterostructure with particularly thin hBN2 and d_{gr} layers is $\eta = \frac{g}{\nu_0} = 0.5 \cdot (0.37694 - 0.066199)/0.15783 = 0.984$, almost entering the deep strong coupling regime where $\eta > 1$

8.3.2 Coupling of Screened and Unscreened (Anti)symmetric Modes

In this section, the coupling mechanism that allows coupling of two modes of different momenta is explained. To do so, let's first consider a heterostructure consisting of a graphene and graphite self-cavity with a homogeneously screened dielectric environment, as depicted in Fig. 8.17. Two types of modes can form in this cavity, screened symmetric (graphite-like) and antisymmetric (graphene-like) modes (as discussed in Sect. 8.2.1.1). Due to the finite size of the self-cavities, only discrete wavelengths of modes can be admitted, which are drawn schematically in Fig. 8.17c with the blue circle, corresponding to the $m = 0$ graphite-like (symmetric) self-cavity mode, and grey and red circles, corresponding to graphene-like $s = 0, 1, 2$ (antisymmetric) modes. Through tuning of the graphene carrier density, two modes of different momenta, such as the $m = 0$ and $s = 2$ mode can be brought into resonance at a frequency ν_{sc} and at a specific graphene carrier density n_{sc}, as shown in Fig. 8.17c. However, because the modes have different, well-defined momenta, they are orthogonal to each other and no energy is transferred from one mode to another in this homogeneously screened heterostructure. As the graphene carrier density is tuned, the screened antisymmetric mode crosses with the screened symmetric mode. The modes become resonant in frequency, and do not form an avoided crossing, as shown in Fig. 8.17d.

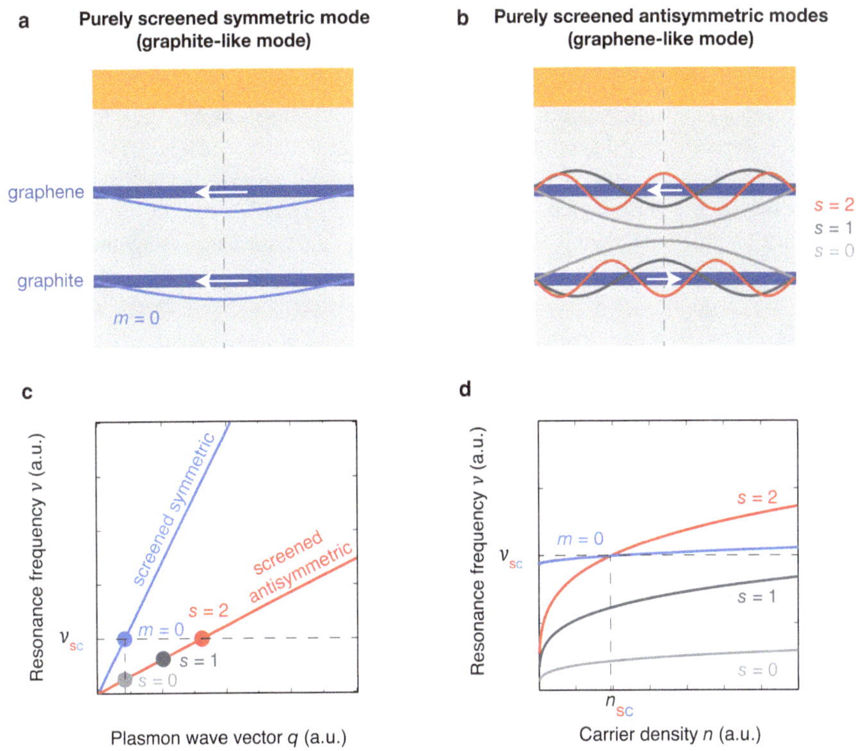

Fig. 8.17 Self-cavities with homogeneous screening. (**a**) Screened symmetric modes can form in the self-cavities, where current densities oscillate in-phase in the graphene and graphite layers. For small graphene carrier densities, the symmetric mode behaves as a screened graphite plasmon as shown in Fig. 8.10. (**b**) Additionally, screened antisymmetric modes can form, where current densities oscillate out of phase. For small graphene carrier densities, these modes behave as graphene plasmons that are screened by the gold and graphite (see Fig. 8.10). (**c**) In the case of homogeneous screening of the self-cavities, only these two modes can form with the shown dispersion relations at a given graphene carrier density. However, due to the finite size of the self-cavities, only discrete modes such as indicated by the four circles for the $m = 0$ and $s = 0, 1, 2$ modes can be excited. Through tuning of the graphene carrier density, the $s = 2$ mode can be brought into resonance with the $m = 0$ mode, indicated by ν_{sc}. However, these two modes have different momenta and can not couple in a heterostructure with a homogeneous dielectric environment. (**d**) This allows the modes to cross in frequency without the formation of an avoided crossing. The $m = 0$ mode is sketched to be slightly increasing in frequency with increasing n to consider the symmetric nature of the graphite cavity mode as discussed in Section 8.2.1.1. (Figure adapted from [1] under a CC BY 4.0 license)

In contrast, a graphene and graphite heterostructure can exhibit an inhomogeneously screened dielectric environment. As discussed in Sect. 6.5.3, the micropatterned dielectric environment of a bare graphite heterostructure already gives rise to a graphite self-cavity mode with both screened and unscreened characteristics. Screened and unscreened modes of different momenta can couple at the dielectric boundaries provided by the metal traces, because the momentum distributions of the individual plasmons that form in screened and unscreened regions are broad (see Fig. 6.29). Thus, only the resonance condition of two modes of different momenta in screened and unscreened regions is required to enable coupling.

While the unscreened symmetric mode (light blue) yields much higher resonance frequencies than the screened symmetric (blue) and antisymmetric (red) modes at a given q, the unscreened antisymmetric mode (purple) can be brought into resonance with the screened modes. Specifically, the resonance frequency of the unscreened antisymmetric modes can be varied relative to the resonance frequency of the screened modes by changing the width W_2 for a fixed metal width W_1 (see Fig. 8.18a), which moves the discrete excitations of unscreened antisymmetric modes (purple stars) along the dispersion line (see Fig. 8.18b). When the purple star aligns with the screened symmetric and antisymmetric modes (blue and red circles) such that $\nu_{\mathrm{unsc},1} = \nu_{\mathrm{sc}}$, a maximum coupling strength can be achieved (see Fig. 8.19). Therefore, the unscreened antisymmetric mode can be thought of as the spring in Fig. 8.18a, which mediates the coupling between screened symmetric and antisymmetric modes.

Thus, increasing W_2 relative to W_1 serves as a coupling strength tuning knob until $\nu_{\mathrm{unsc},1} = \nu_{\mathrm{sc}}$ at a width $W_2 = W_{\eta,\mathrm{max}}$ (see Fig. 8.19). When the coplanar strip gap W_2 is further increased, the situation becomes more complicated. In addition to the unscreened "unsc,1" mode (purple), higher order plasmons of the unscreened region (bright purple "unsc,2" mode, brightest purple "unsc,3" mode in Fig. 8.19) start to appear in the resonance frequency range which also couple to the screened modes. This obscures the coupling as more resonances are squeezed into the specific frequency range.

Figure 8.20 provides a series of analytical theory calculations of a fixed gate-tunable graphene heterostructure, where the normalized coupling strength is tuned as a function of a scaling factor α. α scales the widths of the unscreened regions $W_0 = \alpha \cdot 3\,\mu\mathrm{m}$ and $W_2 = \alpha \cdot 8\,\mu\mathrm{m}$. As plotted in Fig. 8.20, the coupling strength g and the normalized coupling strength η measured between the $m = 0$ and the $s = 2$ mode appears to continuously increase with increasing α, but decreases for $W_2 > 8\,\mu\mathrm{m}$ (for $\alpha > 1$) utilizing the normalized coupling strength calculation protocol of this work, which can be ascribed to the obscured coupling mechanism of higher harmonics of the unscreened antisymmetric modes. Through the presence of these additional resonances, the measurement procedure breaks down as more modes are squeezed into the frequency range of interest, such that the mode splitting is naturally decreased, although the overall coupling strength might indeed increase further.

The break-down of the normalized coupling strength measurement protocol becomes increasingly more prominent when higher order unscreened antisymmetric modes additionally occupy the spectral sensitivity range. However, the general simultaneous contribution of multiple modes to the coupling mechanism complicates the reliable extraction of η, even if only one unscreened antisymmtric mode enhances the coupling between screened modes of different momenta, as further discussed in the following section.

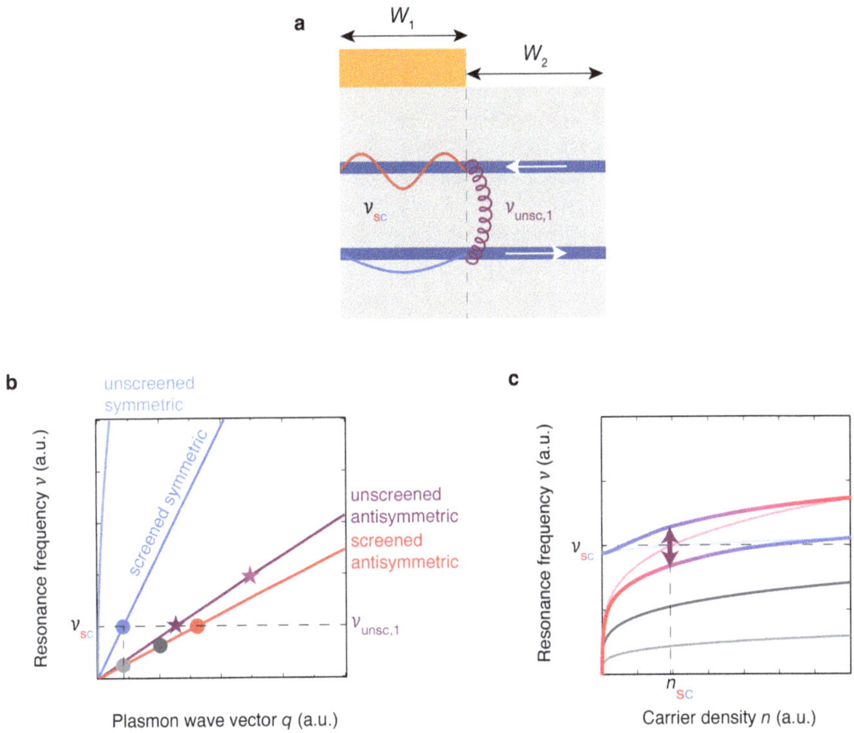

Fig. 8.18 Cavity with inhomogeneous screening. (**a**) The cavity consists of screened and unscreened vdW heterostructure regions. When treating screened and unscreened regions independently, screened symmetric (blue) and antisymmetric (red) modes can form underneath the metal strip, that can couple through the resonant unscreened antisymmetric mode at the boundary. (**b**) Dispersion relation for screened and unscreened, symmetric and antisymmetric modes. The antisymmetric unscreened mode can have a resonant mode at $\nu_{unsc,1}$ as indicated by the purple star, that is close or equal in resonance frequency to the screened antisymmetric mode (red circle). Due to momentum scattering at the screened/unscreened boundaries that provide a wide distribution of available momentum states, screened symmetric and antisymmetric modes of different wavelengths can couple at the boundaries when the unscreened antisymmetric mode approaches similar resonance frequencies. (**c**) An avoided crossing forms as a result of the mode coupling. (Figure adapted from [1] under a CC BY 4.0 license)

8.3.3 *Multimodal Coupling*

This section delves into the complex multimodal hybridization between graphene and graphite plasmonic self-cavity modes, which extends well beyond conventional two-mode couplings, as illustrated in Fig. 8.21.

As sketched in Fig. 8.21a,b, if only two modes are involved in a coupling mechanism, they form an avoided crossing in which the lower polariton branch, that started as an $s = 2$ mode, continues as the $m = 0$ mode at large carrier densities and saturates at the blue dashed line. The upper polariton branch, that started as the

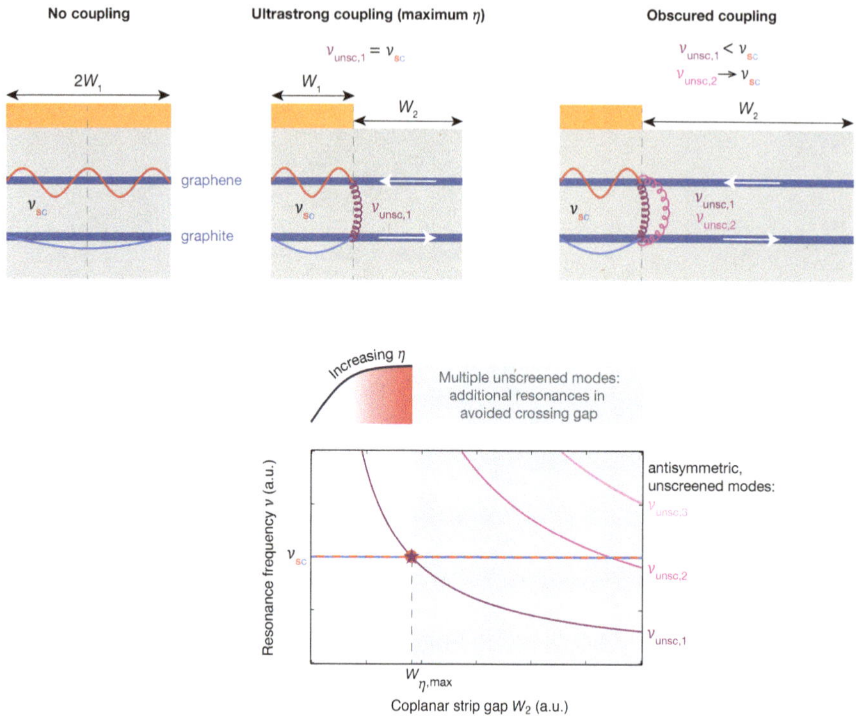

Fig. 8.19 Coupling strength tuning mechanism. A homogeneously screened cavity does not allow coupling of modes of different momenta, as shown in Fig. 8.17, such that $\eta = 0$. Once there is a tiny unscreened gap between the two metal strips, modes of different momenta can couple at the boundaries between screened and unscreened regions, as a wide range of momenta are available through momentum scattering at the screened/unscreened boundaries. The coupling is resonantly enhanced through the resonance frequency of the antisymmetric unscreened mode (purple spring), which can be tuned with the gap width W_2. In the W_2-range between 0 and $W_2 = W_{\eta,\max}$, the purple "unsc,1" mode ($\lambda = W_2$) is pushed continuously closer to ν_{sc}, which increases the normalized coupling strength η (bottom panel). A maximum coupling strength is reached when $\nu_{unsc,1} = \nu_{sc}$, which happens at $W_2 = W_{\eta,\max}$. However, for further increasing W_2, higher harmonics of the unscreened modes (bright purple $\nu_{unsc,2}$ mode with $\lambda = 1/2W_2$ and brightest purple $\nu_{unsc,3}$ mode with $\lambda = 1/3W_2$) approach ν_{sc}, that then additionally couple to the resonant screened modes (top right panel) leading to additional resonances in the unscreened region, which obscures the coupling. (Figure adapted from [1] under a CC BY 4.0 license)

$m = 0$ mode, gains character of the $s = 2$ mode through coupling and continues as the $s = 2$ mode at large graphene carrier densities. Therefore, both branches switch character through the avoided crossing.

When several modes are involved in the coupling mechanism, as shown in Fig. 8.21c,d, the situation gains complexity. As shown in Fig. 8.21c, different antisymmetric s modes are in resonance with the $m = 0$ mode at different graphene carrier densities. But this in turn also means that, when all of these modes couple to the $m = 0$ mode, multiple avoided crossings appear that blur into one another.

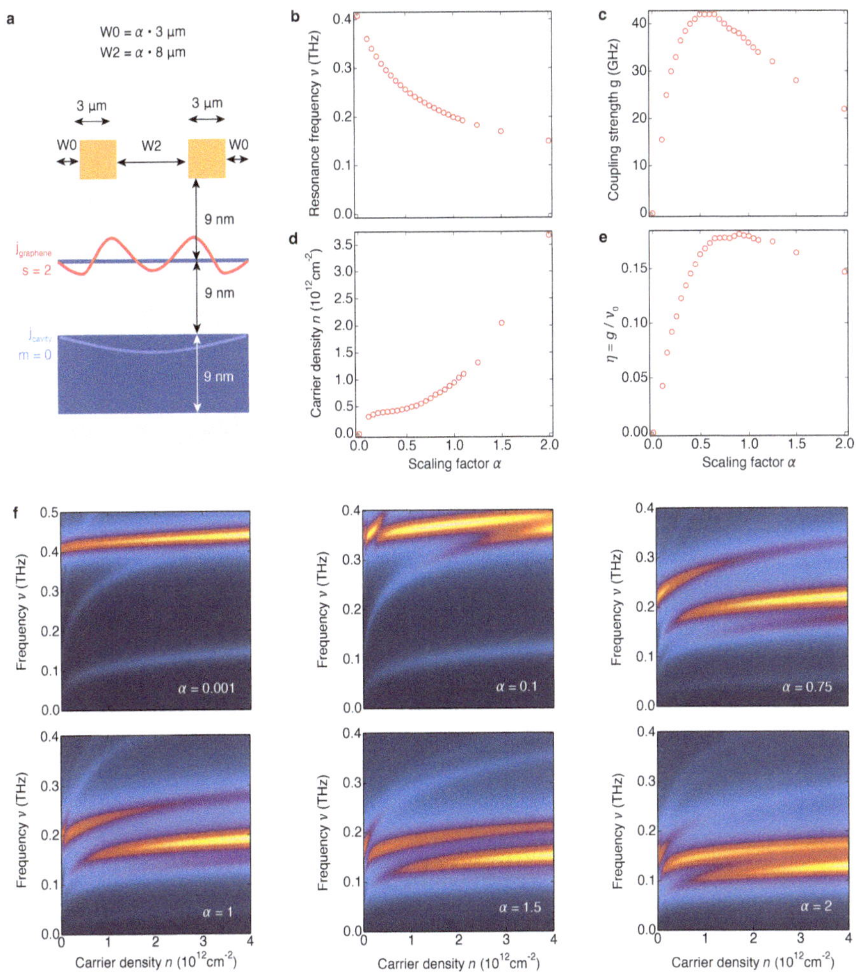

Fig. 8.20 Systematic analytical theory calculations of a heterostructure with varying coplanar strip design. (**a**) Cross-section of the calculated cavity. The widths $W_0 = 3\,\mu$m and $W_2 = 8\,\mu$m are scaled by a factor α. (**b–e**) Resonance frequency ν, coupling strength g, carrier density n and normalized coupling strength η for the avoided crossing between the $s = 2$ and $m = 0$ modes as a function of α. (**f**) Selection of analytical theory calculations for different values of α. Note that the cavity conductivity amplitudes are normalized of the individual panels to the maximum conductivity, but are not comparable between the panels. The maximum conductivity of the panel for $\alpha = 2$ is ~ 200 times larger than of the panel for $\alpha = 0.001$, neglecting filling factors. (Figure adapted from [1] under a CC BY 4.0 license)

Although the $s = 0, 1$ modes are still far away in carrier density to be in resonance with the $m = 0$ mode, they already start forming lower polariton branches and thus show suppressed power law scalings (as measured in GG1, see 8.14). Further, the modes that are already tuned to higher frequencies than the $m = 0$ mode,

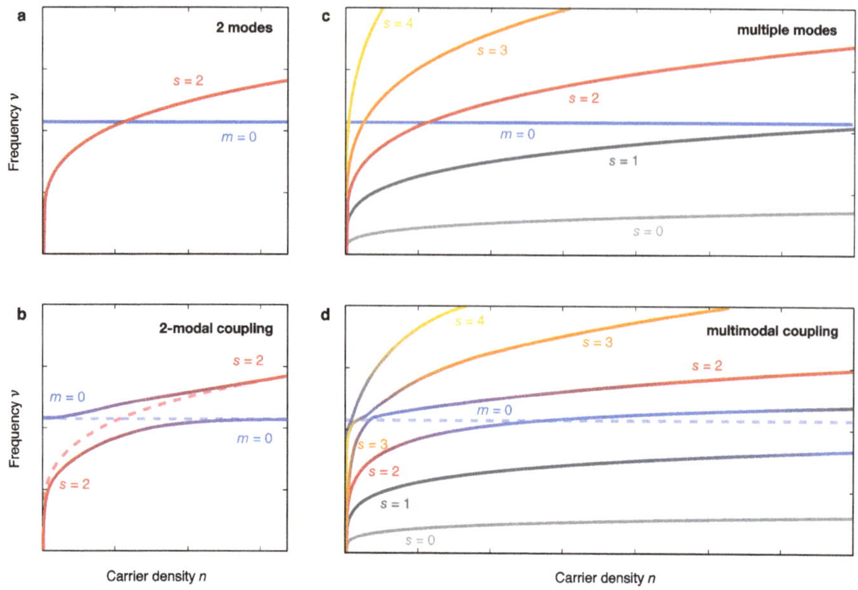

Fig. 8.21 Sketch of 2-modal coupling vs multimodal coupling. (**a**) Uncoupled graphene-like ($s = 2$) and cavity $m = 0$ as a function of graphene carrier density n. (**b**) When the modes hybridise, an avoided crossing appears, and the lower polariton branch saturates to the $m = 0$ frequency. (**c**) Carrier-density dependence of resonance frequency of a cavity mode and multiple graphene-like modes. (**d**) When these modes hybridise, level repulsion is observed around their crossing points, and the power-law scales of graphene-like modes are significantly suppressed. (Figure adapted from [1] under a CC BY 4.0 license)

meaning that they passed the avoided crossing as the lower polariton branch already, start turning into the upper polariton branch of the next s mode that is tuned into resonance with the $m = 0$ mode. For example, when following the path of the branch that starts as the red $s = 2$ mode in Fig. 8.21d, one finds that the mode gains character of the $m = 0$ mode (becomes blue) when being in resonance with the $m = 0$ mode. However, instead of saturating at the frequency of the $m = 0$ mode, it further increases in frequency as it then starts to form the upper polariton branch of the avoided crossing with the dark grey $s = 1$ mode, which occurs at carrier densities slightly extending the sketched range.

The multimodal nature of self-cavity mode couplings adds complexity to the interpretation of cavity conductivity spectra. Accurately extracting the coupling strength g is challenging because the lower polariton branches are shifted to higher resonance frequencies as they begin to form upper polariton branches through hybridization with the next higher momentum mode and the gate mode. This results in an apparent reduction in mode splitting. Therefore, the normalized coupling strengths measured in this thesis should be considered as the lower bound of the actual coupling strength achieved in a given micropatterned vdW heterostructure.

In summary, the previous sections revealed that the multimodal coupling of two self-cavities is enabled through two different mechanisms: Current-current hybridization (Sect. 8.3.1) and coupling of screened and unscreened modes (Sect. 8.3.2). Having discussed how each of the two mechanisms can be maximized or suppressed, let's next aim for spectrally resolving ultrastrong coupling in a graphene-graphite heterostructure using on-chip THz spectroscopy.

8.4 Spectrally Resolving Ultrastrong Coupling

Having identified the two mechanisms driving the multimodal self-cavity coupling in gate-tunable vdW heterostructures within on-chip THz spectroscopy, let's now turn to the design, fabrication, and measurement of a graphene–graphite heterostructure engineered for enhanced coupling (surpassing that of Device GG1 shown in Sects. 8.1 and 8.2.5) with the aim of potentially entering the ultrastrong coupling regime (Sect. 7.4.3).

Based on the design principles for maximizing both current-current hybridization (Sect. 8.3.1) and the coupling between screened and unscreened symmetric and antisymmetric modes (Sect. 8.3.2), an ideal next step might appear as the fabrication and measurement of a heterostructure with a thin graphite layer positioned just a few nanometers from the graphene, embedded in a coplanar strip with $W_2 > W_1$. However, such a configuration would yield a highly congested cavity conductivity spectrum, where numerous graphite and graphene self-cavity modes of finite linewidths overlap and hybridize simultaneously, making it virtually impossible to disentangle and interpret the resulting broad spectrum.

A balance between coupling strength and spectral clarity was found using the device geometry shown in Fig. 8.22. By making both the graphite and the hBN between the graphene and graphite layers thinner as compared to the Device GG1 (see Fig. 8.22), the graphite self-cavity mode could be centered at 0.43 THz while the three graphene modes, $s = 0, 1, 2$, overlap in frequency within the experimental bandwidth.

The experimental results from both DC transport and on-chip THZ spectroscopy measurements on Device GG2 are discussed in the following sections.

8.4.1 DC Transport Measurement

As Device GG2 features four working edge contacts to the graphene, a DC transport measurement could be performed. A current of 100 nA was applied across the device and the voltage drop was measured between contacts A and B (Fig. 8.23a). Fig. 8.23b plots the measured voltage drop as a function of the back gate voltage that was applied between the graphene and graphite layer. The plot shows a very

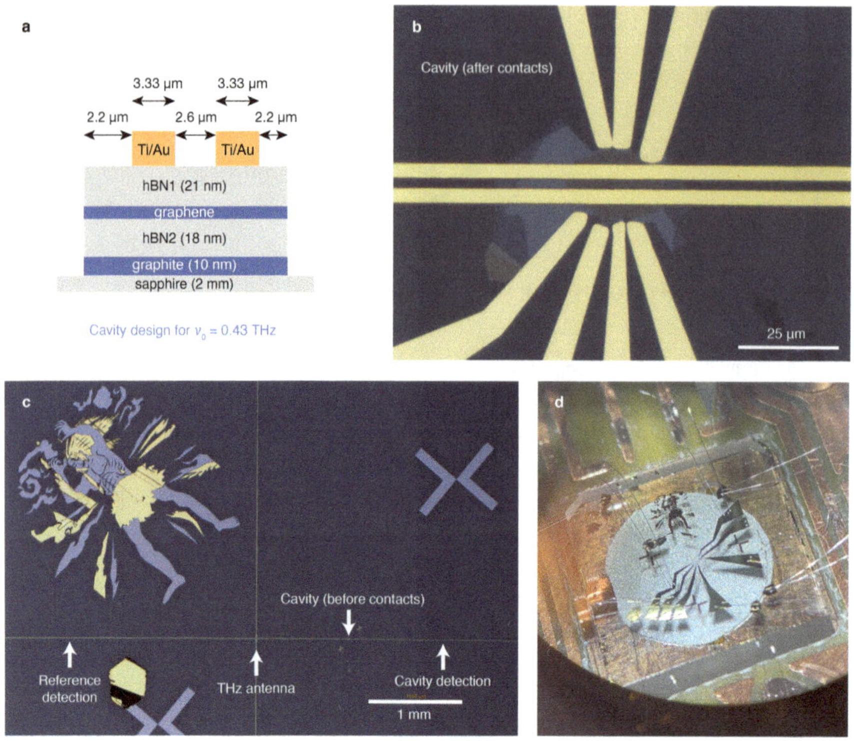

Fig. 8.22 Design of Device GG2. Device GG2 consists of a thin hBN layer between graphene and graphite to center the graphite self-cavity mode at 0.43 THz. (**a**) Cross-section with dimensions. (**b**) Optical micrograph of the integrated and contacted heterostructure. (**c**) Optical micrograph of the full on-chip THz spectroscopy circuitry. (**d**) Picture of the device mounted and wirebonded to a sample holder. (Parts of the figure adapted from [1] under a CC BY 4.0 license)

symmetric and narrow voltage drop curve indicating a good device quality. The maximum voltage drop corresponds to the back gate voltage at which the resistance of the graphene is maximum, marking the charge neutrality point.

The voltage drop could not be accurately converted into a precise resistance value due to the irregular shape of the channel used for the DC transport measurement. Three edge contacts were positioned on one side of the coplanar strip and one on the opposite side. Consequently, although the applied current flows across the entire coplanar strip, the voltage drop was measured only on one side, where the actual current is less than 100 nA. This setup prevents a direct comparison between the measured voltage drop and the applied current, making it not possible to reliably determine the resistance. However, by more carefully designing the placement of the edge contacts, DC resistance measurements can be performed and compared to the on-chip THz spectroscopy measurements in the future.

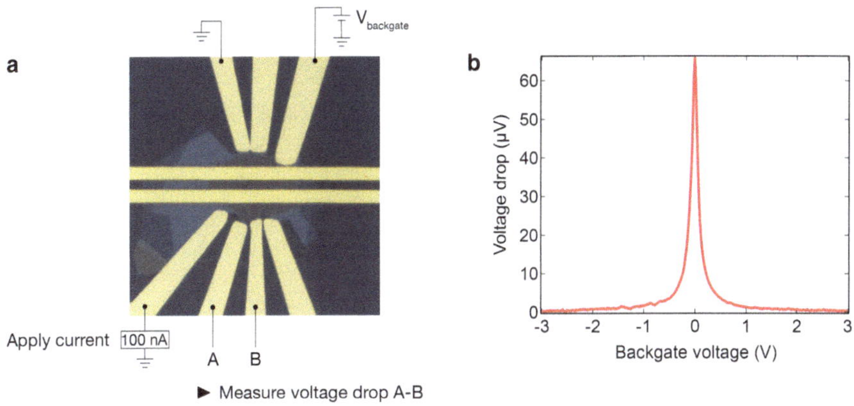

Fig. 8.23 DC transport measurement of Device GG2. (**a**) Measurement configuration in which a current of 100 nA was applied across the device and the voltage drop was measured between contacts A and B. (**b**) Measured voltage drop in graphene as a function of the applied back gate voltage. The resistance of the graphene as a function of back gate voltage (or carrier density) could not be reliably extracted from the voltage measurements due to the irregular positioning of the edge contacts on the graphene layer. As the two top left contacts to the graphene were electrically connected, a current was applied across the device, while the voltage drop was detected on the bottom, only on one side of the coplanar strip

8.4.2 Time Domain Data and Extracted Cavity Conductivity

Next, on-chip THz spectroscopy is performed on Device GG2 (Fig. 8.24a). The Fourier transforms were taken (Fig. 8.24b), and complex cavity conductivity (real and imaginary parts are shown in Fig. 8.24c, d) were calculated in the same procedure as Device GG1. A reference trace measured on a different circuit of identical design was used for the analysis of Device GG2 due to an imperfection in the coplanar strip on the reference side of Device GG2. The back gate voltage corresponding to the charge neutrality point was selected as the back gate voltage which led to the maximum voltage drop across the sample in the 4 point DC transport measurement (Fig. 8.23).

8.4.3 Data Fitting

The cavity conductivity spectra were then fit with a sum of four Drude-Lorentz models to capture the four expected self-cavity modes (see Fig. 8.25). Due to the congested nature of the spectrum in Device GG2, the data were fit using a global fit procedure across all carrier densities. Each carrier density was fit using the same fit function as used for Device GG1 (a sum of four Drude-Lorentz models and a sloped line), where the resonance frequency of the lowest three frequency modes were

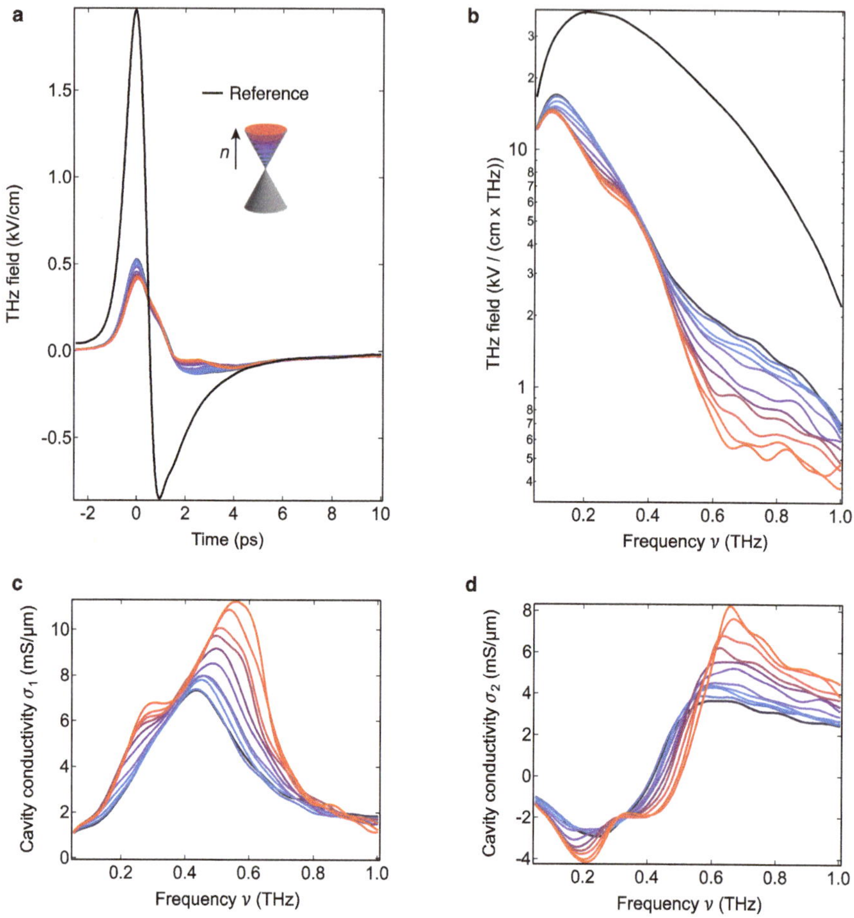

Fig. 8.24 Experimental data for Device GG2. (**a**) Measured time-domain traces taken on the reference coplanar strip (black trace) and on the cavity side for different graphene carrier densities (colored traces). (**b**) Real part of the Fourier transforms of the transients shown in panel (**a**). (**c**) Real and (**d**) imaginary parts of the cavity conductivity obtained from the Fourier transforms in (**b**). The maximum graphene carrier density is $n = 7.6 \times 10^{12}\,\text{cm}{-}2$. (Figure adapted from [1] under a CC BY 4.0 license)

assumed to scale with power-law behavior (the best approximation that could be used to capture the increase in resonance frequency as a function of carrier density). The powers of each of these modes were varied to find the best global fit over all carrier densities.

As the carrier density increases, all modes shift to higher frequencies, even when the graphene $s = 2$ mode exceeds the resonance frequency of the uncoupled $m = 0$ mode of the graphite self-cavity, that is expected to be at $\nu_0 = 0.43\,\text{THz}$ from the analytical theory. Thus, rather than merely crossing in frequency, as would be

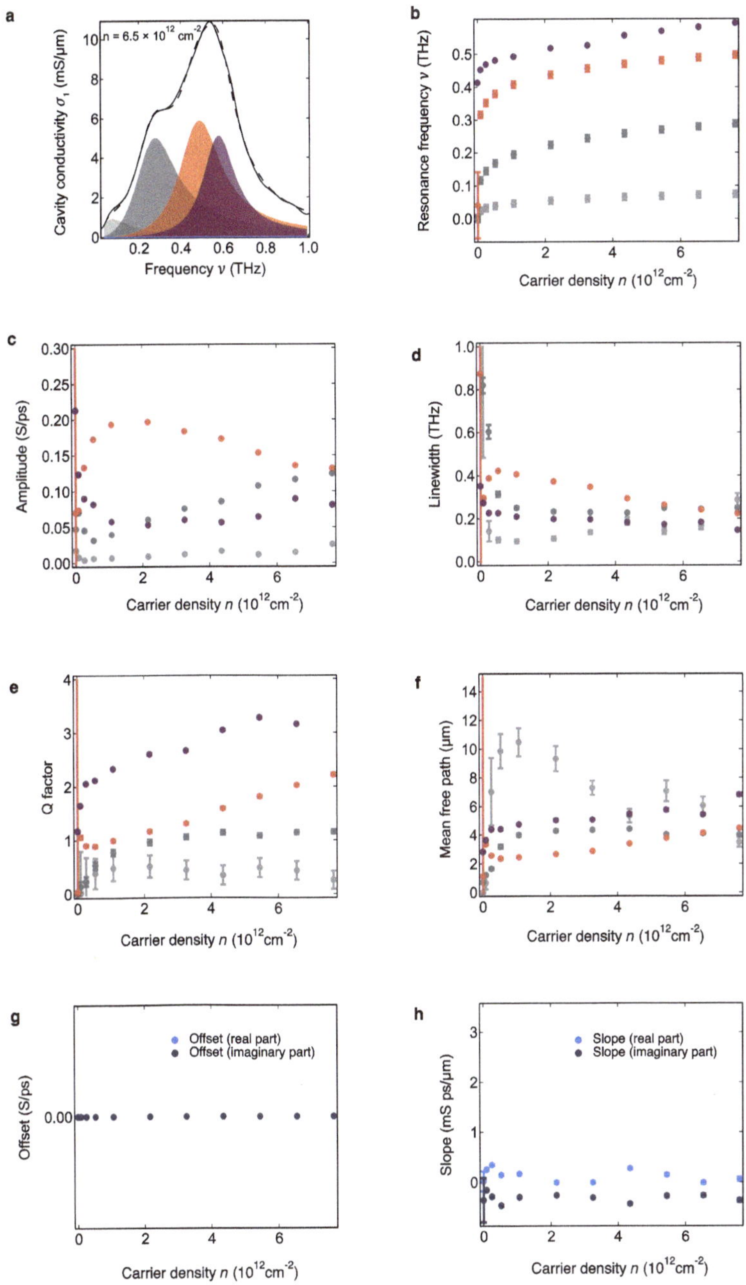

Fig. 8.25 Fit parameters from Device GG2. (**a**) Example fit result at a graphene carrier density of $n = 6.5 \times 10^{12}$ cm^{-2}. Real and imaginary parts of the cavity conductivity were fit to a sum of four Drude-Lorentz models and an offset with a slope. (**b**) Resonance frequencies, (**c**) amplitudes, (**d**) linewidths, (**e**) quality factors, (**f**) mean free paths, (**g**) offsets and (**h**) slopes as a function of graphene carrier density. (Figure adapted from [1] under a CC BY 4.0 license)

expected given the different momenta of the $m = 0$ and $s = 2$ modes, they hybridize and split. The multimodal splitting of the modes therefore appears slightly different than expected for two coupled modes, as discussed in Sect. 8.3.3.

The other parameters extracted from the fits such as linewidths, quality factors and mean free paths (see Fig. 8.25) suggest, that the scattering in Device GG2 is generally increased when compared to Device GG1 (see Fig. 8.3).

8.4.4 Comparison to Analytical Theory Calculations

Figure 8.26 shows the experimentally detected resonance frequencies of the four self-cavity modes as a function of carrier density in comparison to calculations of the same device geometry with the analytical theory. The lower scattering rate

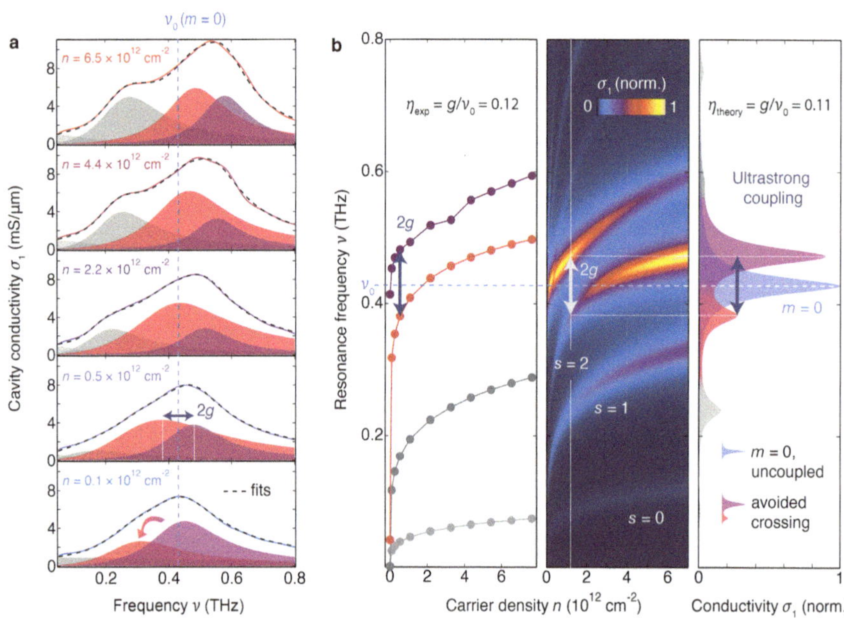

Fig. 8.26 Ultrastrong coupling between plasmonic self-cavity modes of graphene and graphite. (**a**) Experimentally-obtained real part of the near-field THz conductivity spectra of a selection of measured graphene carrier densities (see Fig. 8.24). The data were fit to a sum of four Drude-Lorentz models as indicated by the shaded peaks (see Fig. 8.25). (**b**) Left panel: Detected resonance frequencies as a function of carrier density for the four Drude-Lorentz models fit to the experimental data. The experimentally extracted normalized coupling strength $\eta_{\text{exp}} = 0.12 \pm 0.01$, achieving the ultrastrong coupling regime. The analytical theory for this device (center panel) follows similar trends, with $\eta_{\text{theory}} = 0.11 \pm 0.01$. The right panel shows the calculated conductivity spectra at a carrier density of $n = 0$, for the uncoupled $m = 0$ mode and for a carrier density of $n = 1.24 \times 10^{12}$ cm-2 at which the avoided crossing appears. (Figure adapted from [1] under a CC BY 4.0 license)

($\tau^{-1} = 0.03$ THz) than experimentally detected linewidth ($\Gamma \approx 0.2 - 0.4$ THz, see Fig. 8.25) was chosen in the analytical theory calculations for ease of interpretation and enhanced visibility of the avoided crossing (as shown in Sect. 6.4.9, the scattering rate does not change the resonance frequency of the cavity modes).

The normalized coupling strength η can be calculated from the mode splitting of the avoided crossing. However, there are different approaches to doing so that are elaborated on below, but are found to give approximately the same value for η. The approach chosen in Fig. 8.26 and also previously used to estimate the normalized coupling strength for GG1 (Sect. 8.2.5) calculates η at the carrier density, where the hybrid modes split equally around the unhybridised graphite $m = 0$ mode. This approach has the advantage that the normalized coupling strength can be extracted from the experimental data without comparing the data to the analytical theory.

The experimentally detected symmetric spectral separation of the two hybrid modes around the uncoupled graphite self-cavity $m = 0$ mode (dashed blue line in Fig. 8.26) occurs at a graphene carrier density of $n = 0.5 \times 10^{12}$ cm^{-2}. The mode splitting at this carrier density corresponds to twice the coupling strength g, which is found to be (50 ± 5) GHz (~0.2 meV). Comparing this to the resonance frequency of the uncoupled graphite $m = 0$ mode, (0.43 ± 0.01) THz (~1.8 meV), the normalized coupling strength is

$$\eta_{\text{exp}} = \frac{g}{\nu_0} = \frac{0.05\ \text{THz}}{0.43\ \text{THz}} \approx 0.12, \tag{8.56}$$

characterizing this heterostructure as in the ultrastrong coupling regime.

When determining η from the analytical calculations using the same approach, one finds

$$\eta_{\text{theory}}(n_1) = \frac{g}{\nu_0} = \frac{0.045\text{THz}}{0.43\text{THz}} = 0.105 \approx 0.11, \tag{8.57}$$

marked with the red solid line n_1 in Fig. 8.27f.

A more conventional approach to determine the normalized coupling strength is to evaluate it at the carrier density n where the two uncoupled modes, those that would form the avoided crossing, would intersect in the absence of coupling. According to the analytical theory, the lowest graphite resonance ($m = 0$) without graphene (or with graphene at charge neutrality) lies at $\nu_0 = 0.43$ THz, indicated by the dashed white or blue line in Fig. 8.27e, f. The undistorted graphene resonances, modeled with the graphite gate as a thick perfect metal, are shown in Fig. 8.27c. The $s = 2$ graphene mode crosses the $m = 0$ graphite mode at $n_1 = 1.13 \times 10^{12}$ cm^{-2}, marked by the left red line n_0 in Fig. 8.27f. At this density, the hybridized modes (solid white lines) exhibit an asymmetric splitting of approximately 0.093 THz, yielding a normalized coupling strength

$$\eta_{\text{theory}}(n_0) = \frac{g(n_0)}{\nu_0} = \frac{0.093\text{THz}/2}{0.43\text{THz}} = 0.108 \approx 0.11, \tag{8.58}$$

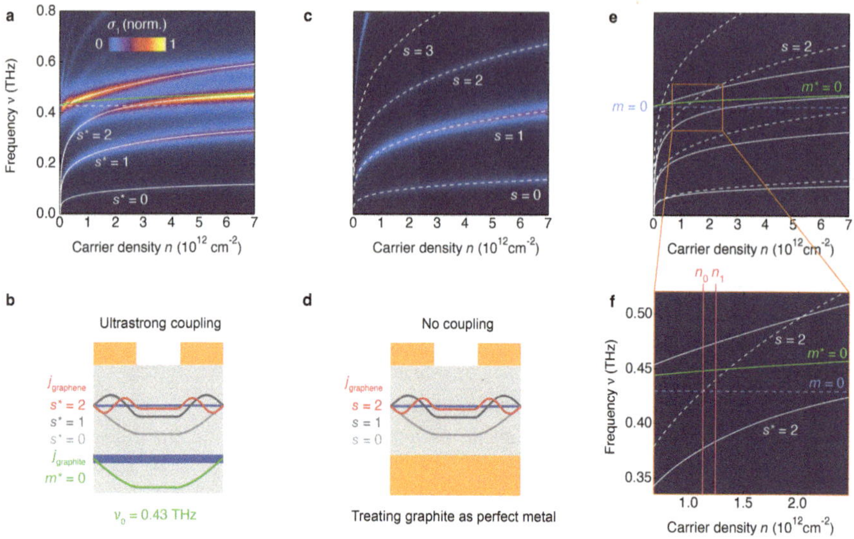

Fig. 8.27 Analytical theory calculations for Device GG2: Coupling strength calculation approaches. (**a**) Cavity conductivity spectra when the graphene modes couple to the graphite, $m = 0$ mode (**b**). Through the coupling with the graphite self-cavity, the s modes are suppressed in resonance frequency, forming the hybridized $s^* = 0, 1, 2$ modes and an avoided crossing. **c** Undistorted cavity conductivity spectra for the graphene $s = 0, 1, 2, 3$ modes obtained from the analytical theory when treating the graphite layer as a thick perfect metal layer (**d**). The cavity conductivity spectra in (**a**) and (**c**) are normalized and comparable between the two panels. For better comparison, the coupled modes (solid lines) of panel (**a**) and the uncoupled modes (dashed lines) of panel (**c**) are put on top of one another in panel (**e**), in addition to the $m = 0$ mode, which shows the resonance frequency of the graphite in the absence of graphene and the $m^* = 0$ mode, which gives the resonance frequency of the graphene carrier density dependent graphite-like symmetric mode. (**f**) Zoomed-in carrier density and frequency range of the avoided crossing shown in (**e**) with the carrier density markers n_0 and n_1 for different normalized coupling strength calculation approaches discussed in the main text. (Figure adapted from [1] under a CC BY 4.0 license)

consistent with the value obtained using the symmetric splitting method above. While the asymmetry observed in the avoided crossing has little effect on the extracted η, it can be seen as additional evidence for entering the non-perturbative ultrastrong coupling regime (see Figure 3a in Ref. [16]). Moreover, it is important to note that all measured η-values presented in this section represent lower bounds of the true coupling strengths within the heterostructure, owing to the multimodal nature of the coupling discussed in Sect. 8.3.3. Thus, it is valid to say that in Device GG2, the graphene and graphite self-cavities are hybridized in the *ultrastrong coupling* regime.

While the coupling of screened and unscreened modes becomes directly clear through the formation of self-cavity modes of different momenta, that would otherwise not have been possible, the amount of frequency shifting through the current-current hybridization becomes only visible when plotting the hybridized, screened symmetric mode, as shown in green in Fig. 8.27e, f. The green $m^* = 0$ curve was calculated with the dispersion relation found by solving Maxwell's equations for a purely screened symmetric mode in a vdW heterostructure consisting of the same graphene, hBN and graphite layers as Device GG2. By excluding the coupling to unscreened regions, this curve isolates the effect of current–current hybridization, which can be quantified by comparing it to the blue dashed line representing the uncoupled $m = 0$ graphite mode. To ensure matching resonance frequencies with the inhomogeneous Device GG2, the plasmon wavevector was set to $q = \frac{2\pi}{5.375 \cdot W_1}$, where $W_1 = 3.33$ µm is the same width as for Device GG2. The factor 5.375 reflects the elongation of the plasmon wavelength caused by unscreened regions, increasing it from the standard quarter-wave $\lambda = 4 \cdot W_1$ to $\lambda = 5.375 \cdot W_1$, as revealed in Sect. 6.5.3.2 through current density calculations for varying coplanar strip gaps.

Thus, Device GG2 exhibits an avoided crossing between the $s = 2$ and $m = 0$ modes that is characterized by the ultrastrong coupling regime independent of the normalized coupling strength analysis protocol, which is further evidenced by significantly modified electrodynamic properties of the system. Tracking the resonance frequencies of the four modes as a function of carrier density (seen in Fig. 8.26b) reveals that the lowest order graphene mode $s = 0$ is again primarily uncoupled from the graphite self-cavity as it scales with $\nu_{s=0} \propto n^{0.24 \pm 0.01}$. However, the $s = 1$ and $s = 2$ modes are significantly suppressed when compared to their uncoupled nature (see Fig. 8.27b).

In the following two sections, two further phenomena of ultrastrong self-cavity coupling are explored: Spectral weight transfer (Sect. 8.4.5) and linewidth broadening (Sect. 8.4.6).

8.4.5 Spectral Weight Transfer Through Mode Hybridization

Another effect of enhanced mode coupling is spectral weight transfer. As previously shown in Fig. 6.18c, the conductivity amplitude of uncoupled graphene plasmonic modes is found to scale with $A \propto n^{0.5}$. In case that the scattering rate is constant as a function of graphene carrier density, the spectral weight of an unhybridized graphene plasmonic mode should therefore scale with $n^{0.5}$. In contrast, the graphite self-cavity mode, when being uncoupled, should only be weakly affected when the graphene carrier density n is increased.

The spectral weight of each experimentally detected mode was calculated as the area of each Drude-Lorentz model shown in Fig. 8.26a and plotted as a function of carrier density in Fig. 8.28a, which is found to be in good agreement with analytical theory calculations (Fig. 8.28b). While the rather uncoupled graphene plasmonic $s = 0$ and $s = 1$ modes indeed show increasing spectral weight with increasing graphene carrier density, the hybridizing $s = 2$ and $m = 0$ modes (magenta and purple) are showing significant deviations from the expected trend. As the graphene becomes metallic, the spectral weight of the upper branch of

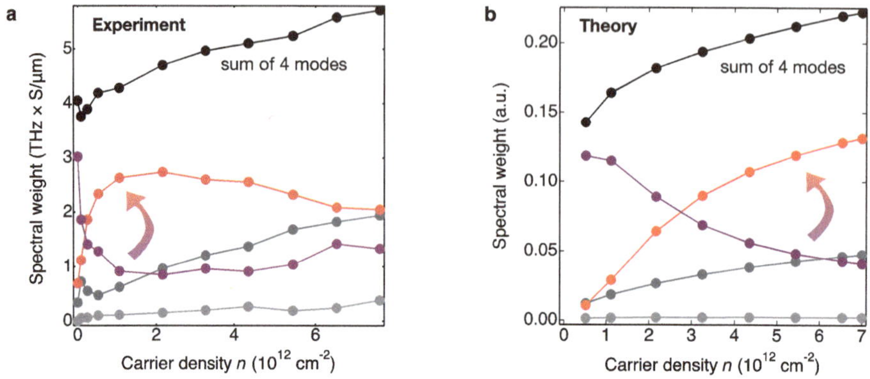

Fig. 8.28 Spectral weight transfer of the hybrid modes of Device GG2 shown in Fig. 8.26 from (**a**) experimental data and from (**b**) analytical theory calculations. In the absence of coupling, all four modes are expected to have increasing spectral weight with increasing graphene carrier density. In contrast, both experimental and calculated data show that this only applies for the $s = 0$ (bright grey) and $s = 1$ (darker grey) modes. The magenta and purple modes exchange spectral weight. (Figure (a) adapted from [1] under a CC BY 4.0 license)

the avoided crossing (purple) decreases, as the lower branch (magenta) rapidly increases in spectral weight. This finding stands in sharp contrast to the expected spectral weight behavior of graphene and graphite self-cavity modes in the absence of coupling. However, the overall spectral weight of the four modes (see Fig. 8.28), is still increasing with increasing graphene carrier density.

The detected transfer in spectral weight across the avoided crossing is thus a consequence of the hybridization and a further signature of ultrastrong coupling. Does this also apply for the detected broadened linewidths of Device GG2 in comparison to Device GG1?

8.4.6 *Linewidth Broadening Through Mode Hybridization*

As seen in Figs. 8.25, 8.26, the linewidths of the graphene plasmonic modes detected in the ultrastrong coupling heterostructure (Device GG2) are clearly broadened when compared to Device GG1 (Figs. 8.2, 8.3). Furthermore, the maximum scattering rate of the $s = 2$ mode in the ultrastrong coupling heterostructure (Device GG2), evidenced by the broadest linewidth, is detected at the carrier density ($n = 0.5 - 1.36 \times 10^{12}\ \mathrm{cm}^{-2}$) at which the coupling of the $s = 2$ and $m = 0$ mode is resonant (see Fig. 8.25d). This finding suggests that the detected linewidth broadening of the ultrastrong coupling Device GG2 in comparison to the less coupled Device GG1 is a feature of ultrastrong coupling. To investigate this phenomenon further, the analytical theory is used in this section to simulate the effect of linewidth broadening through enhanced self-cavity hybridization.

To obtain a better understanding of the linewidth broadening, let's investigate the effect of the graphite self-cavity linewidth on the linewidth of the graphene self-cavity modes using the analytical theory. As shown in Fig. 8.29, an increasing graphite self-cavity linewidth does not affect the linewidth of the graphene modes in a heterostructure with small W_2 and with graphene and graphite layers that are separated by a rather thick hBN flake (see Fig. 8.29a). In contrast, if the layers are only separated by a ~20 nm hBN flake (see Fig. 8.29b), the graphene-like (antisymmetric) modes broaden slightly through the formation of the hybrid symmetric and antisymmetric modes. This happens as both symmetric and antisymmetric modes have contributions of the scattering rates of both layers, and the effect is most prominent if the distance between the layer is small. However, the effect of mode broadening of graphene plasmonic modes is additionally enhanced in an ultrastrong coupling heterostructure (where $W_2 \gtrapprox W_1$) with an inhomogeneous dielectric environment (see Fig. 8.29c).

Thus, the broadening of the graphene modes observed in Device GG2 results from both coupling mechanisms: current-current hybridization (Sect. 8.3.1) and the interaction between screened and unscreened modes at the dielectric boundaries formed by the coplanar strip (Sect. 8.3.2). Moreover, Refs. [17–19] report that linewidth broadening is a well-established signature of ultrastrong coupling. While additional factors, such as the slightly irregular shape of the graphite flake (see Sect. 6.5.1, Fig. 6.23c), make it difficult to quantify the exact contribution, the linewidth broadening observed in Device GG2 nonetheless provides clear and consistent evidence pointing to an additional signature of ultrastrong coupling.

An additional effect accompanying linewidth broadening in the ultrastrong coupling regime is the emergence of lineshape asymmetry (see Ref. [17]). As shown in Fig. 8.30a, when the graphene and graphite scattering rates are assumed to be identical, the THz spectrum of the ultrastrong coupling cavity can be accurately fitted using a sum of multiple Drude-Lorentz models. However, when the graphite scattering rate is expected to be higher than that of graphene, and the graphene and graphite plasmonic cavity modes are coupled in the ultrastrong regime, the graphene plasmons broaden (as discussed above). More notably, the resulting THz spectrum can no longer be satisfactorily fitted by a sum of Drude-Lorentz models if the resonance frequencies are fixed to those obtained from the equal-linewidth case (see Fig. 8.30b, c). This suggests that, in addition to broadening, the plasmonic lineshapes become asymmetric in the ultrastrong coupling regime, as also discussed in Ref. [17]. Subtle indications of such asymmetry are visible in Figs. 8.3 and 8.25, where slight mismatches appear between the measured data and the symmetric fit curves. Nevertheless, fitting the data with symmetric Drude-Lorentz profiles still yields parameter values that lie well within the experimental error margins. This is particularly valid as the linewidth of the graphite plasmon is only marginally broader, by only a factor of two, compared to the graphene plasmon (comparable to the plot in Fig. 8.30b).

Interestingly, in the opposite limit, if the mode with the stronger dipole strength (graphite self-cavity) has a narrower linewidth than the coupling collective mode of a 2D material, the resonance from the collective mode of the 2D material can be

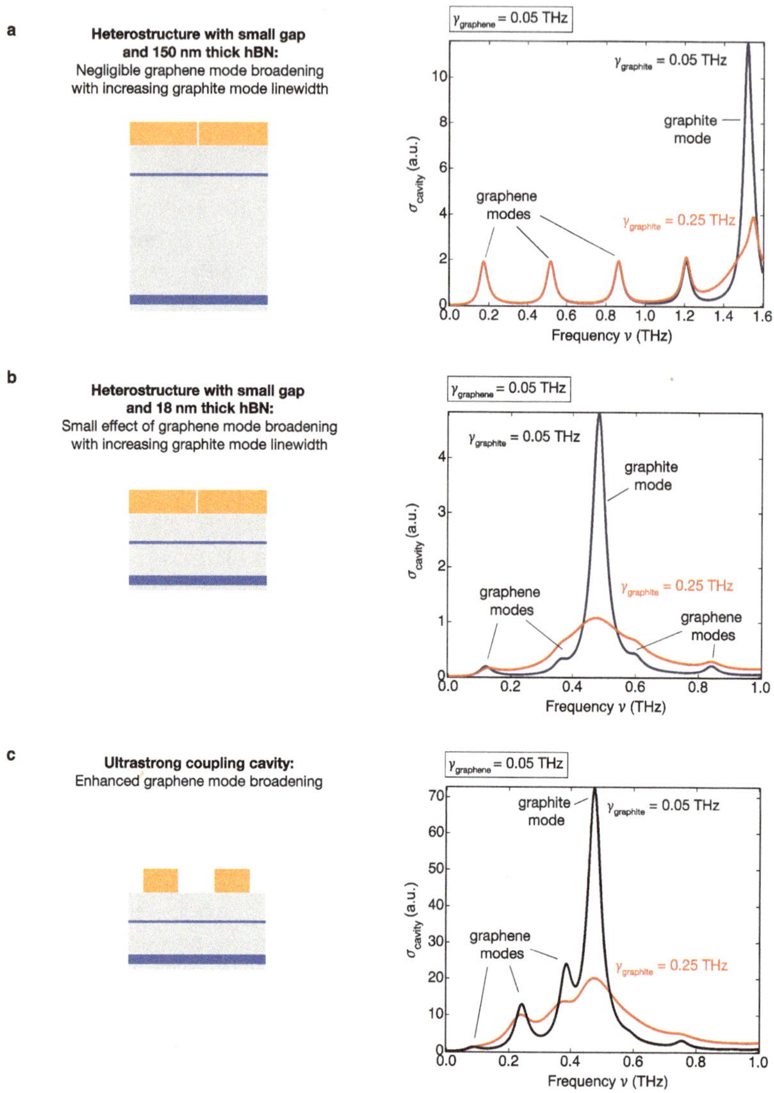

Fig. 8.29 Increased linewidth of graphene plasmonic modes through ultrastrong coupling. (**a**) In a device with small W_2, the broadening of the graphene modes through an increased scattering rate of the graphite mode is negligible, if the hBN layer between graphene and graphite is thick. Cavity parameters: $W_0 = 0\,\mu\text{m}$, $W_1 = 4\,\mu\text{m}$ and $W_2 = 0.1\,\mu\text{m}$, $d_{\text{gr}} = 30\,\text{nm}$, $d_{\text{hBN1}} = 10\,\text{nm}$, $d_{\text{hBN2}} = 150\,\text{nm}$ and $n = 4 \times 10^{12}\,\text{cm}^{-2}$. (**b**) Similar cavity geometry as in (**a**) with reduced hBN2 layer thickness. Current-current hybridization leads to broadening of the graphene modes if the graphite layer has a larger scattering rate. (**c**) In an ultrastrong coupling heterostructure (Device GG2) with $W_1 \approx W_2$, the effect of graphene mode broadening is even stronger. Cavity parameters: $W_0 = 2.2\,\mu\text{m}$, $W_1 = 3.33\,\mu\text{m}$ and $W_2 = 2.6\,\mu\text{m}$, $d_{\text{gr}} = 10.5\,\text{nm}$, $d_{\text{hBN1}} = 21.1\,\text{nm}$, $d_{\text{hBN2}} = 18.3\,\text{nm}$ and $n = 1.24 \times 10^{12}\,\text{cm}^{-2}$. (Parts of the figure adapted from [1] under a CC BY 4.0 license)

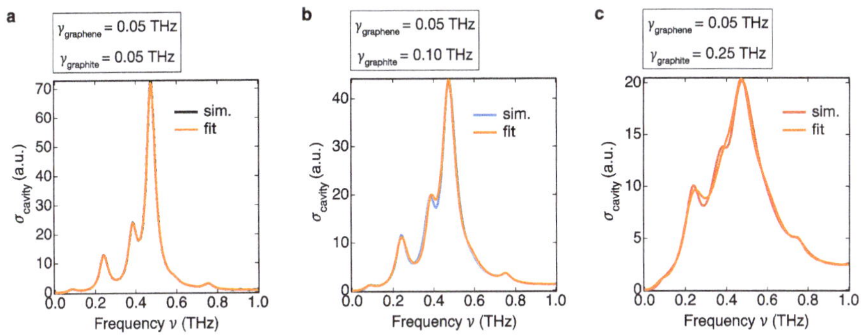

Fig. 8.30 Asymmetric lineshapes through hybridization of modes with different linewidths. (**a**) If graphene and graphite modes have the same linewidths in an ultrastrong coupling cavity, the cavity THz spectrum can be perfectly fit with a sum of multiple Drude-Lorentz models. (**b**, **c**) If the graphite mode has a larger linewidth than the graphene modes, the spectrum can't be fit precisely with a sum of multiple Drude-Lorentz models anymore, through the formation of asymmetric lineshapes in the ultrastrong coupling regime. Cavity parameters match those of Device GG2, and those shown in Fig. 8.29c

enhanced and narrowed down by (ultra)strong coupling to the gate cavity mode. This principle could be used to enhance weak or broad collective modes of a 2D material via ultrastrong coupling to a strong and narrow self-cavity mode of its electrostatic gate in the future, using the graphite thereby as a "sensing layer" (see Fig. A.11 in the Appendix).

Thus, similar to avoided crossings, spectral weight transfer and a reduced power law of graphene plasmonic modes, altered linewidths are another reporter of (ultra)strong coupling.

8.5 Summary

The present chapter discussed the first experimental evidence of the multimodal, ultrastrong coupling between gate-tunable plasmonic self-cavity modes in graphene and its graphite gate, in the deep subwavelength regime using on-chip THz spectroscopy (Sects. 8.1 and 8.4). The detected avoided crossings, along with significantly suppressed power laws of the graphene plasmonic modes, spectral weight transfer, broadened linewidths and a normalized coupling strength of $\eta \approx 0.12$ all serve as reporters of enhanced hybridization.

The analytical theory developed for capturing the THz electrodynamics of vdW heterostructures consisting of a single conductive layer (Sect. 6.4) was extended to capture the multimodal interaction of two conductive self-cavities (graphene and graphite) in a gate-tunable heterostructure (Sect. 8.2). The theory reveals two different coupling mechanisms (Sect. 8.3):

1. **Current-current hybridization:** When two conductive layers, such as graphene and a graphite gate, are separated by only a few tens of nanometers, as is often the case in high-quality DC transport vdW heterostructures, the oscillating currents in each layer can hybridize, forming symmetric (in-phase) and antisymmetric (out-of-phase) plasmonic modes. This hybridization can significantly weaken the power-law behavior of graphene plasmonic modes ($\nu_0 = n^x$, with $x \ll 0.25$) because the symmetric and antisymmetric modes combine characteristics from both layers. The hybridization effect is especially strong when the 2D conductivities of the two layers are similar, and when the gate layer is separated by only a thin hBN layer from the gated 2D material. In such an extreme case, hybrid modes with the **same momentum** can lead to avoided crossings, approaching coupling strengths that nearly enter the "deep-strong coupling" regime ($\eta > 1$). However, due to the multimodal nature of the coupling, the spectrum becomes too congested to accurately capture the electrodynamics when hybrid graphene and graphite plasmonic modes of the same momentum overlap in frequency. To avoid this, the graphite layer should not be too thin (≥ 10 nm), so it can further serve effectively as a back gate, too.
2. **Coupling of screened and unscreened plasmonic modes:** The metallic transmission traces impose sharp electrodynamic boundary conditions, leading to the formation of screened symmetric and antisymmetric modes beneath the metal traces, as well as unscreened modes within the coplanar strip gap and outside the metal regions. While momentum is generally conserved in these sub-wavelength heterostructures (see Sect. 6.5.3.3), the coupling between modes of **different momenta** only depends on the resonance alignment of screened and unscreened modes. The unscreened symmetric mode, with a high resonance frequency, contrasts with the unscreened antisymmetric mode, which can resonate at frequencies comparable to the screened graphene and graphite modes of different momenta. This resonance matching allows the unscreened antisymmetric mode to resonantly enhance the coupling at the dielectric boundaries between screened and unscreened regions. The coupling strength can be tuned by adjusting the relative widths of the unscreened regions (in the coplanar strip gap and outside the metal traces) and the screened regions (metal trace widths). When the unscreened regions are much narrower than the metal traces, the unscreened antisymmetric mode has a much higher resonance frequency than the screened modes, suppressing the coupling. Conversely, when the gap is wider than the metal traces, resonance frequencies align, allowing ultrastrong coupling between modes of different momenta.

8.6 Discussion

This thesis chapter revealed two different coupling mechanisms in gate-tunable vdW heterostructures integrated to on-chip THz spectroscopy circuitry. However, the formation of symmetric and antisymmetric modes due to current-current hybridization in two spatially separated layers is a known phenomenon in literature [13] and appears in many different contexts. Screened plasmons in a heterostructure consisting of a single conductive layer, such as graphene, are also called "acoustic" plasmons [20]. This type of 2D plasmons can be interpreted as an antisymmetric plasmon oscillation between the carriers in a graphene layer and the image charges that form in the screening metal. There are also "gap plasmons", which are surface plasmons that form in two metal layers that are only separated by a thin dielectric spacer [21]. Similarly, the appearance of (anti)symmetric modes is also known from metal-insulator-metal waveguides, where both surface plasmonic and photonic (anti)symmetric modes can form [22]. While the plasmonic modes define the scenario in which light couples to the free electrons in the metal layers, the photonic modes involve the interaction of light with the dielectric spacer. In optical phonon spectroscopy, it is also possible to detect analogies to symmetric and antisymmetric modes. While adjacent layers oscillate in-phase in case of optical phonon polaritons, they oscillate out-of-phase in case of shear optical phonons [23].

While the formation of symmetric and antisymmetric modes is well-established in the literature, and edge reflections of plasmonic modes in vdW heterostructures have been observed using techniques such as scanning probe techniques [24], this thesis presents several additional key advancements beyond those outlined in Sect. 6.7:

1. **Full understanding of gate—sample interactions in on-chip THz spectroscopy**

Previous on-chip THz experiments, such as [7, 12, 25–27], did not have a separate reference and thus were measuring relative changes in conductivity, based on the assumption that their electrostatic gates are transparent and do not contribute noticeably to the THz signal. However, this work shows that the plasmonic modes that form in an electrostatic graphite gate and in a graphene layer can interact strongly and hybridize. This interaction can be cleanly resolved for metallic graphite gates as they, for a given momentum, feature higher resonance frequencies than the 2D material they probe. In contrast, for some semiconducting gates, the 2D conductivity of the gate can become comparable to that of the gated 2D layer. Such a scenario can lead to significantly enhanced coupling between the plasmonic modes of graphene and gate, as seen in Sect. 8.3.1, and further results in a highly congested cavity conductivity spectrum where various hybridized self-cavity modes of different momenta of both gate and gated 2D material overlap in frequency. This makes it difficult to disentangle individual modes from one another, and while the resulting broad spectrum may look like a Drude response, a Drude fit would yield incorrect material properties (Sect. 6.6).

A further disadvantage of semiconducting gates is, that the exact carrier density cannot always be precisely calibrated, so effects are qualitative [28], and gating over a wide carrier-density and displacement field range, particularly at low temperatures, remains challenging [29]. On the other hand, graphite gates are the standard for DC transport due to their reliability and minimal disorder (see Fig. S11 in Ref. [30]). In developing a robust framework for the analysis of graphite-gated heterostructures in on-chip THz spectroscopy, this thesis offers a practical alternative design framework.

2. **Delivery of distinct design principles for “cavity sensing” or “cavity control”**

While the measured devices highlighted in this work exhibited coupling between sample and graphite self-cavity modes, the analytical theory further provides design principles to construct heterostructures where the coupling between gated 2D material and gate is minimized. The theory predicts that a heterostructure built with a thick hBN, thick graphite gate, and with a coplanar strip patterned to cover all but a small gap between the traces minimises both coupling mechanisms described above.[3] This also includes the hybridization induced linewidth broadening discussed in Sect. 8.4.6, which can occur if the plasmon with the larger oscillator strength has a larger scattering rate than the plasmon that forms in the layer with lower scattering rate. If the gate layer is made thick enough ($\gtrsim$20 nm) such that its 2D conductivity is sufficiently larger than that of the 2D material it gates, hybridization induced linewidth broadening can be suppressed. This enables the simple subtraction of the graphite back gate contribution to the overall cavity conductivity spectrum, even if the graphite self-cavity mode is located within the spectral sensitivity range (see also Sect. A.7 of the Appendix: full 3D electromagnetic simulations clearly show that such a subtraction is possible). Measurements in such a **“cavity sensing”** geometry thus enables probing the undistorted electrodynamics of a 2D material (see Fig. 8.31a, c), which will provide deeper insight into the dynamics of correlated materials in future experiments [31].

To spectrally resolve the THz electrodynamics of a vdW heterostructure in the ultrastrong coupling regime (**“cavity coupling”**), both coupling mechanisms must be enhanced (see Fig. 8.31b, d). Current–current hybridization is enhanced by separating the 2D material and gate with a thin hBN layer and using a gate of comparable conductivity, such as thin graphite. The second mechanism (hybridization of screened and unscreened modes) is maximized by embedding the heterostructure in a coplanar strip with a gap width $W_2 > W_1$, aligning the unscreened antisymmetric mode with screened modes of varying momenta.

While these design principles for maximum self-cavity coupling appear straightforward, there are multiple complexities to consider. Firstly, the higher the normalized coupling strength, the lower the resonance frequencies of the coupling self-cavity modes. As the self-cavity modes have finite linewidths, measuring such high coupling strengths is challenging as many hybrid modes are congested in the spectral sensitivity range that overlap in frequency, making it impossible to disentangle individual modes from a measured spectrum (see additional data sets of Devices GG3 and GG4 in the Appendix Sect. A.3).

[3] As seen in Sects. 2.2.3 and 6.7, the coplanar strip gap can only be narrowed down locally around the sample beneath the transmission lines, not over the course of a few mm, as otherwise a significant amount of the signal gets attenuated (see also Fig. A.10 in the Appendix).

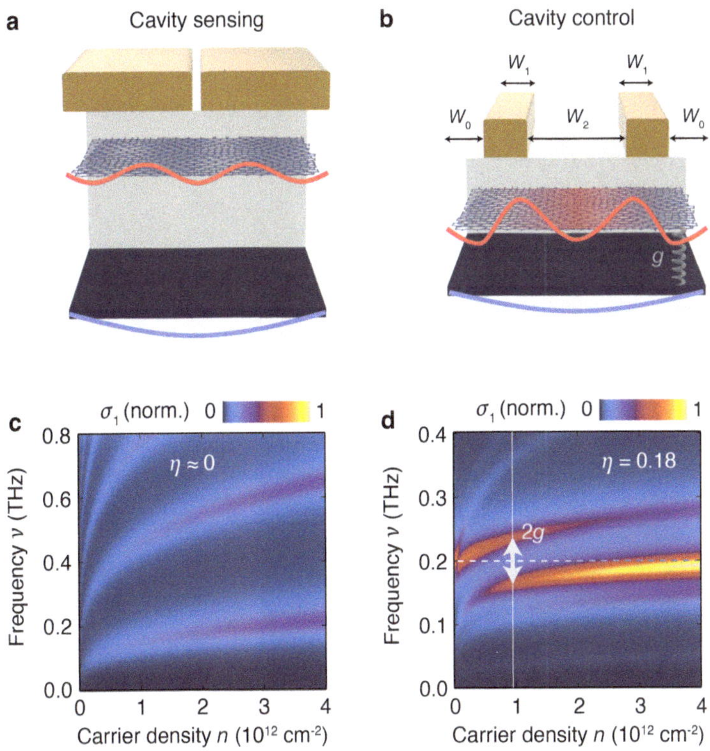

Fig. 8.31 Coupling strength tunability through geometry: From cavity sensing to cavity control. (**a**) A cavity with a small-gap coplanar strip, no extended sample regions and thick graphite and thick hBN2 layer behaves as a sensing tool, meaning the properties of a sample of interest can be probed without the additional influence of resonant self-cavity coupling to the graphite gate. (**b**) A heterostructure with extended 2D material region W_0 and gap W_2, and thin graphite and hBN layers can be used to control collective modes of 2D materials. (**c**) calculation of a device as in (**a**) with a graphene monolayer. Multiple undisturbed graphene plasmonic modes can be obtained that all scale as $\nu \propto n^{0.25}$. (**d**) Calculation of a device as in (**b**) with a graphene monolayer. An avoided crossing with $\eta = 0.18$ is observed. (Figure adapted from [1] under a CC BY 4.0 license)

Secondly, while particularly thin graphite gates $\lesssim$10 nm maximize the hybridization, they also can't be made too thin to still enable reliable electrostatic gating. Since the charge carrier density in such thin graphite gates is comparable to that of the material they control (graphene), both the Fermi levels of the graphene and that of the graphite gate are simultaneously affected when tuning the system. This effect can indeed be seen in the data sets of two additional gate-tunable graphene heterostructures with thin graphite gate (compare Devices GG2, GG3 and GG4 in the Appendix Sect. A.3) when not looking only at positive, but also at negative graphene dopings. While the graphene plasmon yields minimum spectral weight at the charge neutrality point of graphene, the graphite plasmon has minimum spectral

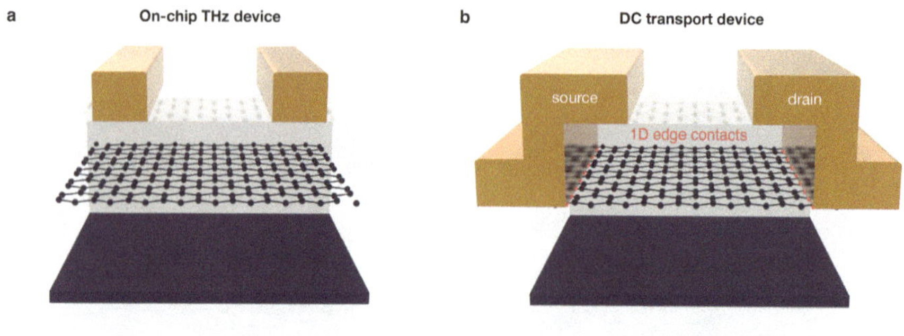

Fig. 8.32 On-chip THz device in comparison to a typical DC transport device. (**a**) The dielectric environment of a vdW heterostructure is micropatterned due to the coplanar strips, that allow the coupling of plasmonic modes of different momenta that form in both the 2D material and the gate at the boundaries between screened and unscreened regions. (**b**) The edge contacts (source and drain) in typical DC transport devices also lead to micropatterning of the dielectric environment of a vdW heterostructure. This suggests that self-cavity modes and cavity-hybridization effects may also play a role in devices other than on-chip THz spectroscopy samples

weight at a negative graphene doping.[4] This finding aligns with literature that shows that thin graphite is slightly hole doped [32]. The analytical theory can be built upon in the future by incorporating graphite band structure calculations to obtain a more complete picture of the electron-hole asymmetries.

3. **Exploration of intrinsic, deterministic cavity control of gate-tunable vdW heterostructures**

Having demonstrated that cavity electrodynamics and hybridization between plasmonic modes in both the gate and gated 2D material are essential considerations for vdW heterostructures of typical dimensions, this becomes particularly significant given that the plasmonic self-cavity resonances of graphite flakes lie within the same energy ($\sim$0.1–10 m eV) or frequency ($\sim$25 GHz–2.5 THz) range as the phenomena they electrically tune. While this work demonstrates ultrastrong coupling between plasmonic graphite self-cavity modes and graphene plasmonic modes, resulting in avoided crossings, suppressed power laws, and spectral weight transfer (and potentially increased scattering, as discussed in Sect. 8.4.6), it suggests that cavity-coupled effects from proximitized graphite gates could enable deterministic cavity control of various phenomena in future vdW heterostructures. The analytical theory further reveals that such cavity electrodynamics are photon-number-independent, implying that even quantum fluctuations may be sufficient to induce self-cavity

[4] The electron-hole asymmetry of the investigated devices GG1 (Sect. 8.1) and GG5 (Sect. A.3) that consist of a thick graphite cannot be evaluated as the lowest frequency modes of the graphite layers (symmetric modes) lie outside the spectral sensitivity range. However, as shown for Device GG5 in Fig. A.3 in the Appendix, the graphene plasmonic modes are centered relatively symmetric around the charge neutrality point.

effects. Finally, because microstructured gates and contacts in typical vdW devices inherently create inhomogeneous dielectric environments, similar to the metallic coplanar strips used here (see Fig. 8.32), these hybridization effects may also be crucial for the general understanding of strongly correlated quantum phases in vdW heterostructures.

By combining simultaneous DC transport and on-chip THz spectroscopy measurements, the chip-scale platform developed in this thesis provides a powerful avenue for probing emergent phases in 2D materials ("cavity sensing") and for experimentally testing theoretical proposals of cavity-controlled phenomena ("cavity control") through tailored heterostructure designs and precise measurements.

References

1. Kipp G, Bretscher HM, Schulte B, Herrmann D, Kusyak K, Day MW, Kesavan S, Matsuyama T, Li X, Langner SM, Hagelstein J, Sturm F, Potts AM, Eckhardt CJ, Huang Y, Watanabe K, Taniguchi T, Rubio A, Kennes DM, Sentef MA, Baudin E, Meier G, Michael MH, McIver JW (2025) Cavity electrodynamics of van der Waals heterostructures. Nat Phys 21:1926–1933. https://doi.org/10.1038/s41567-025-03064-8
2. Michael MH, Kipp G, Potts AM, Day MW, Matsuyama T, Meier G, Bretscher HM, McIver JW (2025) Resolving self-cavity effects in two-dimensional quantum materials. arXiv preprint arXiv:2505.12799
3. Kesavan S (2024) Engineering THz plasmonic cavity modes in microstructured van der Waals heterostructures. Master's thesis. University of Hamburg
4. Langner SM (2024) On-chip THz spectroscopy of van der Waals heterostructures. Bachelor thesis. University of Hamburg
5. Graef H, Mele D, Rosticher M, Banszerus L, Stampfer C, Taniguchi T, Watanabe K, Bocquillon E, Fève G, Berroir J et al (2018) Ultra-long wavelength Dirac plasmons in graphene capacitors. J Phy: Mater 1:01LT02. https://doi.org/10.1088/2515-7639/aadd8c
6. Yoshioka K, Bernard G, Wakamura T, Hashisaka M, Sasaki K-I, Sasaki S, Watanabe K, Taniguchi T, Kumada N (2024) On-chip transfer of ultrashort graphene plasmon wave packets using terahertz electronics. Nat Electron 7:537–544. https://doi.org/10.1038/s41928-024-01197-x
7. Zhao W, Wang S, Chen S, Zhang Z, Watanabe K, Taniguchi T, Zettl A, Wang F (2023) Observation of hydrodynamic plasmons and energy waves in graphene. Nature 614:688–693. https://doi.org/10.1038/s41586-022-05619-8
8. Fei Z, Goldflam M, Wu J-S, Dai S, Wagner M, McLeod A, Liu M, Post K, Zhu S, Janssen G et al (2015) Edge and surface plasmons in graphene nanoribbons. Nano Lett 15:8271–8276. https://doi.org/10.1021/acs.nanolett.5b03834
9. Alonso-González P, Nikitin AY, Gao Y, Woessner A, Lundeberg MB, Principi A, Forcellini N, Yan W, Vélez S, Huber AJ et al (2017) Acoustic terahertz graphene plasmons revealed by photocurrent nanoscopy. Nat Nanotechnol 12:31–35. https://doi.org/10.1038/nnano.2016.185
10. Xu S, Li Y, Vitalone RA, Jing R, Sternbach AJ, Zhang S, Ingham J, Delor M, McIver J, Yankowitz M, Queiroz R, Millis AJ, Fogler MM, Dean CR, Hone J, Liu M, Basov D (2023) Electronic interactions in Dirac fluids visualized by nano-terahertz spacetime mapping. arXiv preprint arXiv:2311.11502
11. Kim BS, Sternbach AJ, Choi MS, Sun Z, Ruta FL, Shao Y, McLeod AS, Xiong L, Dong Y, Chung TS et al (2023) Ambipolar charge-transfer graphene plasmonic cavities. Nat Maters 22:838–843. https://doi.org/10.1038/s41563-023-01520-5

12. Gallagher P, Yang C-S, Lyu T, Tian F, Kou R, Zhang H, Watanabe K, Taniguchi T, Wang F (2019) Quantum-critical conductivity of the Dirac fluid in graphene. Science 364:158–162. https://doi.org/10.1126/science.aat8687
13. Hwang E, Sarma SD (2009) Plasmon modes of spatially separated double-layer graphene. Phys Rev B 80:205405. https://doi.org/10.1103/PhysRevB.80.205405
14. Sarma SD, Madhukar A (1981) Collective modes of spatially separated, two-component, two-dimensional plasma in solids. Phys Rev B 23:805. https://doi.org/10.1103/PhysRevB.23.805
15. Christensen J, Manjavacas A, Thongrattanasiri S, Koppens FHL, García de Abajo FJ (2011) Graphene plasmon waveguiding and hybridization in individual and paired nanoribbons. ACS Nano 6:431–440. https://doi.org/10.1021/nn2037626
16. Zhang Q, Lou M, Li X, Reno JL, Pan W, Watson JD, Manfra MJ, Kono J (2016) Collective non-perturbative coupling of 2D electrons with high-quality-factor terahertz cavity photons. Nat Phys 12:1005–1011. https://doi.org/10.1038/nphys3850
17. Canales A, Karmstrand T, Baranov DG, Antosiewicz TJ, Shegai TO (2023) Polaritonic linewidth asymmetry in the strong and ultrastrong coupling regime. Nanophotonics 12:4073–4086. https://doi.org/10.1515/nanoph-2023-0492
18. Houdré R, Stanley RP, Oesterle U, Weisbuch C (2002) Strong coupling regime in semiconductor microcavities. Comptes Rendus Physique 3:15–27. https://doi.org/10.1016/S1631-0705(02)01299-9
19. Törmä P, Barnes WL (2014) Strong coupling between surface plasmon polaritons and emitters: a review. Rep Progr Phys 78:013901. https://doi.org/10.1088/0034-4885/78/1/013901
20. Lee I-H, Yoo D, Avouris P, Low T, Oh S-H (2019) Graphene acoustic plasmon resonator for ultrasensitive infrared spectroscopy. Nat Nanotechnol 14:313–319. https://doi.org/10.1038/s41565-019-0363-8
21. Ding F, Yang Y, Deshpande RA, Bozhevolnyi SI (2018) A review of gap-surface plasmon metasurfaces: fundamentals and applications. Nanophotonics 7:1129–1156.
22. Park J, Kim K-Y, Lee I-M, Na H, Lee S-Y, Lee B (2010) Trapping light in plasmonic waveguides. Opt Express 18:598–623. https://doi.org/10.1364/OE.18.000598
23. Jia G (2023) Hyperbolic shear polaritons and their potential applications. Nano TransMed 2:100019. https://doi.org/10.1016/j.ntm.2023.100019
24. Chen J, Badioli M, Alonso-González P, Thongrattanasiri S, Huth F, Osmond J, Spasenovic M, Centeno A, Pesquera A, Godignon P, Zurutuza Elorza A, Camara N, de Abajo FJG, Hillenbrand R, Koppens FHL (2012) Optical nano-imaging of gate-tunable graphene plasmons. Nature 487:77–81. https://doi.org/10.1038/nature11254
25. Parker-Jervis R, Park S, Zonetti S, Sydoruk O, Wu J, Wood C, Li L, Davies A, Linfield E, Cunningham J (2020) On-chip terahertz spectroscopy of magnetoplasmons in a two-dimensional electron gas. In: 2020 45th international conference on infrared, millimeter, and terahertz waves (IRMMW-THz). https://doi.org/10.1109/irmmw-thz46771.2020.9370779
26. Chen S-D, Feng Q, Zhao W, Qi R, Zhang Z, Abeysinghe D, Uzundal C, Xie J, Taniguchi T, Watanabe K, Wang F (2024) Direct measurement of terahertz conductivity in a gated monolayer semiconductor. arXiv preprint arXiv:2409.17633
27. Seo J, Lu Z, Park S, Yang J, Xia F, Ye S, Yao Y, Han T, Shi L, Watanabe K et al (2024) On-chip terahertz spectroscopy for dual-gated van der waals heterostructures at cryogenic temperatures. Nano Lett. https://doi.org/10.1021/acs.nanolett.4c04137
28. Hesp NC, Svendsen MK, Watanabe K, Taniguchi T, Thygesen KS, Torre I, Koppens FH (2022) WSe2 as transparent top gate for infrared near-field microscopy. Nano Letters 22:6200–6206. https://doi.org/10.1021/acs.nanolett.2c01658
29. Sterbentz R, Kim B, Flores-Garibay A, Haley KL, Pereira NT, Watanabe K, Taniguchi T, Island JO (2024) Gating graphene with a semiconductor. arXiv preprint arXiv:2410.06432
30. Zibrov AA, Kometter C, Zhou H, Spanton E, Taniguchi T, Watanabe K, Zaletel M, Young A (2017) Tunable interacting composite fermion phases in a half-filled bilayergraphene Landau level. Nature 549:360–364. https://doi.org/10.1038/nature23893

31. Van Loon EGCP, Hafermann H, Lichtenstein AI, Rubtsov AN, Katsnelson MI (2014) Plasmons in strongly correlated systems: spectral weight transfer and renormalized dispersion. Phys Rev Lett 113:246407. https://doi.org/10.1103/PhysRevLett.113.246407
32. Novoselov KS, Geim AK, Morozov SV, Jiang DE, Zhang Y, Dubonos SV, Grigorieva IV, Firsov AA (2004) Electric field effect in atomically thin carbon films. Science 306:666–669 (2004). https://doi.org/10.1126/science.1102896

Chapter 9
Conclusion

Abstract In this chapter, I summarize the key findings of this thesis, including the self-cavity behavior of van der Waals materials under THz excitation, the ultrastrong coupling between graphene and graphite plasmons, as well as the different contributing coupling mechanisms.

On-chip THz spectroscopy is a promising tool for investigating the THz response of vdW heterostructures [2–8]. By confining THz light to the near-field in metallic transmission lines interfaced to these micron-sized materials, the discrepancy between the free-space THz wavelength and small sample size can be overcome [9–11]. This approach has been used to measure gate voltage-dependent changes in transmitted THz fields, from which estimates of the sample conductivity have been made. This thesis introduced an upgraded circuit architecture containing a separate reference arm, enabling in situ referencing for the first time along with spectral readout from 0.05 − 1 THz [12]. The updated circuitry was employed to investigate various graphene and graphite heterostructures of different geometries, all of which exhibited one or multiple resonances in the detected near-field optical conductivity spectra. These findings are particularly intriguing as they deviate from the expected Drude response typical of metals, which is commonly observed in far-field THz spectroscopy of large-area graphene samples [13, 14], and previously reported to be observed in on-chip THz spectroscopy [2, 7, 8].

Based on the experimental results and full 3D electromagnetic simulations presented in this thesis, an analytical theory was developed to describe the near-field optical conductivity spectra and associated resonances in vdW heterostructures. The analytical theory revealed that subwavelength-sized conductive 2D materials, such as graphene and graphite flakes, naturally form plasmonic *self-cavities* for THz light (see Fig. 9.1 and Chap. 6). Similar to the mirrors of optical cavities, the edges of the 2D material provide sharp boundaries for excited currents, thereby enabling the generation of a series of standing waves of current density with discrete momenta.

G. Kipp, *Self-Cavity Electrodynamics of Van der Waals Heterostructures*, Springer Theses, https://doi.org/10.1007/978-3-032-31979-1_9

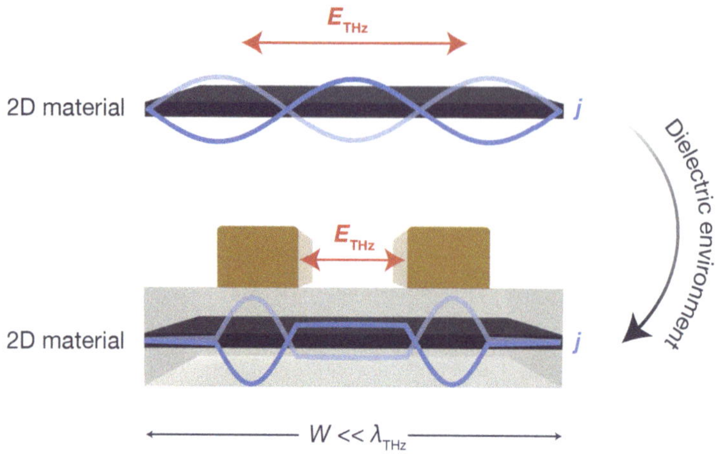

Fig. 9.1 Self-cavity formation of a 2D material for THz light. (**a**) Due to their subwavelength-size ($W \ll \lambda_{THz}$), 2D materials naturally form plasmonic self-cavities, where the edges of the flake act as mirrors to excited currents, allowing the formation of standing waves of current density j of discrete momenta. The dielectric environment thereby majorly influences the current density profile and thus momenta of the self-cavity modes. (Figure adapted from [15] under a CC BY 4.0 license)

These momenta, and thus the resonance frequencies of the self-cavity modes, are governed not only by the width of the 2D material but also significantly by the design of the micropatterned dielectric environment created by the on-chip THz spectroscopy circuitry. For example, the resonance frequencies of plasmonic self-cavity modes in a 2D material can be decreased by increasing the width of the material or by enhancing the screening. The latter can be achieved by reducing the thickness of the insulating layer between the 2D material and the metal traces, or by increasing the width of the metal traces (Sect. 6.5).

Subsequently, Chap. 7 explored the promising field of cavity engineering, the design of cavities to control light-matter interactions [16–26]. Conventionally, cavity engineering is performed by placing an atom, a molecule or a material ("matter") inside a cavity. If a discretised cavity mode energetically overlaps with some characteristic energy scale of the matter inside the cavity, energy can be exchanged, leading to the formation of hybrid modes. These hybrids can drastically influence the macroscopic properties of the material itself [16, 17]. As the light-matter coupling strength g increases, the system enters the ultrastrong coupling regime, characterized by non-perturbative light-matter interactions. In this regime, historically defined by $\eta = g/\nu_0 \gtrsim 0.1$, where ν_0 defines the frequency at which cavity and matter are resonant, even low-amplitude drives or photon vacuum fluctuations could lead to the formation of new thermodynamic ground states [25, 27–29].

The intriguing finding of the plasmonic self-cavity formation of conductive 2D materials opens a new avenue for cavity engineering. Rather than placing a 2D material inside a cavity, the 2D material itself acts as the cavity. Thus, if more than one self-cavity is comprised in a vdW heterostructure, as is typically the case in gate-tunable heterostructures, the possibility is given to perform cavity engineering via coupling of two self-cavities, such as gate and gated 2D material. Notably, the plasmonic cavity resonances of prototypical graphite gates (~25 GHz–2.5 THz) coincidentally fall on the same energy scale of many phenomena in vdW heterostructures (~0.1–10 meV) (see Fig. 1.4). This raised the question: can plasmonic self-cavity modes in graphite gates modify or control the THz electrodynamics of a gated vdW material?

This thesis shows that the answer is yes and provides the first experimental evidence that a plasmonic cavity mode intrinsic to a microstructured graphite electrostatic gate can hybridize with multiple plasmonic cavity modes in a gated graphene layer (see Fig. 9.2). This interaction reaches the ultrastrong coupling regime ($\eta \gtrsim 0.1$), as evidenced by the formation of avoided crossings between graphene and graphite modes of differing momenta ($s = 2$ and $m = 0$ modes, see Fig. 9.2), the transfer of spectral weight from graphite to graphene modes (Sect. 8.4.5), and the quantification of hybridization through the normalized coupling strength of $\eta_{\text{exp}} = 0.12 \pm 0.01$. If the resonance condition is not met within the measured carrier density range ($s = 0, 1$ modes for Device GG2, see Fig. 9.2) or frequency range ($s = 0, 1, 2$ modes for Device GG1, see Sect. 8.2.5), a complete avoided crossing cannot be detected. However, as these modes begin forming avoided crossings within the measured range, suppressed power-law scalings of resonance frequencies as a function of carrier density can still be observed, serving as additional reporters of strong hybridization (with uncoupled graphene plasmons scaling as $\nu \propto n^{0.25}$, while coupled graphene plasmons scale as $\nu \propto n^x$, with $x < 0.25$).

The analytical theory was extended to capture the THz electrodynamics of heterostructures comprising two self-cavities, such as back gate and gated graphene (see right panel of Fig. 9.2b), and enabled the identification of two distinct coupling mechanisms. The first mechanism, named as current-current hybridization (see Fig. 9.3a or Sect. 8.3.1), manifests through the formation of symmetric (in-phase current oscillations) and antisymmetric (out-of-phase current oscillations) modes in the two conductive layers. This hybridization becomes particularly strong ($\eta \gg 0.1$), when the 2D conductivities of the layers are comparable and if the self-cavities are separated by only a thin insulating layer (Sect. 8.3.1). If the gate layer has a much higher 2D conductivity than the 2D material, the antisymmetric mode still behaves like a pure 2D material plasmon and the symmetric mode still behaves like a pure gate plasmon. Otherwise, the distinct 2D material and gate modes merge into hybridised modes, mixing their characteristics. While this coupling occurs between modes of the same momentum, modes of different momenta can couple through the second coupling mechanism.

The second coupling mechanism involves the interaction between screened and unscreened (anti)symmetric modes, which arises when a vdW heterostructure is placed in an inhomogeneous dielectric environment (see Fig. 9.3b). Because parts of the heterostructure are screened by the metal traces of the circuitry and others

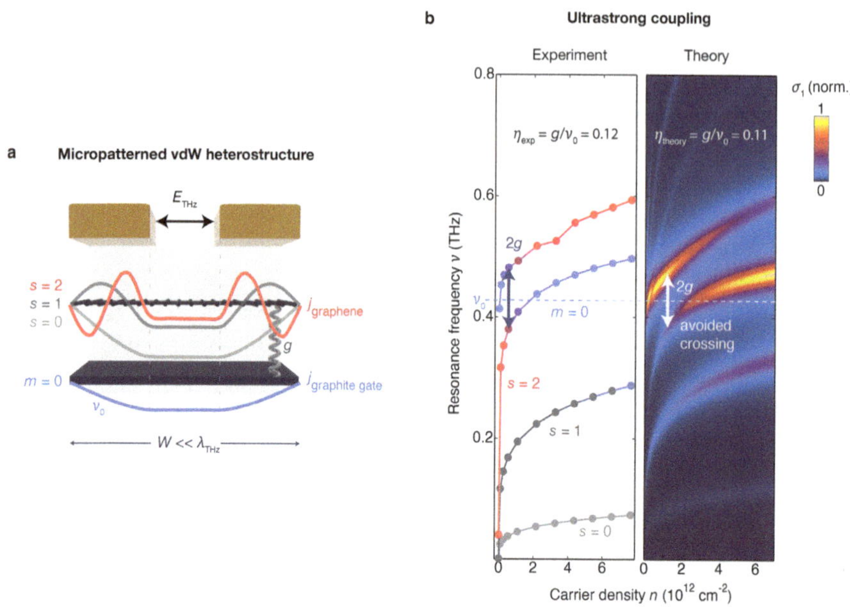

Fig. 9.2 VdW heterostructure illustrating ultrastrong coupling between plasmonic self-cavity modes. (**a**) Schematic of a vdW heterostructure consisting of a graphene layer and a few-nm-thick graphite gate embedded in a micropatterned dielectric environment provided by a metallic transmission line. Due to their subwavelength-size ($W \ll \lambda_{THz}$), both graphene and graphite gate naturally form plasmonic self-cavities. Unscreened currents between the metal strips enable the hybridization of screened graphene and graphite modes of different momenta at the dielectric boundaries provided by the metal traces, such that multiple screened plasmonic modes of graphene ($s = 0, 1, 2$) can hybridise with a screened plasmonic mode of the graphite gate ($m = 0$). (**b**) The frequencies of the screened plasmonic modes in graphene can be tuned with carrier density into resonance with the screened graphite mode, ν_0. For the given carrier density range, the resonance condition can be achieved between the $s = 2$ (magenta) and $m = 0$ (blue) mode. Both experimental data (left panel) and simulated data from the analytical theory (right panel) reveal that avoided crossings appear, with energy splittings of twice the coupling strength, g. The experimentally extracted normalized coupling strength $\eta_{exp} = 0.12 \pm 0.01$, and the analytical theory delivers $\eta_{theory} = 0.11 \pm 0.01$, both marking the entry to the ultrastrong coupling regime. (Figure adapted from [1] under a CC BY 4.0 license)

remain unscreened, screened and unscreened plasmonic modes can hybridize at the dielectric boundaries provided by the metal traces. This type of coupling is already present in heterostructures consisting of a single self-cavity (Sect. 6.5.3). However, in case of a heterostructure hosting two self-cavities, this effect becomes more pronounced as the unscreened antisymmetric mode has a resonance frequency that is comparable to the resonance frequencies of screened symmetric and antisymmetric modes, allowing to resonantly enhance the coupling of the two screened modes (Sect. 8.3.2).

The analytical theory thus provides generalisable design principles for enhancing or minimising the coupling strength. As the resonance frequencies of the

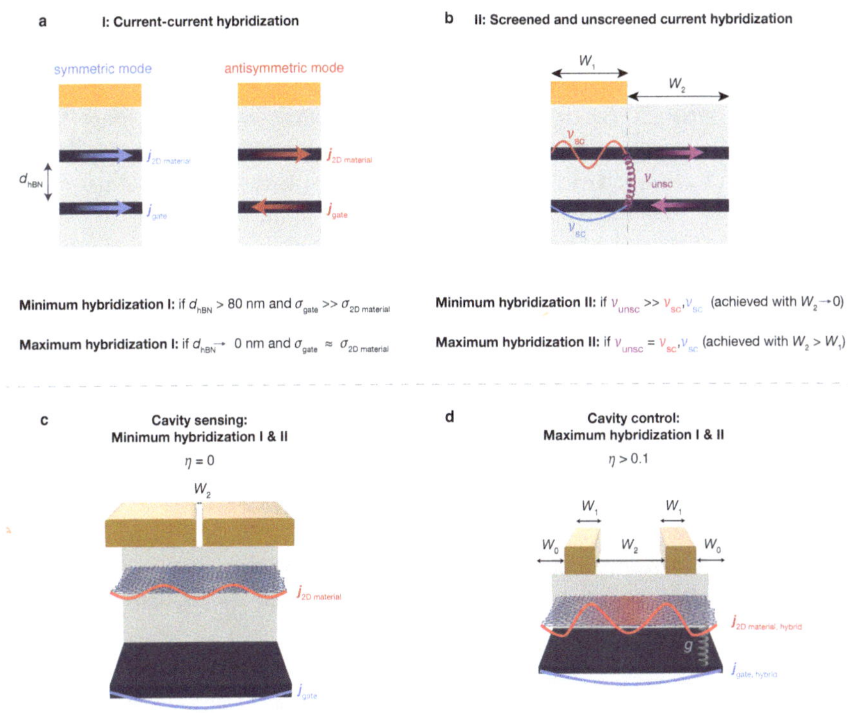

Fig. 9.3 Self-cavity coupling mechanisms and designs for cavity sensing or cavity control. (**a**) The first coupling mechanism is the current-current hybridization that leads to the formation of symmetric (currents oscillate in-phase) and antisymmetric (currents oscillate out-of-phase) modes. If this type of hybridization is minimised (as achieved by thick hBN layers and by using a gate with much larger conductivity than the 2D material, such as a thick graphite gate), the antisymmetric mode still behaves like a pure 2D material plasmonic mode and the symmetric mode still behaves like a pure gate plasmonic mode. Otherwise, this type of hybridization can lead to significantly altered resonance frequencies of 2D material and gate with $\eta \gg 0.1$. (**b**) The second coupling mechanism includes the hybridization of symmetric and antisymmetric modes of different momenta that form in either screened or unscreened regions of a micropatterned vdW heterostructure. This hybridization is maximised if symmetric (ν_{sc}) and antisymmetric (ν_{sc}) modes in the screened region and the antisymmetric mode in the unscreened region ($\nu_{unsc,1}$) are in resonance ($\nu_{sc} = \nu_{sc} = \nu_{unsc}$), as the unscreened antisymmetric mode mediates and resonantly enhances the coupling of screened modes of different momenta at the dielectric boundaries (grey dashed line). As ν_{sc} and ν_{unsc} are defined via the widths of the screened and unscreened regions, tuning W_2 with a fixed W_1 serves as a tuning knob for the second coupling mechanism. (**c**) A cavity with a small-gap transmission line, no extended sample regions, thick graphite gate (~20 nm) and thick hBN layer (>80 nm) between the 2D material and graphite gate suppresses both coupling mechanisms, enabling the sensing of the THz electrodynamics of a 2D material without the additional influence of resonant cavity coupling to the graphite gate. (**d**) A cavity with extended 2D material region W_0 and gap W_2, and rather thin graphite gate (~10 nm) and hBN layers (~20 nm) can be used to control collective modes of 2D materials. (Parts of the figure adapted from [1] under a CC BY 4.0 license)

(un)screened (anti)symmetric modes are determined by the widths of the corresponding regions, adjusting the ratio of the metal trace widths (screened regions) to the coplanar strip gap width (unscreened region) provides a tuning knob for the coupling strength. By designing the heterostructure with a thick graphite gate and thick hBN layer between the gate and gated material, to suppress the first coupling mechanism, and by embedding the heterostructure in a transmission line geometry with small gap, to suppress the second coupling mechanism, the overall coupling strength can be minimised (see Fig. 9.3c). Such an approach enables the precise sensing of the intrinsic THz electrodynamics of a 2D material. In contrast, the coupling strength can be maximised in thin, heavily micropatterned vdW heterostructures, where the conductivities of the gate and gated layer are comparable (see Fig. 9.3c).

Previous on-chip THz spectroscopy experiments did not report on the self-cavity and cavity-coupling electrodynamics of vdW heterostructures [2, 7, 8].[1] This may be attributed to two facts: first, earlier studies observed broad near-field conductivity spectra resembling a Drude response, consistent with expectations from far-field THz spectra of large-area metallic samples [13, 14]. It was thus reasonable to interpret these data using a Drude model, without considering potential resonances. However, the analytical theory developed in this thesis reveals that these broad spectra likely resulted from a series of low-frequency self-cavity modes within the measurement bandwidth (see Sect. 6.5). In irregularly shaped samples, the self-cavity modes experience inhomogeneous broadening, producing an accumulation of a series of overlapping modes that appeared as a broad Drude-like response. However, fitting such data with a Drude model yields parameters, such as Drude weights and scattering rates, that can differ significantly from the intrinsic material properties (Sect. 6.6).

Second, prior works primarily measured relative changes in THz transmission as a function of gate voltage to estimate the conductivity of the 2D material, based on the assumption that the semiconducting gate is transparent and did not contribute to the transmitted THz signal. Only with the introduction of a separate reference arm in this thesis was it possible to fully characterize back gate materials, such as graphite flakes, which revealed mode hybridization with the gated 2D material. Furthermore, the analytical theory in this work demonstrated that electrostatic gates with low conductivity, comparable to that of the gated 2D material, do not eliminate but rather enhance the current-current hybridization (see Fig. 9.3a), which may influence the interpretation of previous experiments.

In summary, this thesis shows that all vdW heterostructures of typical dimensions act as THz self-cavities, and reveals the general relevance of finite-size effects in vdW heterostructures for both investigating and tailoring macroscopic properties. Studying the THz electrodynamics of a vdW heterostructure, therefore, directly involves exploring its self-cavity electrodynamics.

[1] Recent work from Zhao et al. [4] did report the observation of hydrodynamic plasmons in graphene.

In addition, this thesis presents a theoretical framework for describing the THz conductivity of vdW heterostructures under a near-field probe, which is the first to explicitly account for the interaction between the sample and the measurement apparatus. This represents a significant advancement, addressing a long-standing challenge in THz near-field probes, namely the inability to disentangle probe-induced perturbations, whether from scanning tips, transmission lines, or emitters [30]. Moreover, the analytical theory does not only account for the influence of the metal traces of the THz circuitry, but furthermore enables the active design of both the metal traces and integrated heterostructure to enhance the sensitivity to specific features of the material's response.

By combining this analytical theory with an upgraded circuit architecture, this work unlocks the full potential of on-chip THz spectroscopy as a chip-scale platform for quantitative, finite-momentum measurements of the complex THz cavity conductivity of 2D samples. The platform enables precise extraction of material parameters, tuning and spectrally resolved read-out of self-cavity coupling strengths, and simultaneous DC transport measurements. As a result, on-chip THz spectroscopy now emerges as a powerful tool for both "cavity sensing" and "cavity control", uniquely positioned to probe and manipulate a wide range of emergent phases in vdW heterostructures.

References

1. Kipp G, Bretscher HM, Schulte B, Herrmann D, Kusyak K, Day MW, Kesavan S, Matsuyama T, Li X, Langner SM, Hagelstein J, Sturm F, Potts AM, Eckhardt CJ, Huang Y, Watanabe K, Taniguchi T, Rubio A, Kennes DM, Sentef MA, Baudin E, Meier G, Michael MH, McIver JW (2025) Cavity electrodynamics of van der Waals heterostructures. Nat Phys 21:1926–1933. https://doi.org/10.1038/s41567-025-03064-8
2. Gallagher P, Yang C-S, Lyu T, Tian F, Kou R, Zhang H, Watanabe K, Taniguchi T, Wang F (2019) Quantum-critical conductivity of the Dirac fluid in graphene. Science 364:158–162. https://doi.org/10.1126/science.aat8687
3. Parker-Jervis R, Park S, Zonetti S, Sydoruk O, Wu J, Wood C, Li L, Davies A, Linfield E, Cunningham J (2020) On-chip terahertz spectroscopy of magnetoplasmons in a two-dimensional electron gas. In: 2020 45th international conference on infrared, millimeter, and terahertz waves (IRMMW-THz). https://doi.org/10.1109/irmmw-thz46771.2020.9370779
4. Zhao W, Wang S, Chen S, Zhang Z, Watanabe K, Taniguchi T, Zettl A, Wang F (2023) Observation of hydrodynamic plasmons and energy waves in graphene. Nature 614:688–693. https://doi.org/10.1038/s41586-022-05619-8
5. Sterbentz RM, Island JI (2023) On-chip sub-diffraction THz spectroscopy of materials and liquids.In: Meziani YM, Velázquez-Pérez J (eds) IntechOpen. https://doi.org/10.5772/intechopen.110877
6. Potts AM, Nayak AK, Nagel M, Kaj K, Stamenic B, John DD, Averitt RD, Young AF (2023) On-chip time-domain terahertz spectroscopy of superconducting films below the diffraction limit. Nano Lett 23:3835–3841. https://doi.org/10.1021/acs.nanolett.3c00412
7. Chen S-D, Feng Q, Zhao W, Qi R, Zhang Z, Abeysinghe D, Uzundal C, Xie J, Taniguchi T, Watanabe K, Wang F (2024) Direct measurement of terahertz conductivity in a gated monolayer semiconductor. arXiv preprint arXiv:2409.17633

8. Seo J, Lu Z, Park S, Yang J, Xia F, Ye S, Yao Y, Han T, Shi L, Watanabe K et al (2024) On-chip terahertz spectroscopy for dual-gated van der waals heterostructures at cryogenic temperatures. Nano Lett. https://doi.org/10.1021/acs.nanolett.4c04137
9. Graef H, Mele D, Rosticher M, Banszerus L, Stampfer C, Taniguchi T, Watanabe K, Bocquillon E, Fève G, Berroir J et al (2018) Ultra-long wavelength Dirac plasmons in graphene capacitors. J Phy: Mater 1:01LT02. https://doi.org/10.1088/2515-7639/aadd8c
10. Yoshioka K, Bernard G, Wakamura T, Hashisaka M, Sasaki K-I, Sasaki S, Watanabe K, Taniguchi T, Kumada N (2024) On-chip transfer of ultrashort graphene plasmon wave packets using terahertz electronics. Nat Electron 7:537–544. https://doi.org/10.1038/s41928-024-01197-x
11. Fei Z, Goldflam M, Wu J-S, Dai S, Wagner M, McLeod A, Liu M, Post K, Zhu S, Janssen G et al (2015) Edge and surface plasmons in graphene nanoribbons. Nano Lett 15:8271–8276. https://doi.org/10.1021/acs.nanolett.5b03834
12. Kusyak K, Schulte B, Matsuyama T, Kipp G, Bretscher HM, Day MW, Meier G, Potts AM, McIver JW (2025) Monolithic optoelectronic circuit design for on-chip terahertz applications. APL Photon 10:076117. https://doi.org/10.1063/5.0278183
13. Horng J, Chen C-F, Geng B, Girit C, Zhang Y, Hao Z, Bechtel HA, Martin M, Zettl A, Crommie MF et al (2011) Drude conductivity of Dirac fermions in graphene. Phys Rev B—Condensed Matter Mater Phys 83:165113. https://doi.org/10.1103/PhysRevB.83.165113
14. Razanoelina M, Bagsican FR, Kawayama I, Zhang X, Ma L, Murakami H, Vajtai R, Ajayan PM, Kono J, Tonouchi M (2016) Probing low-density carriers in a single atomic layer using terahertz parallel-plate waveguides. Opt Express 24:3885–3893. https://doi.org/10.1364/OE.24.003885
15. Michael MH, Kipp G, Potts AM, Day MW, Matsuyama T, Meier G, Bretscher HM, McIver JW (2025) Resolving self-cavity effects in two-dimensional quantum materials. arXiv preprint arXiv:2505.12799
16. Schlawin F, Kennes DM, Sentef (MA) Cavity quantum materials. Appl Phys Rev 9. https://doi.org/10.1063/5.0083825
17. Hübener H, De Giovannini U, Schäfer C, Andberger J, Ruggenthaler M, Faist J, Rubio A (2021) Engineering quantum materials with chiral optical cavities. Nat Mater 20:438–442. https://doi.org/10.1038/s41563-020-00801-7
18. Eckhardt CJ, Chattopadhyay S, Kennes DM, Demler EA, Sentef MA, Michael MH (2023) Theory of resonantly enhanced photo-induced superconductivity. Nat Commun 15:2300. https://doi.org/10.1038/s41467-024-46632-x
19. Sentef MA, Ruggenthaler M, Rubio A (2018) Cavity quantum-electrodynamical polaritonically enhanced electron-phonon coupling and its influence on superconductivity. Sci Adv 4:eaau6969. https://doi.org/10.1126/sciadv.aau6969
20. Curtis JB, Michael MH, Demler E (2023) Local fluctuations in cavity control of ferroelectricity. Phys Rev Res 5:043118. https://doi.org/10.1103/PhysRevResearch.5.043118
21. Masuki K, Ashida Y (2024) Cavity moiré materials: controlling magnetic frustration with quantum light-matter interaction. Phys Rev B 109:195173. https://doi.org/10.1103/physrevb.109.195173
22. Bloch J, Cavalleri A, Galitski V, Hafezi M, Rubio A (2022) Strongly correlated electron–photon systems. Nature 606:41–48. https://doi.org/10.1038/s41586-022-04726-w
23. Boström EV, Sriram A, Claassen M, Rubio A (2023) Controlling the magnetic state of the proximate quantum spin liquid $\alpha - RuCl_3$ with an optical cavity. NPJ Comput Mater 9:202. https://doi.org/10.1038/s41524-023-01158-6
24. Wang X, Ronca E, Sentef MA (2019) Cavity quantum electrodynamical Chern insulator: towards light-induced quantized anomalous Hall effect in graphene. Phys Rev B 99:235156. https://doi.org/10.1103/PhysRevB.99.235156
25. Riolo R, Tomadin A, Mazza G, Asgari R, MacDonald AH, Polini M (2024) Tuning Fermi liquids with sub-wavelength cavities. arXiv preprint arXiv:2403.20067

26. Andberger J, Graziotto L, Sacchi L, Beck M, Scalari G, Faist J (2024) Terahertz chiral sub-wavelength cavities breaking time-reversal symmetry via ultrastrong light-matter interaction. Phys Rev B 109:L161302. https://doi.org/10.1103/PhysRevB.109.L161302
27. Forn-Díaz P, Lamata L, Rico E, Kono J, Solano E (2019) Ultrastrong coupling regimes of light-matter interaction. Rev Mod Phys 91:025005. https://doi.org/10.1103/RevModPhys.91.025005
28. Frisk Kockum A, Miranowicz A, De Liberato S, Savasta S, Nori F (2019) Ultrastrong coupling between light and matter. Nat Rev Phys 1:19–40. https://doi.org/10.1038/s42254-018-0006-2
29. Bayer A, Pozimski M, Schambeck S, Schuh D, Huber R, Bougeard D, Lange C (2017) Terahertz light–matter interaction beyond unity coupling strength. Nano Lett 17:6340–6344. https://doi.org/10.1021/acs.nanolett.7b03103
30. Hillenbrand R, Abate Y, Liu M, Chen X, Basov D (2025) Visible-to-THz near-field nanoscopy. Nat Rev Mater. https://doi.org/10.1038/s41578-024-00761-3

Chapter 10
Outlook: Cavity Sensing and Control of Quantum Phases in 2D Materials

Abstract This concluding chapter presents an outlook on a set of experiments inspired by the findings of this thesis, some of which are already underway.

The experimental and theoretical findings presented in this thesis are pivotal for establishing on-chip THz spectroscopy as a powerful technique to probe the low-energy electrodynamics of 2D vdW heterostructures, while simultaneously offering a novel platform to explore the effects of cavity coupling on these materials (see Fig. 10.1). These results have already deepened the understanding of self-cavity-mediated THz emission from WTe_2 [3] and cavity conductivity spectra of superconducting NbN films [4], and now lay the groundwork for a series of follow-up experiments. These efforts focus on the tailored design of heterostructures and coplanar strip geometries to selectively capture intrinsic ("cavity sensing") or engineered ("cavity control") phenomena of target vdW materials within the accessible measurement bandwidth.

To give a taste of what's to come in the near future, the following sections introduce a newly implemented measurement setup, built to enable on-chip THz experiments at millikelvin temperatures, a crucial condition for investigating strongly correlated phenomena in vdW heterostructures at temperatures below 1 K (Sect. 10.1). Additionally, several experimental proposals supported by analytical theory calculations are presented, demonstrating that the principles outlined in this thesis extend beyond plasmons to a broad array of other phenomena and collective modes (Sect. 10.2).

G. Kipp, *Self-Cavity Electrodynamics of Van der Waals Heterostructures*,
Springer Theses, https://doi.org/10.1007/978-3-032-31979-1_10

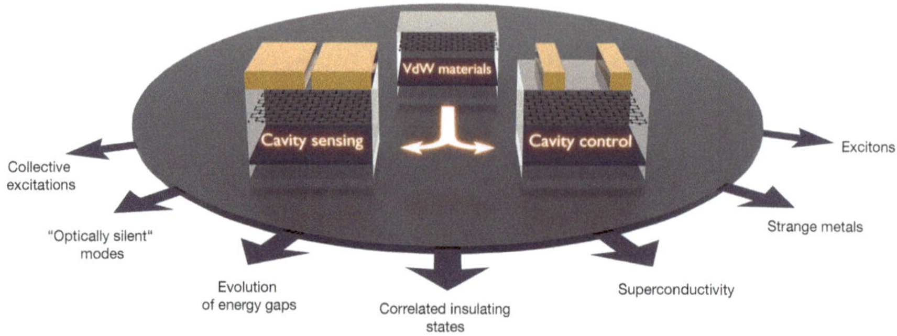

Fig. 10.1 Novel platform for cavity sensing and control through geometric design of diverse quantum phases in 2D materials

10.1 On-chip THz Spectroscopy at Millikelvin Temperatures

The on-chip THz spectroscopy measurements in this work were conducted in an optical cryostat with a base temperature of 4 K. However, many vdW heterostructures, such as magic-angle twisted bilayer graphene, exhibit their most intriguing strongly correlated phases below 1 K (see Fig. 7.4). These phases can be tuned using external magnetic fields to explore phenomena such as both integer and fractional quantum anomalous Hall effects [6, 7]. As a result, accessing the full suite of emergent phenomena in vdW heterostructures requires an experimental setup that enables measurements at millikelvin temperatures and in strong magnetic fields.

To meet these requirements, a custom magnet millikelvin cryostat with optical access was specifically designed and built together with the company *JanisULT* (Fig. 10.2). The cryostat is designed with direct optical access to the sample via five specially coated windows and heat shields, essential for minimizing blackbody radiation and maintaining low temperatures. It also features a tunable magnetic field up to 9 T. Optics can be positioned within 8 cm of the sample, allowing laser beams and camera light to focus onto the optoelectronic circuits for operating photoconductive switches and imaging the device. This He3-based cryostat achieves a base temperature of 300 mK without laser illumination and provides 1 mW of cooling power at 500 mK. Considering blackbody radiation, laser-induced heating, and ohmic heating from voltage biases, the estimated total heat load is approximately 0.5 mW. This heat load enables a base temperature of 450 mK or lower during on-chip THz measurements, which enables investigating many emergent phenomena in vdW heterostructures [5].

In February 2025, after 5 years of designing, building and preparing the optical cryostat for operation, the first THz signal was detected at a base temperature of 480 mK and in a 7 T magnetic field by Dr. Benedikt Schulte and Dr. Matthew W. Day. Thus, this mK cryostat is now operable for performing on-chip THz spectroscopy and enables a plethora of future experiments.

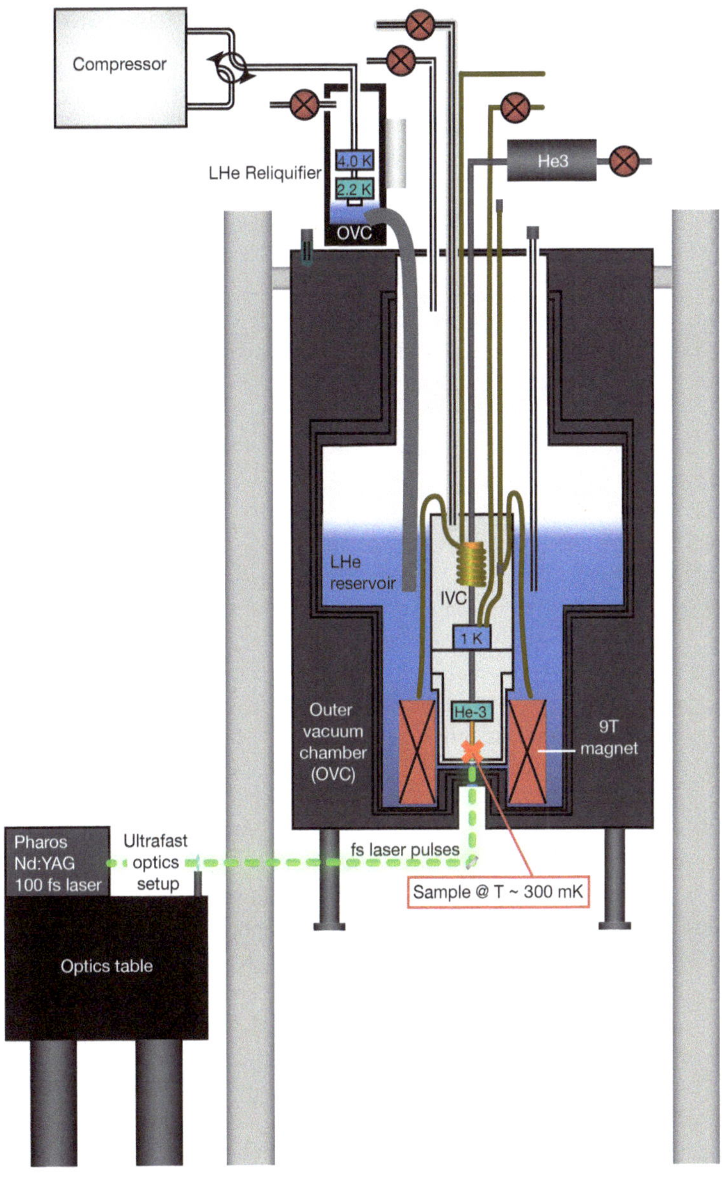

Fig. 10.2 On-chip THz spectroscopy setup for ∼300 mK temperature and 9 T magnetic field measurements. A newly built ultrafast optics setup (similar to the one discussed in Fig. 4.1) that is operated by a 100 fs pulsed Nd:YAG laser system by *Light conversion* is sending pump and probe beams towards the optical mK cryostat next to the optics table. The cryostat operates by using He-4 to cool a 9 T magnet and to pre-cool the inner vacuum chamber (IVC) down to 4 K. The "1 K pot" placed in the IVC is then pumped to suck liquid 4-He into the pot, cooling it down to 1.6 K via evaporation. The "1 K pot" is then used to condense the He-3 gas, making it drip into the "He-3 pot". By pumping on the liquid He-3 with a charcoal, a base temperature of 300 m K is reached. The sample is connected to the "He-3 pot" and mounted in the center of the 9 T magnet [5]

10.2 Exploring Emergent Phases in 2D Quantum Materials

The principles developed for cavity sensing and control extend beyond plasmons to other emergent phenomena and fundamental excitations. For example, on-chip THz spectroscopy can access collective modes with no net dipole moment that are typically rendered "optically silent" with respect to far-field probes. This is possible because on-chip THz spectroscopy confines electromagnetic fields to the near field, enabling coupling to finite momentum modes and breaking spatial symmetries that protect these excitations from far-field detection. As a result, even charge-neutral, but current-carrying modes with no net dipole moment can become accessible.

Two examples of such optically silent modes are explored for their self-cavity responses. The first example are demon modes [8, 9], which correspond to in-phase oscillations of electrons and holes, and the second example is the shear optical phonon [10], involving antisymmetric oscillations of dipole active modes in each layer of a bilayer.

To theoretically investigate the demon mode, the analytical theory is used to model the electron and hole bands as two separate 2D conductivities, each confined to slightly offset layers along the z-direction. This simplified picture captures the physical situation where different electronic bands are localized on different atoms or orbitals that are spatially separated by only Ångströms to nanometers. Such subtle vertical displacements enable charge-neutral oscillations, even in a monolayer like graphene, to be effectively described as interlayer collective excitations. For simplicity, both bands are assumed to follow identical Drude conductivities. The resulting cavity conductivity spectrum (Fig. 10.3a) displays a bright, symmetric plasmon mode near 0.38 THz (Fig. 10.3b). Additionally, distinct out-of-phase current oscillations appear, resembling in-phase motion of electrons and holes, which are characteristic of **optically silent demon modes** (Fig. 10.3c). The resonance frequencies of the demon modes exhibit pronounced sensitivity to layer separation, analogous to orbital spacing in real materials, presenting measurable changes to the cavity conductivity spectrum under even Ångström-scale variations.

To explore the sensitivity of on-chip THz spectroscopy towards dipole-active phonons in bilayers, the Drude model used for plasmonic self-cavity responses is replaced by a Drude–Lorentz model (see Sect. 5.2.2, with $\omega_0 = 2\pi\nu_{\mathrm{ph}}$ in Eq. 5.19). As shown in Fig. 10.3d, the resulting cavity conductivity spectrum bears close resemblance to its plasmonic counterpart (Fig. 10.3a), but is shifted to higher frequencies due to the phonon resonance at ν_{ph}, below which no modes are supported. Above this threshold, the spectrum exhibits a prominent resonance arising from symmetric ion displacements in each layer, **optical phonons** (Fig. 10.3e), in addition to side peaks corresponding to their antisymmetric counterparts, **shear optical phonons** (Fig. 10.3f). Much like the plasmonic demon mode, the shear phonon is optically silent in the far field, yet its interlayer self-cavity response can be probed with on-chip THz spectroscopy with Ångström-level resolution.

These two examples demonstrate that on-chip THz spectroscopy provides access to the self-cavity response of collective excitations beyond purely plasmonic modes,

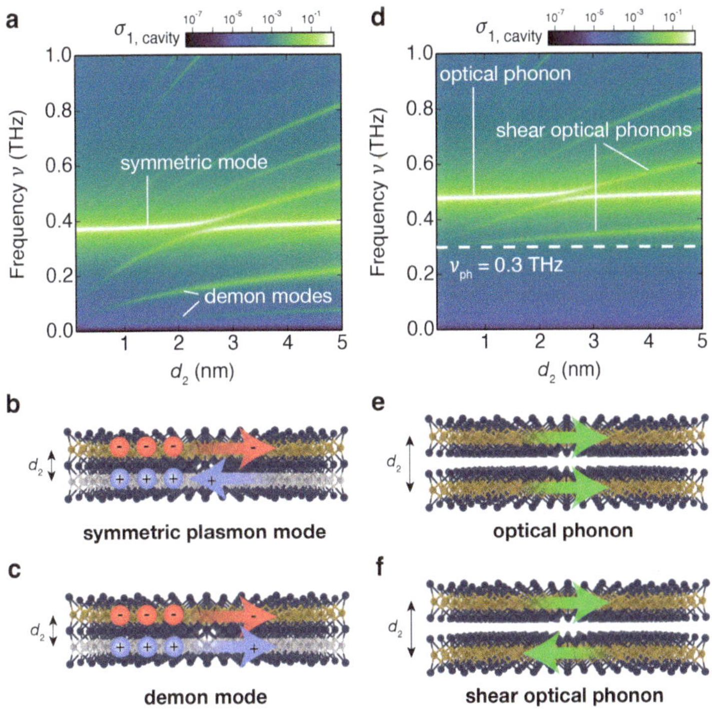

Fig. 10.3 Detecting demon modes and phonons with on-chip THz spectroscopy. (**a**) Cavity conductivity spectrum of a bilayer system (each layer has the same conductivity) as a function of layer separation d_2. Demon modes (lower amplitude resonances) and a symmetric plasmon mode (largest resonance at 0.38 THz) are detected. (**b**) Electrons and holes oscillate out-of-phase in a symmetric plasmon mode. (**c**) Electrons and holes oscillate in-phase in a demon mode. (**d**) Optical phonon spectroscopy of a bilayer detecting optical phonon polaritons (oscillations of adjacent layers in-phase) and shear optical phonons (oscillations out-of-phase). (**e**) If adjacent layers of the crystal lattice move in the same direction, an optical phonon is excited. (**f**) If adjacent layers of the crystal lattice move in opposite directions, a shear optical phonon is excited. Cavity parameters: $W_0 = 0\,\mu$ m, $W_1 = 3\,\mu$ m and $W_2 = 10\,\mu$ m, $d_{\text{material}} = 20$ nm, $d_{\text{hBN}} = 10$ nm, $\nu_{3D} = 100$ THz. (Figure adapted from [2] under a CC BY 4.0 license)

enabling sensitive detection of electronic band separations and insights into the electronic band structure of bilayer systems through the dispersions of optically silent modes. Simulated avoided crossings between optically silent plasmonic or phononic modes and their symmetric counterparts (Fig. 10.3a, d) further show that self-cavity coupling of these modes can likewise be achieved through dielectric micropatterning of the surrounding environment, analogous to the approach employed for plasmonic modes in this work.

The analytical theory is thus capable of predicting cavity conductivity spectra from virtually any input 2D conductivity. In Fig. 10.4, a range of representative conductivities are used as inputs to simulate the self-cavity response of a series of emergent phenomena in vdW heterostructures. Comparing the intrinsic 2D

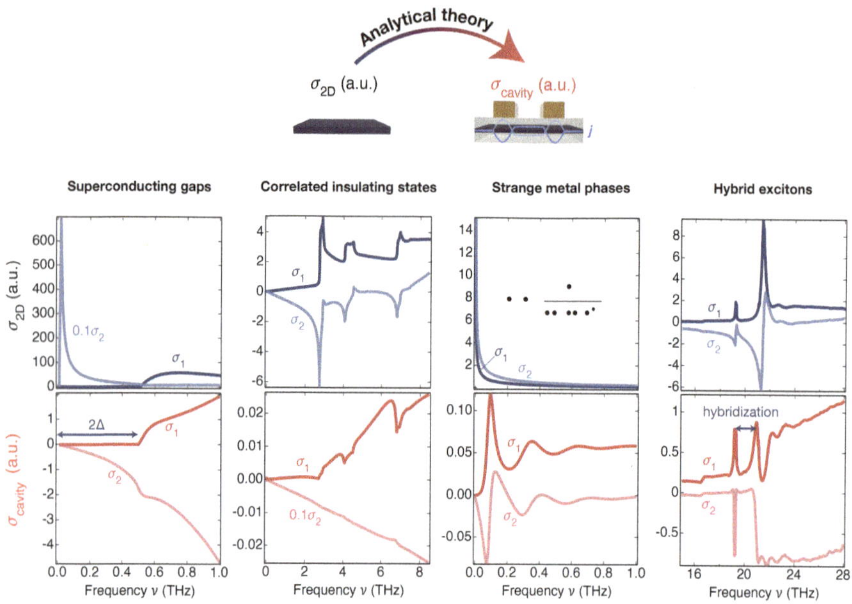

Fig. 10.4 Cavity engineering of other emergent phases with on-chip THz spectroscopy. The analytical theory of this work can be used to simulate how the THz cavity conductivity spectrum looks like (red) for a given intrinsic, 2D conductivity of a phenomenon or collective mode in a 2D quantum material (blue), such as for a superconductor, a correlated insulating state, a strange metal phase or for hybrid excitons. The real and imaginary parts of the input 2D conductivities (blue) were taken from measured or simulated data. The input curves for the superconducting gaps were kindly provided by A. M. Potts using the Zimmermann model (BCS theory). Calculations were performed for the following cavity geometry: $W_0 = 0\,\mu$ m, $W_1 = 0.005\,\mu$ m, $W_2 = 40\,\mu$ m and $d_{hBN1} = 24$ nm. The input curves for the correlated insulating states were extracted from measured data on twisted bilayer graphene from Ref. [11], using the following cavity geometry: $W_0 = 3\,\mu$ m, $W_1 = 3\,\mu$ m, $W_2 = 3\,\mu$ m and $d_{hBN1} = 24$ nm. The input curves for the strange metal phases were simulated with the formula provided in Ref. [12] with $\alpha = 0.7$, and in the following cavity geometry: $W_0 = 0\,\mu$ m, $W_1 = 20.33\,\mu$ m, $W_2 = 2.56\,\mu$ m, $d_{hBN1} = 24$ nm and $d_{gr} = 5$ nm, resulting in $A = 2.89 \cdot 10^{19}$. The input data for the hybrid excitons was extracted from measured data on bilayer graphene (Ref. [13]) using the following cavity dimensions: $W_0 = 3\,\mu$ m, $W_1 = 3\,\mu$ m, $W_2 = 3\,\mu$ m and $d_{hBN1} = 20$ nm. (Figure adapted from [2] under a CC BY 4.0 license)

conductivity spectra (dark blue) with the corresponding cavity conductivity spectra (red) reveals how finite-size effects and the surrounding dielectric environment shape the material's electrodynamics. While certain spectral features are preserved, on-chip THz spectroscopy is able to capture more subtle signatures that are hidden in the far-field.

For instance, in **superconductors**, the real part of the cavity conductivity distinctly captures the superconducting gap at 2Δ, while the imaginary part deviates from the conventional $1/\omega$ divergence, instead exhibiting a pronounced kink near the gap edge.[1] Similarly, the cavity conductivity of **correlated insulating states**, such as those in twisted bilayer graphene [15], reveals the associated energy gap. By spectroscopically tracking gap magnitudes as a function of a tuning parameter, like carrier density or temperature, one can uncover the underlying symmetry of an emergent phases (see Chap. 1) [16]. On-chip THz spectroscopy can thus facilitate the exploration of the microscopic origin of both insulating and superconducting states.

Moreover, on-chip THz spectroscopy allows for unlocking new insights into mysterious **strange metals**, which are a phase of matter with a series of anomalous properties, such as linear temperature-dependent resistivity [17–19] and non-Fermi liquid behavior, defying traditional metallic theories. For frequencies above the temperature of the system, their conductivity is expected to follow a power law, $\sigma = \frac{A}{(-i\omega)^\alpha}$, with a scaling exponent α [12]. This yields the following fractal[2] dispersion relation for screened strange metal plasmons (see Eq. 6.42):

$$q_{\text{screened}} = \frac{i(-i\omega)^{\frac{\alpha+1}{2}}}{\sqrt{d_{\text{hBN}} A/\epsilon_1}}, \tag{10.1}$$

that is intriguingly purely defined by the exponent α.

While the intrinsic 2D conductivity of strange metals shows relatively featureless real and imaginary parts, the cavity conductivity reveals a rich spectrum: the exponent α emerges through the number and frequency positions of oscillations in both real and imaginary parts (see Fig. 10.4). Given that strange metal phases frequently appear near unconventional superconductors or quantum critical points [20], on-chip THz spectroscopy provides an avenue to probe their finite-frequency response, potentially enabling new insights into the optical fingerprints of quantum phase transitions.

Lastly, the response of **excitons**, ubiquitous quasiparticles in 2D systems due to reduced screening, is likewise accessible. Using experimentally measured absorption spectra of bilayer graphene as an input [13], the resulting cavity conductivity reveals clear signatures of exciton-polariton formation and mode hybridization, with frequency shifts that reflect the effects of self-cavity coupling.

These examples demonstrate the potential of on-chip spectroscopy to provide a deeper spectroscopic view of quantum phases and light–matter interactions, not only for conventional plasmons or polaritons but also for emergent quasiparticles and optically silent modes that are inaccessible to far-field techniques. At the heart

[1] To isolate these features from competing superconducting plasmons, the THz circuitry was purposefully designed with narrow metal traces, shifting the plasmonic response outside the explored frequency range. Nonetheless, superconducting plasmons themselves present an intriguing avenue for on-chip THz studies, as demonstrated in Refs. [14].

[2] In this context, "fractal" implies that the exponent α simultaneously governs both the resonance frequencies and scattering rates. This contrasts plasmons in conventional Drude metals, where these quantities are determined independently.

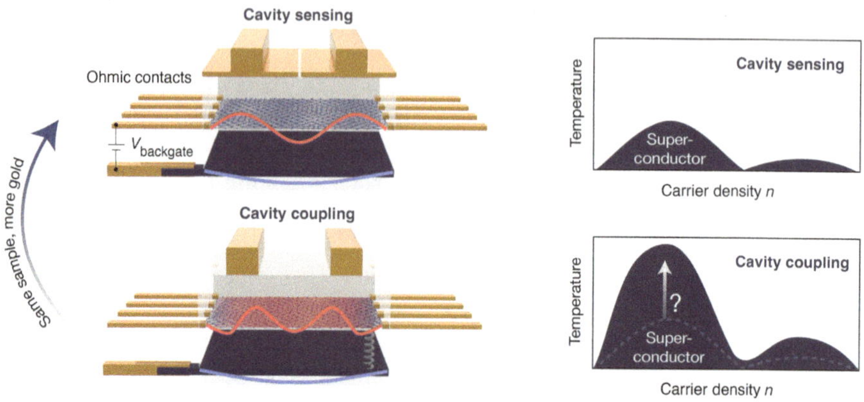

Fig. 10.5 Enhancing superconductivity via cavity-cavity coupling? Experimental proposal for testing the influence of cavity-cavity coupling on macroscopic material properties such as superconductivity: a gate-tunable vdW heterostructure, such as magic-angle twisted bilayer graphene (see Fig. 7.4) can first be measured in a "cavity coupling" geometry, to detect THz spectroscopy *and* DC transport data simultaneously. After this first measurement, additional gold can be evaporated onto the sample to convert the "cavity coupling" geometry into a "cavity sensing" configuration, allowing the THz spectroscopy and DC transport data to be retaken for direct comparison. This approach provides a means to experimentally test theoretical proposals of cavity-enhanced superconductivity (see Sect. 7.4.3) [21–26]

of this approach lies the analytical theory, which disentangles intrinsic material properties from environmental effects and guides the design of device geometries and dielectric environments for the tailored investigation of an emergent phase.[3]

A particularly exciting experimental prospect is to first investigate a system under ultrastrong self-cavity coupling, then reconfigure the same heterostructure by locally widening the coplanar strip metal traces to explore its intrinsic electrodynamics in a "sensing" cavity (see Fig. 10.5). Because DC transport measurements can be performed in parallel to spectroscopic cavity sensing and control measurements, one may directly evaluate the impact of self-cavity coupling on the emergence of a correlated quantum phase.

Altogether, with its multifaceted capabilities now well established, on-chip THz spectroscopy provides a novel window into the electrodynamics of vdW heterostructures, enabling both passive sensing and active control of complex quantum phases.

[3] This may also include impedance matching strategies, such as specifically designing graphite gates, to optimize the spectroscopic sensitivity across a broad conductivity range (see Appendix, Sect. A.8).

References

1. Kipp G, Bretscher HM, Schulte B, Herrmann D, Kusyak K, Day MW, Kesavan S, Matsuyama T, Li X, Langner SM, Hagelstein J, Sturm F, Potts AM, Eckhardt CJ, Huang Y, Watanabe K, Taniguchi T, Rubio A, Kennes DM, Sentef MA, Baudin E, Meier G, Michael MH, McIver JW (2025) Cavity electrodynamics of van der Waals heterostructures. Nat Phys 21:1926–1933. https://doi.org/10.1038/s41567-025-03064-8
2. Michael MH, Kipp G, Potts AM, Day MW, Matsuyama T, Meier G, Bretscher HM, McIver JW (2025) Resolving self-cavity effects in two-dimensional quantum materials. arXiv preprint arXiv:2505.12799
3. Li X, Hagelstein J, Kipp G, Sturm F, Kusyak K, Huang Y, Schulte B, Potts AM, Stensberg J, Cordero VQ, Trovatello C, Peng ZH, Hu C, DeStefano JM, Tanigushi T, Wantanabe K, Schuck PJ, Xu X, Rubio A, Michael MH, Day MW, Bretscher HM, McIver JW (2025) Purcell enhancement of photogalvanic currents in wte_2 plasmonic cavities. arXiv preprint arXiv:2507.07987
4. Potts AM, Michael MH, Kipp G, Langner SM, Bretscher HM, Stensberg J, Kaj K, Matsuyama T, Day MW, Sturm F, Nayak AK, Cohen LA, Zhu X, Young A, McIver J () Finite-momentum Cooper plasmons in superconducting terahertz microcavities. arXiv preprint arXiv:2601.10692
5. Schulte BF (2022) Ultrafast Transport in Tunable Graphene Heterostructures. PhD Thesis. University of Hamburg
6. Park H, Cai J, Anderson E, Zhang Y, Zhu J, Liu X, Wang C, Holtzmann W, Hu C, Liu Z et al (2023) Observation of fractionally quantized anomalous Hall effect. Nature 622:74–79. https://doi.org/10.1038/s41586-023-06536-0
7. Zibrov AA, Kometter C, Zhou H, Spanton E, Taniguchi T, Watanabe K, Zaletel M, Young A (2017) Tunable interacting composite fermion phases in a half-filled bilayergraphene Landau level. Nature 549:360–364. https://doi.org/10.1038/nature23893
8. Husain AA, Huang EW, Mitrano M, Rak MS, Rubeck SI, Guo X, Yang H, Sow C, Maeno Y, Uchoa B, Chiang TC, Batson PE, Phillips PW, Abbamonte P (2023) Pines' demon observed as a 3D acoustic plasmon in Sr22RuO4. Nature 621:66–70. https://doi.org/10.1038/s41586-023-06318-8
9. Pines D (1956) Electron interaction in solids. Canadian Journal of Physics 34:1379–1394. https://doi.org/10.1139/p56-154
10. Jia G (2023) Hyperbolic shear polaritons and their potential applications. Nano TransMed 2:100019. https://doi.org/10.1016/j.ntm.2023.100019
11. Calderón MJ, Bascones E (2020) Correlated states in magic angle twisted bilayer graphene under the optical conductivity scrutiny. NPJ Quantum Mater 5:57. https://doi.org/10.1038/s41535-020-00258-6
12. Michon B, Berthod C, Rischau CW, Ataei A, Chen L, Komiya S, Ono S, Taillefer L, van der Marel D, Georges A (2023) Reconciling scaling of the optical conductivity of cuprate superconductors with Planckian resistivity and specific heat. Nat Commun 14:3033. https://doi.org/10.1038/s41467-023-38762-5
13. Ju L, Wang L, Cao T, Taniguchi T, Watanabe K, Louie SG, Rana F, Park J, Hone J, Wang F, McEuen PL (2017) Tunable excitons in bilayer graphene. Science 358:907–910. https://doi.org/10.1126/science.aam9175
14. Potts AM (2023) On-chip terahertz time domain spectroscopy for sub-diffraction van der Waals heterostructures. PhD Thesis. University of California, Santa Barbara
15. Huang X, Wang T, Miao S, Wang C, Li Z, Lian Z, Taniguchi T, Watanabe K, Okamoto S, Xiao D et al (2021) Correlated insulating states at fractional fillings of the WS2/WSe2 moiré lattice. Nat Phys 17:715–719. https://doi.org/10.1038/s41586-020-2868-6
16. Basov DN, Averitt RD, Van Der Marel D, Dressel M, Haule K (2011) Electrodynamics of correlated electron materials. Rev Mod Phys 83:471. https://doi.org/10.1103/RevModPhys.83.471

17. Cao Y, Chowdhury D, Rodan-Legrain D, Rubies-Bigorda O, Watanabe K, Taniguchi T, Senthil T, Jarillo-Herrero P (2020) Strange metal in magic-angle graphene with near Planckian dissipation. Phys Rev Lett 124:076801. https://doi.org/10.1103/PhysRevLett.124.076801
18. Jindal A, Saha A, Li Z, Taniguchi T, Watanabe K, Hone JC, Birol T, Fernandes RM, Dean CR, Pasupathy AN, Rhodes D (2023) Coupled ferroelectricity and superconductivity in bilayer Td-MoTe2. Nature 613:48–52. https://doi.org/10.1038/s41586-022-05521-3
19. Xia Y, Han Z, Watanabe K, Taniguchi T, Shan J, Mak KF (2025) Superconductivity in twisted bilayer WSe2. Nature 637:833–838
20. Alexandradinata A, Armitage N, Baydin A, Bi W, Cao Y, Changlani HJ, Chertkov E, Da Silva Neto EH, Delacretaz L, Baggari IE et al (2020) The future of the correlated electron problem. arXiv preprint arXiv:2010.00584
21. Eckhardt CJ, Chattopadhyay S, Kennes DM, Demler EA, Sentef MA, Michael MH (2023) Theory of resonantly enhanced photo-induced superconductivity. Nat Commun 15:2300. https://doi.org/10.1038/s41467-024-46632-x
22. Curtis JB, Raines ZM, Allocca AA, Hafezi M, Galitski VM (2019) Cavity quantum Eliashberg enhancement of superconductivity. Phys Rev Lett 122:167002. https://doi.org/10.1103/PhysRevLett.122.167002
23. Allocca AA, Raines ZM, Curtis JB, Galitski VM (2019) Cavity superconductorpolaritons. Phys Rev B 99:020504. https://doi.org/10.1103/PhysRevB.99.020504
24. Chakraborty A, Piazza F (2021) Long-range photon fluctuations enhance photon-mediated electron pairing and superconductivity. Phys Rev Lett 127:177002. https://doi.org/10.1103/PhysRevLett.127.177002
25. Sentef MA, Ruggenthaler M, Rubio A (2018) Cavity quantum-electrodynamical polaritonically enhanced electron-phonon coupling and its influence on superconductivity. Sci Adv 4:eaau6969. https://doi.org/10.1126/sciadv.aau6969
26. Kozin VK, Thingstad E, Loss D, Klinovaja J (2025) Cavity-enhanced superconductivity via band engineering. Phys Rev B 111. https://doi.org/10.1103/physrevb.111.035410

Appendix A

A.1 Device Parameters

The following Tables A.1 and A.2 include all geometric dimensions of the devices discussed in Chap. 6, where W_0 denotes the width of the graphite extending the coplanar strips, W_1 is the width of each metal trace, W_2 is the width of the coplanar strip gap, d_{hBN} is the thickness of the hexagonal boron nitride (subscripts 1 and 2 denote hBN layers above or below the graphene layer, d_{gr} is the thickness of the graphite flake and l defines the length of the device along the propagation direction of the THz pulse.

For the series of test devices (TD series) that did not exhibit noticeable deviations between measured sample and reference signals, the resonance frequency of the lowest in frequency resonances were theoretically determined assuming the excitation of either purely screened or purely unscreened cavity modes with $\lambda = 2W_{1,2}$. Device TD7 consists of a stack of thin palladium gold (PdAu) and graphite, separated by hBN (see Fig. A.1). The expected resonance frequency was determined by assuming that graphite and PdAu layers, both placed in the coplanar strip gap, are screening each other.

The detected THz conductivity spectra of devices BG1 to BG9E were fit to a sum of Drude-Lorentz models (see Sect. 5.2.2) to obtain the resonance frequencies $\nu_{exp,1}$ and $\nu_{exp,2}$, linewidths Γ_1 and Γ_2, and corresponding quality factors Q_1 and Q_2 of the detected resonances (see Table A.3).

A.2 From 3D to 2D Conductivity

This section briefly discusses how the 2D conductivity discussed in this thesis is defined and how it can be obtained from the 3D (bulk) conductivity. First, the 3D conductance G is defined as the inverse of the resistance R:

G. Kipp, *Self-Cavity Electrodynamics of Van der Waals Heterostructures*,
Springer Theses, https://doi.org/10.1007/978-3-032-31979-1

Table A.1 Device parameters for the test devices shown in Figs. 6.3 and 6.4, and expected resonance frequencies of the $m = 0$ mode (assuming a screened (unscreened) 2D plasmon with momentum $q = 2\pi/(2W_1)$ ($q = 2\pi/(2W_2)$)). A bulk plasma frequency of $\nu_{3D} = 118$ THz was assumed for graphite

Device	W_1	W_2	d_{hBN1}	d_{hBN2}	d_{gr}	ν_{theory} (THz)
TD1	3.7 μm	2.4 μm	24.5 nm	—	6.5 nm	0.65
TD2	3.27 μm	2.9 μm	17.9 nm	—	6.3 nm	0.63
TD3	2.33 μm	3.51 μm	22.7 nm	21.6 nm	graphene	0.96
TD4	3.5 μm	2.6 μm	20.4 nm	25.2 nm	graphene	0.63
TD5	3.7 μm	2.7 μm	21+27.4 nm	—	2.9 nm	1.58
TD6	3.3 μm	2.44 μm	16.8 nm	17.1 nm	5 nm	2.56
TD7	5.7 μm	4.6 μm	29.2 nm	24.9 nm	$d_{gr} = 8.5$ nm	0.49
					$d_{PdAU} = 5$ nm	?

Table A.2 Device parameters for the graphite devices shown in Figs. 6.5, 6.6 and 6.7

Device	W_0	W_1	W_2	d_{hBN}	d_{gr}	l
BG1	3-6 μm	5.31 μm	4.47 μm	28 nm	5.9 nm	9.8 μm
BG2	2.05 μm	5.49 μm	2.3 μm	23.7 nm	6.9 nm	12.3 μm
BG3	4.8 μm	3.5 μm	2.31 μm	24.8 nm	4.5 nm	13.2 μm
BG4	1.5 μm	3.67 μm	2.4 μm	24.4 nm	4.6 nm	38 μm
BG5	0.2 μm	3.27 μm	2.9 μm	17.9 nm	6.3 nm	8.2 μm
BG6E	1.9 μm	3.55 μm	2.26 μm	26.0 nm	16 nm	11 μm
BG7	10 μm	3.47 μm	2.64 μm	90 nm	4 nm	16.47 μm
BG8	0-16 μm	2.93 μm	3.03 μm	16.5 nm	5.5 nm	20.4 μm
BG9E	1.2 μm	2.95 μm	4.84 μm	30 nm	7.5 nm	8.8 μm
BGXL	1 μm	16.18 μm	13.85 μm	31.5 nm	6.7 nm	8.8 μm

Table A.3 Fitted plasmon resonance frequencies ν, linewidths Γ and quality factors Q for the graphite devices shown in Figs. 6.5, 6.6 and 6.7

Device	$\nu_{exp,1}$ (THz)	Γ_1 (THz)	Q_1	$\nu_{exp,2}$ (THz)	Γ_2 (THz)	Q_2
BG1	0.123±0.001	0.22±0.01	0.55	0.636±0.003	0.87±0.01	0.72
BG2	0.141±0.001	0.29±0.01	0.48	0.680±0.003	0.8±0.01	0.85
BG3	0.161±0.001	0.296±0.001	0.53	0.75±0.001	0.855±0.007	0.88
BG4	0.230±0.001	0.412±0.004	0.56	0.761±0.001	0.523±0.007	1.46
BG5	0.185±0.001	0.29±0.002	0.66	0.932±0.006	1.12±0.03	0.83
BG6E	0.43±0.01	0.29±0.01	1.48	–	–	–
BG7	0.435±0.003	0.27±0.01	1.63	–	–	–
BG8	0.16±0.03	–	–	–	–	–
BG9E	0.271±0.003	0.43±0.01	0.63	1.00±0.03	1.24±0.17	0.81
BGXL	<0.05 THz	–	–	–	–	–

$$G = \frac{1}{R}. \tag{A.1}$$

The resistance R, in turn, is defined as

$$R = \rho \frac{l}{A}, \tag{A.2}$$

whereby ρ defines the resistivity, typically given in Ω m, l is the length of the conductor and A is the area of the cross-section of the conductor. The inverse of the resistivity is the conductivity σ. The resistance can then be reformulated to obtain the sheet resistance R_s (in Ohms), which is the resistance of thin materials with thickness d (such that $A = d \cdot W$):

$$R = \rho \frac{l}{dW} = \frac{\rho}{d} \frac{l}{W} = R_s \frac{l}{W}. \tag{A.3}$$

Now, the 2D conductance is defined as the inverse of the sheet resistance, such that the 2D conductivity (in Siemens) is given by

$$\sigma_{\mathrm{2D}} = \sigma_{\mathrm{3D}} d. \tag{A.4}$$

The 2D Drude conductivity of a graphite flake is therefore given by (using the definition of the bulk plasma frequency $\omega_{\mathrm{3D}} = \sqrt{ne^2/(m\epsilon_0)}$):

$$\sigma_{\mathrm{2D,graphite}} = \sigma_{\mathrm{3D}} d_{\mathrm{gr}} = \frac{ne^2}{\Gamma m} d_{\mathrm{gr}} = \frac{\epsilon_0 \omega_{\mathrm{3D}}^2 d_{\mathrm{gr}}}{\Gamma}. \tag{A.5}$$

In order to obtain the 2D conductivity in Siemens per meter, as detected in this thesis with on-chip THz spectroscopy, the 2D conductivity is multiplied by the multiplicative filling factor (see Sect. 6.4.7).

A.3 Additional Data Sets

This section provides additional data sets of one test device consisting of a stack of PdAu and graphite purely placed in the coplanar strip gap (Fig. A.1) and of three further gate-tunable graphene devices, providing measured data on both electron and hole doped sides (Figs. A.2 and A.3).

In the main text of this thesis, detected THz spectra of only the electron doped side of graphene were discussed that correspond to positive back gate voltages. However, measurement data was collected also on the electron hole doped side for a series of devices, including the ultrastrong coupling device GG2 (see Figs. A.2 and A.3). Interestingly it was found that the cavity conductivity spectra are relatively symmetric for Devices GG2 and GG5, but the spectra exhibit an

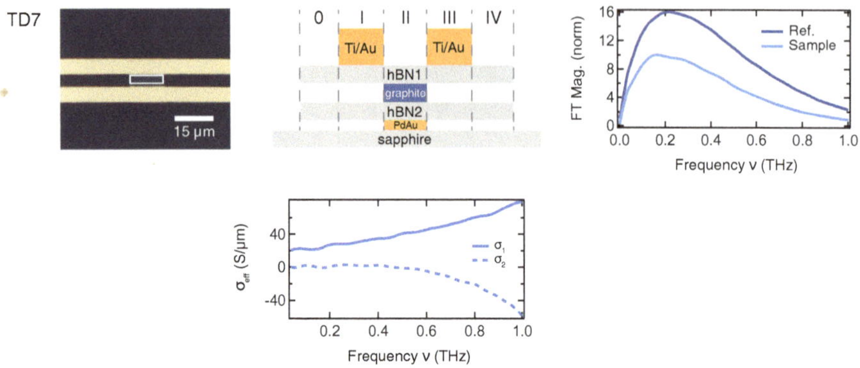

Fig. A.1 Measured effective near-field conductivity spectrum of device TD7. No resonance is observed within the spectral sensitivity range, but the data suggests the presence of a high-amplitude resonance outside this range. This may be attributed to the excitation of a plasmonic resonance in the PdAu layer of the device. Given the significantly higher carrier concentration in PdAu compared to graphite, it is possible that the graphite resonance is obscured by the PdAu resonance, which exhibits much greater spectral weight

electron-hole asymmetry for Devices GG3 and GG4. While the charge neutrality point of graphene of the individual devices was defined as the applied back gate voltage at which the overall spectral weight within the spectral sensitivity range was minimized, the graphite $m = 0$ mode, that is expected to be the resonance with the maximum amplitude in the spectrum, has first a decreasing resonance frequency and then an increasing resonance frequency with increasing negative back gate voltage in the spectra of Devices GG3 and GG4. In contrast, the resonance frequency of the graphite mode is purely increasing with increasing positive back gate voltage on the electron doped side (marked by the black curves).

This phenomenon can be attributed to the varying graphite thickness in the devices. While Devices GG2 and GG5 contain a rather thick graphite layer with a carrier density that far exceeds the one of the (doped) graphene layer, the graphite layer of Devices GG3 and GG4 is rather thin $d_{gr} < 10$ nm. This has the effect that the Fermi level of thin graphite (<10 nm) can be electrostatically tuned by the graphene layer. This finding agrees with literature showing that thin graphite is slightly hole doped when a back gate voltage of 0 V is applied [1], meaning that the lowest 3D bulk plasma frequency would be found when a slightly positive voltage is applied between an electrostatic gate and a thin graphite layer. As graphene serves here as the electrostatic gate, and the minimum resonance frequency of the graphite $m = 0$ mode was observed with a small negative voltage applied from the graphite to the graphene layer, this effect is equivalent to applying a small positive voltage from graphene to the graphite layer.

Thus, both Fermi levels of graphene and graphite can be tuned with the applied back gate voltage. This obscures the detected cavity conductivity spectra and while ultrastrong coupling is also expected for Devices GG3 (b) and GG4 (c) based on

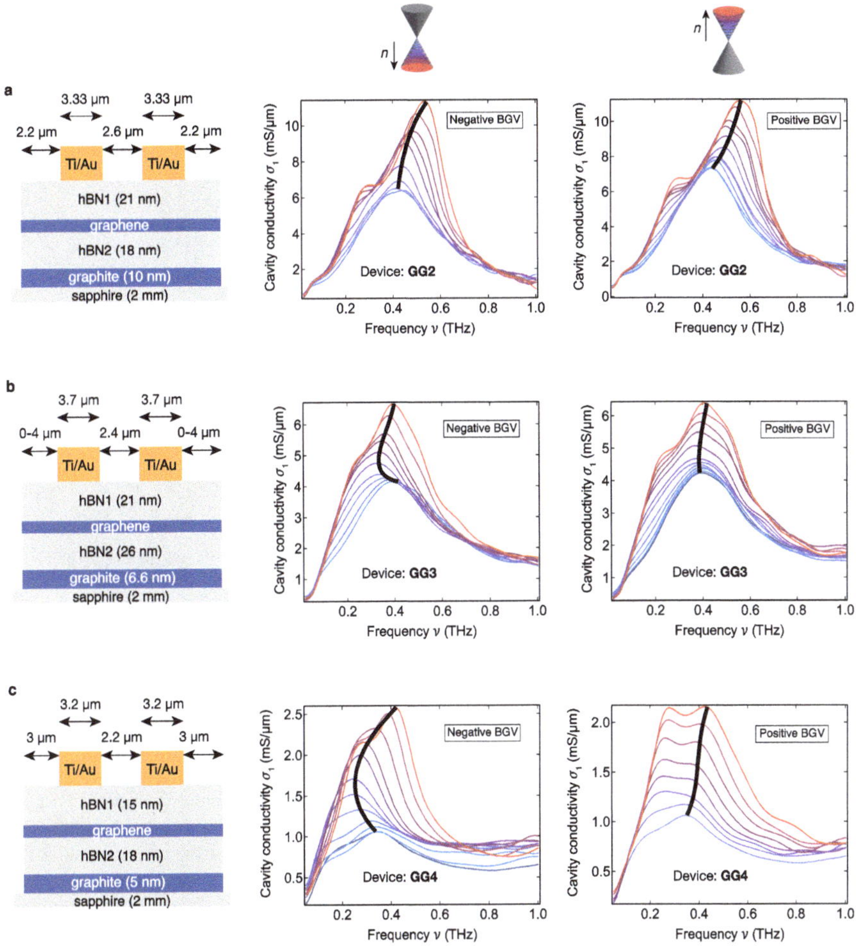

Fig. A.2 Additional on-chip THz spectroscopy data sets of gate-tunable graphene heterostructures for both hole (center panels, negative back gate voltages (BGV)) and electron (right panels, positive BGV) doped sides. (**a**) Data for Device GG2, the ultrastrong coupling device discussed in Sect. 8.4. Only the electron doped data was discussed in the main thesis as the data could be fitted more cleanly. Additionally, the data sets taken on two similar graphene/graphite heterostructures (**b**, **c**) that contained a thinner graphite flake than Device GG2 showed an interesting trend of the graphite plasmon at low negative back gate voltages (marked by the black curve), as explained in the main text of this section

their geometric designs, it is not possible to reliably extract information from the spectra without considering the bandstructure of (thin) graphite in the analytical theory. First graphite bandstructure calculations have already been initiated to better understand the observed electron-hole asymmetries.

The detected cavity conductivity spectra of Device GG5 (similar cavity geometry as Device GG1 discussed in Sect. 8.1) presented in Fig. A.3 show that while the

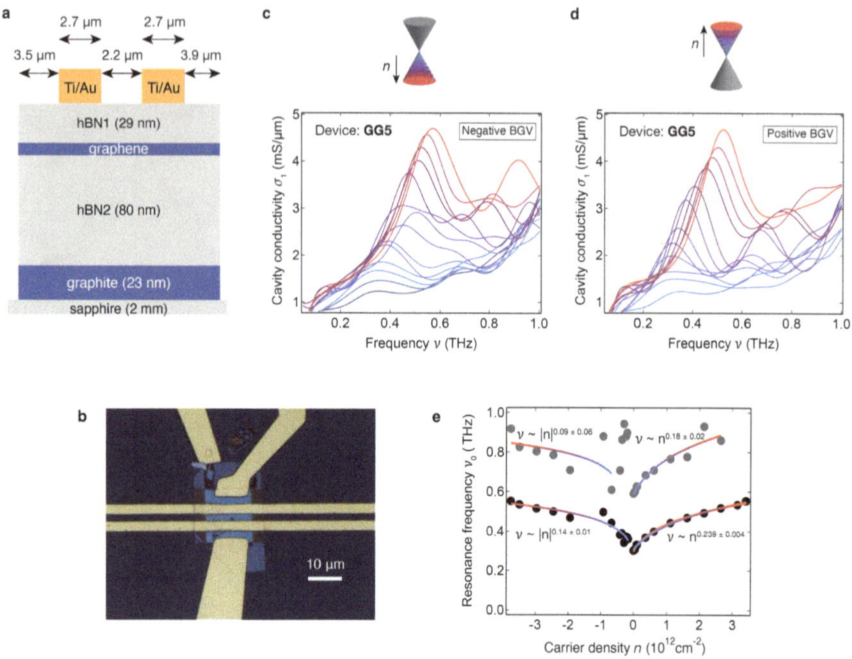

Fig. A.3 Additional on-chip THz spectroscopy data set of a gate-tunable graphene heterostructure with a geometry similar to Device GG1. (**a**) Cross-section of the device with parameters. The hBN2 and graphite layers were designed thick to push the graphite $m = 0$ mode outside the spectral sensitivity range (expected resonance frequency from theory is 1.22 THz). (**b**) Optical micrograph of the device. The top left gold patch is the contact to the graphite gate and the other two gold patches contact the graphene layer. The parallel metal traces are the transmission line. (**c**, **d**) Real parts of the cavity conductivity as a function of frequency for different negative (**c**) and positive (**d**) graphene carrier densities. (**e**) The data of (**c**) and (**d**) was fit to a sum of Drude-Lorentz models to extract the resonance frequencies as a function of graphene carrier density for the $s = 1, 2$ modes. The $s = 0$ mode resonating below 0.2 THz was found difficult to fit and is therefor not shown here. The power scaling of the graphene plasmonic $s = 1, 2$ is found to deviate significantly from the expected $\nu_0 \propto n^{0.25}$ behaviour that is expected for uncoupled graphene modes, indicating strong hybridization of the graphene modes with the graphite $m = 0$ mode outside the spectral sensitivity range

detected resonance frequencies are aligned relatively symmetric around the charge neutrality point of graphene, the noise level on the negative back gate voltage side (hole doped graphene) is much higher compared to the positive back gate voltage side (electron doped graphene). Significantly reduced graphene plasmon power law scalings ($\nu_0 \propto n^x$, $x \ll 0.25$) are detected on both electron and hole doped sides for the $s = 1$ and $s = 2$ modes. Similar to Device GG1 discussed in Sect. 8.1, the $s = 2$ mode is even more suppressed than the $s = 1$ mode as it hybridized more strongly to the $m = 0$ mode located outside the spectral sensitivity range around 1.22 THz.

A.4 Dispersion Relation of Double-Screened 2D Plasmons

The boundary conditions for double-screened 2D plasmon are very similar to the screened 2D plasmon's boundary conditions (see Sect. 6.4.2), but an additional reflection has to be taken into account as the electric field is reflected from the additional metal layer (Fig. A.4).

The electric field is therefore (split into two distinct regions):

$$E_x = \begin{cases} \left(t_1 e^{-kz} + r_1 e^{kz}\right) E_0 e^{iqx-i\omega t} & , 0 < z < d_{\mathrm{hBN1}} \\ \left(t_2 e^{kz} + r_2 e^{-kz}\right) E_0 e^{iqx-i\omega t} & , -d_{\mathrm{hBN2}} < z < 0, \end{cases} \quad \text{(A.6)}$$

where $z = 0$ defines the out-of-plane position of the 2D material and $z < 0$ defined the region in the hBN between the 2D material and the bottom metal layer. From the electric field along the x-direction (in-plane), the electric field along the z-direction (out-of-plane) can be found using Gauss's law: $\nabla \cdot \vec{E} = 0$:

$$E_z = \begin{cases} \left(\frac{iq}{k} t_1 e^{-kz} - \frac{iq}{k} r_1 e^{kz}\right) E_0 e^{iqx-i\omega t} & , 0 < z < d_{\mathrm{hBN1}} \\ -\left(\frac{iq}{k} t_2 e^{kz} - \frac{iq}{k} r_2 e^{-kz}\right) E_0 e^{iqx-i\omega t} & , -d_{\mathrm{hBN2}} < z < 0. \end{cases} \quad \text{(A.7)}$$

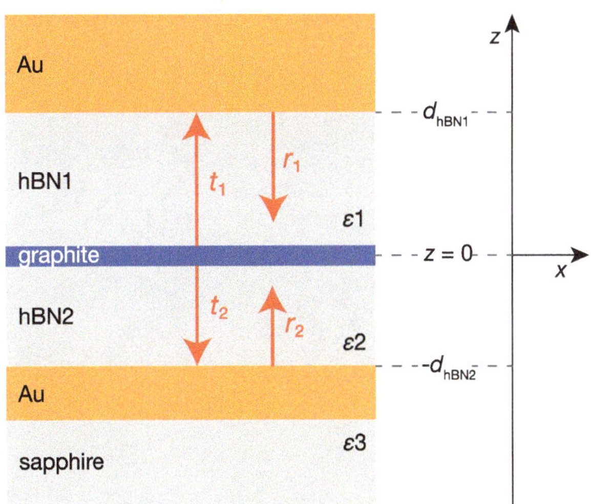

Fig. A.4 Double-screened graphite. Transmission and reflection coefficients of the electric field in a graphite heterostructure that is screened by two metallic gold layers

Similarly, the density distribution in the graphene is given by:

$$\rho_{2D} = \epsilon_0 \left(\epsilon_1 E_z(z = 0^+) - \epsilon_2 E_z(z = 0^-) \right) \\ = \frac{iq\epsilon_0}{k} E_0 \left(\epsilon_1 t_1 - \epsilon_1 r_1 + \epsilon_2 t_2 - \epsilon_2 r_2 \right) e^{iqx - i\omega t}, \tag{A.8}$$

as $\nabla \cdot \vec{E} = \rho/\epsilon_0$ in a material. $\epsilon_1 = \epsilon_2 = \epsilon_{\text{hBN}}$ in the case of graphene.

Next, the magnetic field along the y-direction can be written down that is given by the Maxwell's equation $\partial_x B_y = \frac{\epsilon_i}{c^2} \partial_t E_z$:

$$B_y = \frac{-i\omega}{c^2} \begin{cases} \epsilon_1 \left(\frac{1}{k} t_1 e^{-kz} - \frac{1}{k} r_1 e^{kz} \right) E_0 e^{iqx - i\omega t} & , 0 < z < d_{\text{hBN1}} \\ -\epsilon_2 \left(\frac{iq}{k} t_2 e^{kz} - \frac{iq}{k} r_2 e^{kz} \right) E_0 e^{iqx - i\omega t} & , -d_{\text{hBN2}} < z < 0 \end{cases} \tag{A.9}$$

At the metal interfaces, the electric field parallel to the surfaces vanishes:

$$E_x(z = d_{\text{hBN1}}) = t_1 e^{-qd_{\text{hBN1}}} + r_1 e^{qd_{\text{hBN1}}} \equiv 0, \tag{A.10}$$

$$r_1 = -t_1 e^{-2qd_{\text{hBN1}}}, \tag{A.11}$$

$$E_x(z = -d_{\text{hBN2}}) = t_2 e^{-qd_{\text{hBN2}}} + r_2 e^{qd_{\text{hBN2}}} \equiv 0, \tag{A.12}$$

$$r_2 = -t_2 e^{-2qd_{\text{hBN2}}}, \tag{A.13}$$

while at $z = 0$, the electric and magnetic field parallel to the interface can be matched:

$$t_1 \left(1 - e^{-2qd_{\text{hBN1}}} \right) = t_2 \left(1 - e^{-2qd_{\text{hBN2}}} \right), \tag{A.14}$$

$$- i\omega \frac{\epsilon_1}{c^2 q} t_1 \left(1 + e^{-2qd_{\text{hBN1}}} \right) + i\omega \frac{\epsilon_2}{c^2 q} t_2 \left(1 + e^{-2qd_{\text{hBN2}}} \right) \\ = \frac{1}{c^2 \epsilon_0} \frac{j_{2D}}{E_0} = \frac{\sigma_{2D}(\omega)}{\epsilon_0} \frac{1}{c^2} t_2 \left(1 - e^{-2qd_{\text{hBN2}}} \right). \tag{A.15}$$

By inserting the first equation into the second equation for t_1, the following equation is obtained:

$$- i\omega \frac{\epsilon_{\text{hBN}}}{c^2 q} t_2 \left(\frac{\left(1 - e^{-2qd_{\text{hBN2}}}\right)\left(1 + e^{-2qd_{\text{hBN1}}}\right)}{1 - e^{-2qd_{\text{hBN1}}}} + \left(1 + e^{-2qd_{\text{hBN2}}}\right) \right) \\ = \frac{\sigma_{2D}(\omega)}{\epsilon_0} \frac{1}{c^2} t_2 \left(1 - e^{-2qd_{\text{hBN2}}} \right), \tag{A.16}$$

that can be transformed to:

$$
\begin{aligned}
&- i\omega\frac{\epsilon_{\text{hBN}}}{c^2 q}t_2\left(1+e^{-2qd_{\text{hBN2}}}\right)\left(\frac{\left(1+e^{-2qd_{\text{hBN1}}}\right)\left(1-e^{-2qd_{\text{hBN2}}}\right)}{\left(1-e^{-2qd_{\text{hBN1}}}\right)\left(1+e^{-2qd_{\text{hBN2}}}\right)}+1\right)\\
&= \frac{\sigma_{\text{2D}}(\omega)}{\epsilon_0}\frac{1}{c^2}t_2\left(1-e^{-2qd_{\text{hBN2}}}\right),
\end{aligned}
\tag{A.17}
$$

and then simplified to:

$$
- i\omega\frac{\epsilon_{\text{hBN}}}{c^2 q}t_2\left(1+e^{-2qd_{\text{hBN2}}}\right)\left(\frac{\coth\left(qd_{\text{hBN1}}\right)}{\coth\left(qd_{\text{hBN2}}\right)}+1\right)=\frac{\sigma_{\text{2D}}(\omega)}{\epsilon_0}\frac{1}{c^2}t_2\left(1-e^{-2qd_{\text{hBN2}}}\right).
\tag{A.18}
$$

Using the fact that $qd_{\text{hBN}} << 1$, one finds:

$$
2\epsilon_{\text{hBN}}t_2\left(\frac{d_{\text{hBN2}}}{d_{\text{hBN1}}}+1\right)=\frac{\sigma_{\text{2D}}(\omega)q}{i\omega\epsilon_0}t_2\cdot 2qd_{\text{hBN2}}.
\tag{A.19}
$$

This leads to the dispersion relation of the double screened plasmon:

$$
\omega_{\text{ds}}=\frac{\sigma_{\text{2D}}q^2}{i\epsilon_0\epsilon_{\text{hBN}}\left(\frac{1}{d_{\text{hBN1}}}+\frac{1}{d_{\text{hBN2}}}\right)},
\tag{A.20}
$$

that can be simplified with $qd_{\text{hBN}} \ll 1$, using $\sigma_{\text{2D}} = \epsilon_0\omega_{\text{3D}}^2 d_{\text{gr}}/(-i\omega+\Gamma)$ to the well-known term for screened 2D plasmons in graphite (in case of graphene, replace $\omega_{\text{3D}}^2 d_{\text{gr}}$ with $e^2 v_F\sqrt{n}/(\hbar\sqrt{\pi})$) [2, 3]:

$$
\boxed{\omega_{\text{ds}}=\sqrt{\frac{\omega_{\text{3D}}^2 d_{\text{gr}} q^2}{\epsilon_1\left(\frac{1}{d_{\text{hBN1}}}+\frac{1}{d_{\text{hBN2}}}\right)}}}.
\tag{A.21}
$$

A.5 Computing Charge Densities

Sect. 6.4.4 shows how current densities can be calculated for the individual cavity regions as a function of x. The corresponding charge densities can simply be extracted from the current density equations using the continuity equation (see Eq. 6.8):

$$\begin{aligned} 0 &= \delta_t \rho + \delta_x j_x \\ -i\omega\rho &= -iqj_x^+ + iqj_x^- \\ \rho^+ &= \frac{q}{\omega} j_x^+ \\ \rho^- &= -\frac{q}{\omega} j_x^- , \end{aligned} \tag{A.22}$$

where the plus symbol defines currents/charges propagating in positive x-direction (transmitted waves) and the minus symbols defines currents/charges propagating in negative x-direction (reflected waves). For computing charge densities, all transmission terms given for current densities in Sect. 6.4.4 therefore have to be multiplied by q/ω and all reflection terms by $-q/\omega$.

Both current and charge densities are calculated with the analytical theory for exemplary graphite heterostructures in Sects. 6.5.3.2 and 6.5.3.3.

A.6 Current Densities and Potentials of a Heterostructure with Two Self-cavities

This section explicitly presents the current density and potential terms for a heterostructure composed of two self-cavities, such as a graphene and graphite layer.

A.6.1 Current Densities

The currents in the cavity region 0 ($-W_0 - W_1 - W_2/2 < x < -W_1 - W_2/2$) are in the graphene layer

$$\begin{aligned} j_{0,\text{graphene}}(x,t) &= \sigma_{\text{graphene}} \left(r_{0,\text{as}} E_0 e^{-iq_{\text{un,as}}(\nu)x} + t_{0,\text{as}} E_0 e^{iq_{\text{un,as}}(\nu)x} \right) \\ &+ \sigma_{\text{graphene}} \left(r_{0,\text{sym}} E_0 e^{-iq_{\text{un,sym}}(\nu)x} + t_{0,\text{sym}} E_0 e^{iq_{\text{un,sym}}(\nu)x} \right) \end{aligned} \tag{A.23}$$

and in the graphite layer

$$\begin{aligned} j_{0,\text{graphite}}(x,t) &= \sigma_{\text{graphite}} \frac{E_{\text{x,un,graphite}}}{E_{\text{x,un,graphene}}} (q_{\text{un,as}}) \left(r_{0,\text{as}} E_0 e^{-iq_{\text{un,as}}(\nu)x} + t_{0,\text{as}} E_0 e^{iq_{\text{un,as}}(\nu)x} \right) \\ &+ \sigma_{\text{graphite}} \frac{E_{\text{x,un,graphite}}}{E_{\text{x,un,graphene}}} (q_{\text{un,sym}}) \\ &\times \left(r_{0,\text{sym}} E_0 e^{-iq_{\text{un,sym}}(\nu)x} + t_{0,\text{sym}} E_0 e^{iq_{\text{un,sym}}(\nu)x} \right) \end{aligned} \tag{A.24}$$

where $r_{0,\text{as}}$ and $t_{0,\text{as}}$ are the reflection and transmission coefficients of the antisymmetric mode and $r_{0,\text{sym}}$ and $t_{0,\text{sym}}$ are the reflection and transmission coefficients of the symmetric mode in the unscreened region 0 with the dispersion relations $q_{\text{un,sym}}$ (symmetric mode) and $q_{\text{un,as}}$ (antisymmetric mode), E_0 is the incident electric field and $E_{\text{x,un,graphite}}/E_{\text{x,un,graphene}}$ is Eq. 8.45 as a function of $q_{\text{un,sym}}$ or $q_{\text{un,as}}$.

The currents in the cavity region I ($-W_1 - W_2/2 < x < -W_2/2$) are:

$$\begin{aligned} j_{1,\text{graphene}}(x,t) &= \sigma_{\text{graphene}}\left(r_{1,\text{as}}E_0e^{-iq_{\text{sc,as}}(\nu)x} + t_{1,\text{as}}E_0e^{iq_{\text{sc,as}}(\nu)x}\right) \\ &+ \sigma_{\text{graphene}}\left(r_{1,\text{sym}}E_0e^{-iq_{\text{sc,sym}}(\nu)x} + t_{1,\text{sym}}E_0e^{iq_{\text{sc,sym}}(\nu)x}\right), \end{aligned} \tag{A.25}$$

$$\begin{aligned} j_{1,\text{graphite}}(x,t) &= \sigma_{\text{graphite}}\frac{E_{\text{x,sc,graphite}}}{E_{\text{x,sc,graphene}}}(q_{\text{sc,as}})\left(r_{1,\text{as}}E_0e^{-iq_{1,\text{as}}(\nu)x} + t_{1,\text{as}}E_0e^{iq_{\text{sc,as}}(\nu)x}\right) \\ &+ \sigma_{\text{graphite}}\frac{E_{\text{x,sc,graphite}}}{E_{\text{x,sc,graphene}}}(q_{\text{sc,sym}}) \\ &\times\left(r_{1,\text{sym}}E_0e^{-iq_{\text{sc,sym}}(\nu)x} + t_{1,\text{sym}}E_0e^{iq_{\text{sc,sym}}(\nu)x}\right) \end{aligned} \tag{A.26}$$

where $r_{1,\text{as}}$ and $t_{1,\text{as}}$ are the reflection and transmission coefficients of the antisymmetric mode and $r_{1,\text{sym}}$ and $t_{1,\text{sym}}$ are the reflection and transmission coefficients of the symmetric mode in the screened region I with the dispersion relations $q_{\text{sc,sym}}$ (symmetric mode) and $q_{\text{sc,as}}$ (antisymmetric mode), E_0 is the incident electric field and $E_{\text{x,sc,graphite}}/E_{\text{x,sc,graphene}}$ is Eq. 8.24 as a function of $q_{\text{sc,sym}}$ or $q_{\text{sc,as}}$.

The currents in the cavity region II ($-W_2/2 < x < W_2/2$) are in the graphene layer

$$\begin{aligned} j_{2,\text{graphene}}(x,t) &= \sigma_{\text{graphene}}\left(r_{2,\text{as}}E_0e^{-iq_{\text{un,as}}(\nu)x} + t_{2,\text{as}}E_0e^{iq_{\text{un,as}}(\nu)x}\right) \\ &+ \sigma_{\text{graphene}}\left(r_{2,\text{sym}}E_0e^{-iq_{\text{un,sym}}(\nu)x} + t_{2,\text{sym}}E_0e^{iq_{\text{un,sym}}(\nu)x}\right) \\ &+ \sigma_{\text{graphene}}E_{\text{ext}} \end{aligned} \tag{A.27}$$

and in the graphite layer

$$\begin{aligned} j_{2,\text{graphite}}(x,t) &= \sigma_{\text{graphite}}\frac{E_{\text{x,un,graphite}}}{E_{\text{x,un,graphene}}}(q_{\text{un,as}})\left(r_{2,\text{as}}E_0e^{-iq_{\text{un,as}}(\nu)x} + t_{2,\text{as}}E_0e^{iq_{\text{un,as}}(\nu)x}\right) \\ &+ \sigma_{\text{graphite}}\frac{E_{\text{x,un,graphite}}}{E_{\text{x,un,graphene}}}(q_{\text{un,sym}}) \\ &\times\left(r_{2,\text{sym}}E_0e^{-iq_{\text{un,sym}}(\nu)x} + t_{2,\text{sym}}E_0e^{iq_{\text{un,sym}}(\nu)x}\right) \\ &+ \sigma_{\text{graphite}}\frac{E_{\text{x,un,graphite}}}{E_{\text{x,un,graphene}}}(q_{\text{un,as}})E_{\text{ext}} \end{aligned} \tag{A.28}$$

where $r_{2,\mathrm{as}}$ and $t_{2,\mathrm{as}}$ are the reflection and transmission coefficients of the antisymmetric mode and $r_{2,\mathrm{sym}}$ and $t_{2,\mathrm{sym}}$ are the reflection and transmission coefficients of the symmetric mode in the unscreened region II with the dispersion relations $q_{\mathrm{un,sym}}$ (symmetric mode) and $q_{\mathrm{un,as}}$ (antisymmetric mode), E_0 is the incident electric field and $E_{\mathrm{x,un,graphite}}/E_{\mathrm{x,un,graphene}}$ is Eq. 8.45 as a function of $q_{\mathrm{un,sym}}$ or $q_{\mathrm{un,as}}$.

A.6.2 Potentials

Next, the potentials V have to be found, as both current densities and potentials have to be matched to extract the cavity conductivity. As $j = \sigma E$ and $E = -\frac{\mathrm{dV}}{\mathrm{dx}}$, the potentials can be extracted from the just derived relations for the different currents by dividing the currents through the conductivity and dividing the transmission terms by $-iq$ and the reflection terms by iq.

The potentials in the cavity region 0 $(-W_0 - W_1 - W_2/2 < x < -W_1 - W_2/2)$ are in the graphene layer

$$\begin{aligned} V_{0,\mathrm{graphene}}(x,t) = &\frac{r_{0,\mathrm{as}}}{iq_{\mathrm{un,as}}}E_0 e^{-iq_{\mathrm{un,as}}(\nu)x} + \frac{t_{0,\mathrm{as}}}{-iq_{\mathrm{un,as}}}E_0 e^{iq_{\mathrm{un,as}}(\nu)x} \\ &+ \frac{r_{0,\mathrm{sym}}}{iq_{\mathrm{un,sym}}}E_0 e^{-iq_{\mathrm{un,sym}}(\nu)x} + \frac{t_{0,\mathrm{sym}}}{-iq_{\mathrm{un,sym}}}E_0 e^{iq_{\mathrm{un,sym}}(\nu)x} \end{aligned} \tag{A.29}$$

and in the graphite layer

$$\begin{aligned} V_{0,\mathrm{graphite}}(x,t) = &\frac{E_{\mathrm{x,un,graphite}}}{E_{\mathrm{x,un,graphene}}}(q_{\mathrm{un,as}})\left(\frac{r_{0,\mathrm{as}}}{iq_{\mathrm{un,as}}}E_0 e^{-iq_{\mathrm{un,as}}(\nu)x} + \frac{t_{0,\mathrm{as}}}{-iq_{\mathrm{un,as}}}E_0 e^{iq_{\mathrm{un,as}}(\nu)x}\right) \\ &+ \frac{E_{\mathrm{x,un,graphite}}}{E_{\mathrm{x,un,graphene}}}(q_{\mathrm{un,sym}}) \\ &\times \left(\frac{r_{0,\mathrm{sym}}}{iq_{\mathrm{un,sym}}}E_0 e^{-iq_{\mathrm{un,sym}}(\nu)x} + \frac{t_{0,\mathrm{sym}}}{-iq_{\mathrm{un,sym}}}E_0 e^{iq_{\mathrm{un,sym}}(\nu)x}\right). \end{aligned} \tag{A.30}$$

The potentials in the cavity region I $(-W_1 - W_2/2 < x < -W_2/2)$ are:

$$\begin{aligned} V_{1,\mathrm{graphene}}(x,t) = &\frac{r_{1,\mathrm{as}}}{iq_{\mathrm{sc,as}}}E_0 e^{-iq_{\mathrm{sc,as}}(\nu)x} + \frac{t_{1,\mathrm{as}}}{-iq_{\mathrm{sc,as}}}E_0 e^{iq_{\mathrm{sc,as}}(\nu)x} \\ &+ \frac{r_{1,\mathrm{sym}}}{iq_{\mathrm{sc,sym}}}E_0 e^{-iq_{\mathrm{sc,sym}}(\nu)x} + \frac{t_{1,\mathrm{sym}}}{-iq_{\mathrm{sc,sym}}}E_0 e^{iq_{\mathrm{sc,sym}}(\nu)x}, \end{aligned} \tag{A.31}$$

$$
\begin{aligned}
V_{1,\text{graphite}}(x,t) = {} & \frac{E_{\text{x,sc,graphite}}}{E_{\text{x,sc,graphene}}}(q_{\text{sc,as}}) \left(\frac{r_{1,\text{as}}}{iq_{\text{sc,as}}} E_0 e^{-iq_{\text{sc,as}}(\nu)x} + \frac{t_{1,\text{as}}}{-iq_{\text{sc,as}}} E_0 e^{iq_{\text{sc,as}}(\nu)x} \right) \\
& + \frac{E_{\text{x,sc,graphite}}}{E_{\text{x,sc,graphene}}}(q_{\text{sc,sym}}) \\
& \times \left(\frac{r_{1,\text{sym}}}{iq_{\text{sc,sym}}} E_0 e^{-iq_{\text{sc,sym}}(\nu)x} + \frac{t_{1,\text{sym}}}{-iq_{\text{sc,sym}}} E_0 e^{iq_{\text{sc,sym}}(\nu)x} \right).
\end{aligned}
\tag{A.32}
$$

The potentials in the cavity region II ($-W_2/2 < x < W_2/2$) are in the graphene layer

$$
\begin{aligned}
V_{2,\text{graphene}}(x,t) = {} & \frac{r_{2,\text{as}}}{iq_{\text{un,as}}} E_0 e^{-iq_{\text{un,as}}(\nu)x} + \frac{t_{2,\text{as}}}{-iq_{\text{un,as}}} E_0 e^{iq_{\text{un,as}}(\nu)x} \\
& + \frac{r_{2,\text{sym}}}{iq_{\text{un,sym}}} E_0 e^{-iq_{\text{un,sym}}(\nu)x} + \frac{t_{2,\text{sym}}}{-iq_{\text{un,sym}}} E_0 e^{iq_{\text{un,sym}}(\nu)x}
\end{aligned}
\tag{A.33}
$$

and in the graphite layer

$$
\begin{aligned}
V_{2,\text{graphite}}(x,t) = {} & \frac{E_{\text{x,un,graphite}}}{E_{\text{x,un,graphene}}}(q_{\text{un,as}}) \left(\frac{r_{2,\text{as}}}{iq_{\text{un,as}}} E_0 e^{-iq_{\text{un,as}}(\nu)x} + \frac{t_{2,\text{as}}}{-iq_{\text{un,as}}} E_0 e^{iq_{\text{un,as}}(\nu)x} \right) \\
& + \frac{E_{\text{x,un,graphite}}}{E_{\text{x,un,graphene}}}(q_{\text{un,sym}}) \\
& \times \left(\frac{r_{2,\text{sym}}}{iq_{\text{un,sym}}} E_0 e^{-iq_{\text{un,sym}}(\nu)x} + \frac{t_{2,\text{sym}}}{-iq_{\text{un,sym}}} E_0 e^{iq_{\text{un,sym}}(\nu)x} \right).
\end{aligned}
\tag{A.34}
$$

A.7 Simulations of "Cavity Sensing" and "Cavity Control" Designs

A.7.1 Cavity Sensing of Gated 2D Materials

Full 3D electromagnetic simulations (shown in Fig. A.5) were performed by Dr. Toru Matsuyama for a "sensing cavity" design with small coplanar strip gap and $W_0 = 0$ to demonstrate that this geometry indeed can be used to probe the THz conductivity of vdW heterostructures, without the influence of coupling between screened and unscreened graphene and graphite plasmons. The full 3D electromagnetic simulations were performed using a thin graphite layer (called

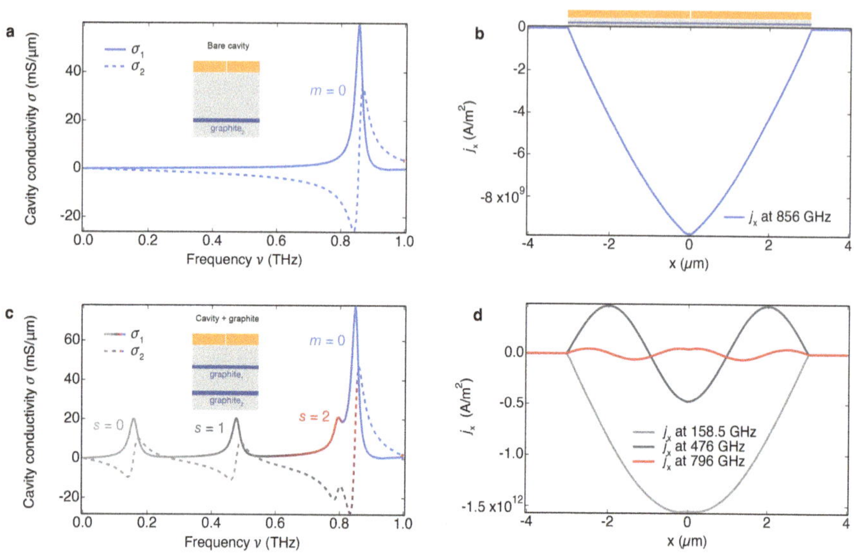

Fig. A.5 Full 3D electromagnetic simulations for a sensing cavity design. (**a**) Design of the simulated bare graphite heterostructure. (**b**) Real and imaginary parts of the cavity conductivity as a function of frequency. The $m = 0$ mode of the graphite$_2$ self-cavity hosts a resonance centered at 856 GHz. (**c**) Current density profile in the graphite$_2$ layer at the resonance frequency. (**d**) An additional graphite layer, graphite$_1$ is introduced into the heterostructure as a sample. (**e**) Cavity conductivity spectrum for the heterostructure with the graphite$_1$. Three additional resonances appear below 1 THz, that correspond to the $s = 0$ (158.5 GHz), $s = 1$ (476 GHz) and $s = 2$ (796 GHz) modes of the graphite$_1$ layer. (**f**) Current density profiles obtained inside the graphite$_1$ layer at the three resonance frequency positions of the $s = 0, 1, 2$ modes. Cavity parameters: $W_0 = 0\,\mu$ m, $W_1 = 3\,\mu$ m and $W_2 = 0.01\,\mu$ m, $d_{gr} = 9$ nm, $d_{hBN1} = 9$ nm, $d_{hBN2} = 93$ nm and $n = 1.1 \cdot 10^{12}\text{cm}^{-2}$. (Figure adapted from [4] under a CC BY 4.0 license)

graphite$_1$) whose simulation parameters were optimized to match the experimental results obtained from the bare graphite self-cavity devices.

The full 3D electromagnetic simulations were performed both for the bare graphite$_2$ heterostructure, and for the heterostructure with an additional, 3 nm thick graphite$_1$ layer assuming a 3D plasma frequency of $\nu_{3D} = 118$ THz and a scattering rate of $\tau^{-1} = 0.03$ THz. The graphite self-cavity mode is centered at a frequency of 856 GHz (see Fig. A.5b) with a current density distribution in the graphite$_2$ layer as shown in Fig. A.5c.

When the additional graphite$_1$ is placed inside the heterostructure, the simulations show three additional resonances ($s = 0, 1, 2$) from the graphite$_1$ layer, while the resonance of the graphite$_2$ self-cavity mode does not change significantly when compared to the heterostructure without the graphite$_1$ layer. The current density distributions of the $s = 0, 1, 2$ modes in the graphite$_1$ layer are shown in Fig. A.5f, exhibiting mostly undistorted modes with wavelengths $\lambda = 4W_1, (4/3)W_1, (4/5)W_1$.

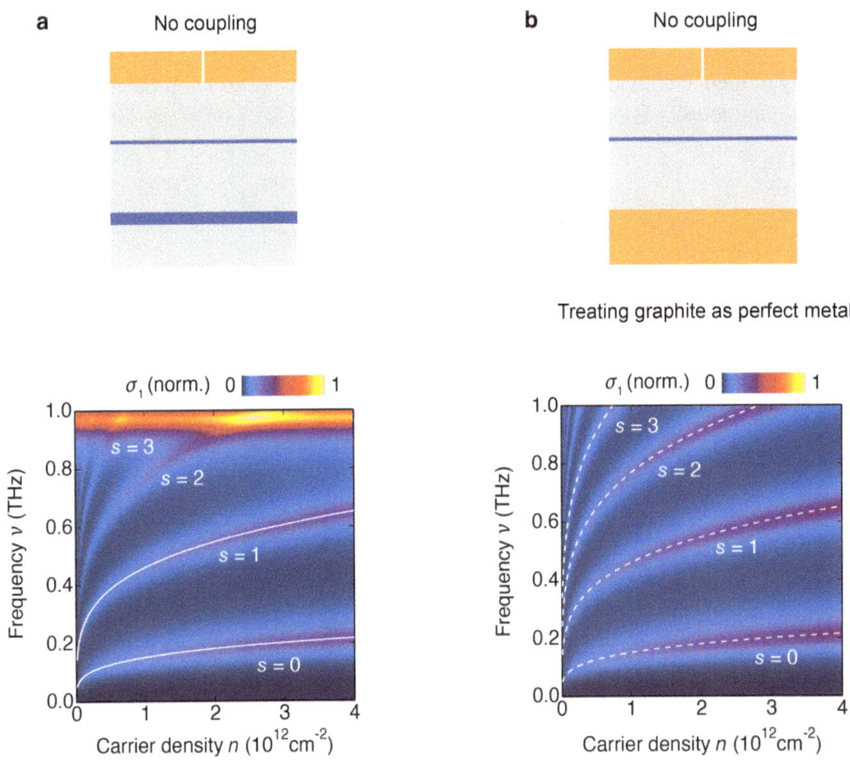

Fig. A.6 Analytical theory calculations for a sensing cavity design as in Fig. A.5c. (**a**) Calculation of the graphene-graphite heterostructure as a function of graphene carrier density. (**b**) Calculation for a heterostructure where the graphite layer is replaced by a perfect metal layer. As the resonances of the graphene plasmonic modes behave identical in both geometries, the sensing nature of a heterostructure with small coplanar strip gap is verified

When the simulations of the heterostructure shown in Fig. 8.20f for $\alpha = 0$ (analytical model) are compared to additional analytical theory simulations where the graphite was replaced by a perfect, thick metal, one can see that both heterostructures exhibit qualitatively similar behavior (see Fig. A.6). The typical resonance frequency scaling of $\nu \sim n^{0.25}$ is observed. In addition, the amplitude of each graphene mode scales with $A_{\text{graphene}} \sim n^{0.5}$. Because the scattering rate τ^{-1} is fixed in the simulations, then the spectral weight of an uncoupled graphene mode scales with $\sim n^{0.5}$.

I would like to note here that while the full 3D electromagnetic simulations were performed for a heterostructure with a 9 nm thick graphite layer, it is advisable to increase the graphite layer thickness to >20 nm to fully suppress current-current hybridization (see Sect. 8.3.1).

Thus, both full 3D electromagnetic simulations and analytical theory simulations illustrate that for a "sensing cavity" geometry, the device can be used to probe the

THz conductivity of vdW heterostructures, without the influence of enhanced self-cavity coupling. The contribution of the graphite gate is independent on the carrier density of the gated 2D material and can thus be directly substracted from the overall cavity conductivity spectrum.

A.7.2 *Cavity Control of Gated 2D Materials*

Full 3D electromagnetic simulations were performed on a heterostructure with geometric parameters equal to the heterostructure with $\alpha = 1$ (see Fig. 8.20) both for the heterostructure with just a graphite$_2$ layer (see Fig. A.7a) and for the heterostructure with an additional graphite$_1$ layer, placed at the same position as the graphene layer in the heterostructure discussed in Fig. 8.20. A thickness of 0.5 nm and a 3D plasma frequency of 276.7 THz (1.144 eV) were chosen to

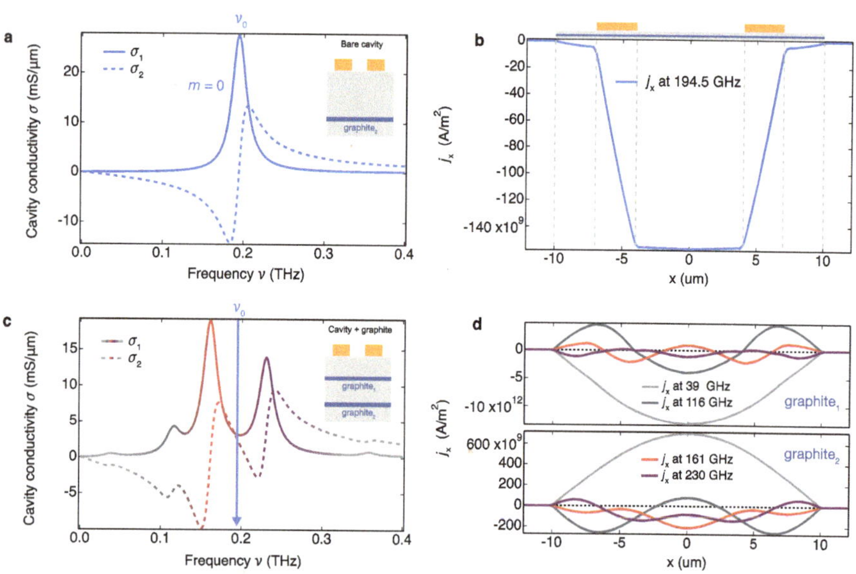

Fig. A.7 Full 3D electromagnetic simulations for an ultrastrongly coupled cavity design as in Fig. 8.20f for $\alpha = 1$. (**a**) Design of the simulated graphite$_2$ self-cavity. (**b**) Real and imaginary part of the cavity conductivity as a function of frequency. The $m = 0$ mode of the graphite$_2$ self-cavity hosts a resonance centered at 194.5 GHz. (**c**) Current density profile in the graphite$_2$ layer at the resonance frequency. (**d**) The heterostructure is filled with an additional graphite$_1$ layer, while all other heterostructure geometry parameters are kept constant. (**e**) Cavity conductivity spectrum for the heterostructure with the graphite$_1$. Instead of a resonance at 194.5 GHz, an avoided crossing appears, where the magenta and purple resonances split equally around the graphite$_2$ self-cavity resonance frequency ν_0. (**f**) Current density profiles obtained inside the graphite$_1$ (upper panel) layer and inside the graphite$_2$ layer (bottom panel) at the frequencies at which the resonances in (**e**) appear. (Figure adapted from [4] under a CC BY 4.0 license)

have the graphite$_1$ layer comparable to a graphene layer with a carrier density of $\sim n = 0.95 \times 10^{12}$ cm^{-2}, the carrier density at which the avoided crossing occurs. The graphite$_2$ layer has the same thickness of 9 nm and typical 3D plasma frequency of 118 THz (0.49 eV). A scattering rate of $\tau^{-1} = 0.02$ THz was chosen for both graphite layers.

The full 3D electromagnetic simulation for the bare graphite$_2$ heterostructure shows that the resonance of the $m = 0$ mode is centered at $\nu_0 = 194.5$ GHz (see Fig. A.7b). The current density distribution j at this resonance frequency is plotted in Fig. A.7c, showing that the current density goes to zero at the edges of the dual graphite heterostructure. When the additional graphite$_1$ layer is placed in the heterostructure (see Fig. A.7d), the cavity conductivity spectrum shown in Fig. A.7e reveals that the magenta and purple modes form a large avoided crossing and are split equally around ν_0, with $g = 34.5$ GHz, $\nu_0 = 194.5$ GHz, giving $\eta = g/\nu_0 \approx 0.18$, equal to the value obtained from the analytical model for this heterostructure.

The current density distributions in both the graphite$_1$ and in the graphite$_2$ layer are shown in Fig. A.7f. All resonances appear in both graphite layers. The signs of the current densities are opposite, as expected from image charges in the two layers.

A.8 Characteristic Impedance Engineering

This section discusses how an optimum signal-to-noise ratio in detected cavity conductivity spectra can be obtained based on comparing the characteristic impedance Z_0 of a transmission line to the impedance Z_s of a sample, that is given by:

$$Z_s = \frac{1}{Y_s} \approx \frac{1}{\sigma_{\text{cavity}} l} = \frac{1}{\sigma_{\text{analytical theory}} C_{\text{filling}} l}, \tag{A.35}$$

where Y_s is the admittance of the sample, l the sample's length, W the sample's width, σ_{cavity} is the cavity conductivity, C_{filling} is the filling factor (see Sect. 6.4.7) and $\sigma_{\text{analytical theory}}$ is the output conductivity of the analytical theory.

Figure A.8 shows two scenarios of signal distortion from discontinuities in a transmission line. Firstly, parts of a signal would reflect if there is a transition from a transmission line with characteristic impedance Z_0 to a transmission line with characteristic impedance Z_1 (see Fig. A.8a). This scenario can be described by Fresnel's equations, as discussed in Sect. 5.3, and the reflection r and transmission t coefficients are [5]:

$$t = \frac{2Z_1}{Z_1 + Z_0} \tag{A.36}$$

$$r = \frac{Z_1 - Z_0}{Z_1 + Z_0}. \tag{A.37}$$

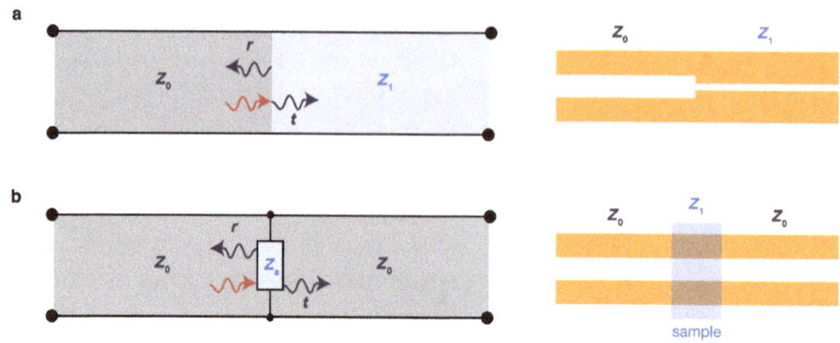

Fig. A.8 Reflections in transmission lines due to discontinuities. (**a**) A signal partially reflects in a transmission line if the characteristic impedance changes from Z_0 to Z_1. The transmission t and reflection r coefficients are described by Eqs. A.36 andA.37. (**b**) If a sample is embedded in a transmission line, the sample can also be modeled as a lumped element with impedance Z_s. The transmission t and reflection r coefficients are described by Eqs. A.38 and A.39

These coefficients were applied in Sect. 5.3 for both transitions from empty transmission line to transmission line with sample and transmission line with sample to empty transmission line to extract the near-field optical conductivity per sample length.

A different approach is to assume that a transmission line with characteristic impedance Z_0 is distorted by a lumped element with impedance Z_s, the sample (see Fig. A.8b). In that case, the transmission and reflection coefficients are [6]:

$$t = \frac{2}{2 + \frac{Z_0}{Z_s}} \tag{A.38}$$

$$r = \frac{-Z_0}{2Z_s + Z_0}. \tag{A.39}$$

Such a "shunt-configuration" is a good approximation for a sample embedded in on-chip THz spectroscopy transmission lines.

To determine the characteristic impedance Z_0 of an on-chip THz spectroscopy transmission line (see Fig. A.8b) that maximizes the signal-to-noise ratio in the conductivity spectrum of a sample with impedance Z_s, let's first plot Eq. A.38 as a function of Z_0/Z_s. As shown in Fig. A.9, most of the pulse is transmitted ($t \approx 1$) if $Z_s \gg Z_0$, whereas almost no transmission occurs ($t \approx 0$) if $Z_s \ll Z_0$. If $Z_0 = 2Z_s$, exactly 50% gets transmitted and 50% gets reflected (see Fig. A.9a).

In order to then find the Z_0/Z_s ratio at which the sensitivity towards small changes in the conductivity is maximized, two aspects have to be considered. First, the transmission should respond strongly to a small variation in conductivity (the derivative of Eq. A.38 with respect to Z_0/Z_s should be large). Second, the overall reflection from a sample must be sufficiently high for changes to be detectable above the noise floor, which generally favors larger Z_0/Z_s ratios. However, excessive reflection can become detrimental: if too little signal reaches the detector switch,

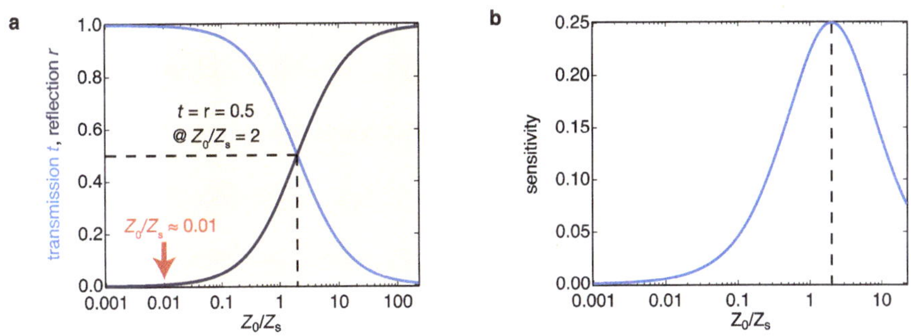

Fig. A.9 Signal transmission and reflection coefficients as a function of impedance contrast. (**a**) Transmission t and reflection r of a signal as a function of the ratio of characteristic impedance Z_0 of a transmission line and impedance of a sample Z_s plotted with Eq. A.38. 50% of a signal gets transmitted and 50% gets reflected at $Z_0/Z_s = 2$. The yellowish shaded area roughly defines the Z_0/Z_s range that is detectable based on the signal-to-noise level of the measurements conducted in this thesis, indicating that $Z_0/Z_s < 0.01$ is not possible to detect. (**b**) Sensitivity towards small changes in conductivity as a function of Z_0/Z_s. Maximum sensitivity is yielded at $Z_0/Z_s = 2$ (see also Ref. [6])

it may struggle to resolve small changes reliably. This means that the product of the absolute value of the derivative of Eq. A.38 with respect to Z_0/Z_s times Z_0/Z_s should be maximum for the optimal sensitivity:

$$\text{sensitivity} = \frac{dt}{d\left(\frac{Z_0}{Z_s}\right)} \times \frac{Z_0}{Z_s}. \tag{A.40}$$

As shown in Fig. A.9b, the sensitivity is maximized at $Z_0/Z_s = 2$, at which exactly 50% of the signal gets transmitted and 50% gets reflected (see Fig. A.9a).

For graphite heterostructures measured in a transmission line with $W_1 = W_2 = 3\,\mu\text{m}$ (metal trace width W_1 and coplanar strip gap size W_2), the detected THz signals on the sample side showed approximately 50% lower amplitudes compared to the reference signals (see Fig. 6.2a for example), indicating that $Z_0/Z_s \approx 2$. In case of the gate-tunable graphene heterostructure, the lowest change in transmission that was still detectable occurred at a relative transmission between the graphene charge neutrality point and the lowest measured graphene conductivity of $\Delta t \sim 0.005$ (see time traces of Device GG1 in Fig. 8.2). This defines a measurable range of $t = 0.5$ to 99.5% with the current signal-to-noise ratio, highlighted in yellow in Fig. A.9a. Although the exact Z_0/Z_s ratio is difficult to determine due to the frequency and filling factor-dependent impedance, these estimates suggest that for low-conductive samples, where $Z_0/Z_s \ll 2$ (as $Z_s = 1/Y \propto 1/\sigma_{2D}$), extracting the near-field optical conductivity becomes progressively more challenging.

However, these estimates overlook an important contribution: the graphite gate. The frequency dependent cavity conductivity is the sum of the near-field conductivity of a 2D material and the near-field conductivity of the gate, such that

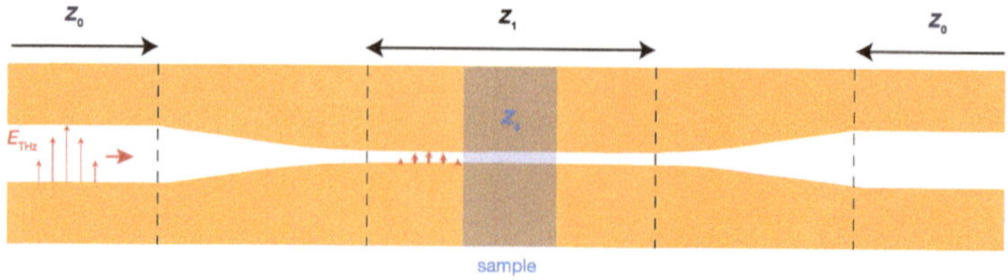

Fig. A.10 Impedances in the sensing cavity geometry. To perform cavity sensing of a sample, a coplanar strip with a $W_1 = W_2 = 3\,\mu$m geometry, that has a characteristic impedance Z_0, is smoothly narrowed down to $W_1 = W_2 = 0.1\,\mu$m in the sample region. The resulting characteristic impedance, Z_1, for this narrower transmission line is lower than Z_0. Since the coplanar strip on the reference side is similarly tapered, the sample impedance Z_s should be compared to Z_1, not to Z_0. For optimal resolution, the ratio $Z_1/Z_s \approx 2$ should be maintained

Eq. A.35 transforms to:

$$Z_s(\nu) = \frac{1}{Y_s(\nu)} \approx \frac{1}{(\sigma_{\text{2D material, cavity}}(\nu) + \sigma_{\text{graphite gate, cavity}}(\nu))l}. \tag{A.41}$$

Thus, the impedance of a sample can be engineered with the graphite gate: dependent on how thick the graphite gate is and where the resonance of the lowest-frequency mode is positioned via the coplanar strip and dielectric environment design, the sample impedance can be tailored to match $Z_0/Z_s \approx 2$ over a possibly large frequency range.

As a resonance in the cavity conductivity spectrum automatically produces a highly frequency-dependent impedance (a minimum impedance is found at the resonance position and a decreasing impedance is found when slowly moving away in frequency from the resonance frequency), a good approach is to center the graphite $m = 0$ mode towards higher frequencies either within or just outside the spectral sensitivity range. This way, the low-frequency tail of the resonance provides a more uniform conductivity offset, reducing $Z_s(\nu)$. By optimizing the graphite thickness to achieve $Z_0/Z_s \approx 2$ at the center of the spectral sensitivity range, even small changes in the conductivity of a 2D material can be detected with maximum sensitivity.

An alternative approach to tuning the ratio Z_0/Z_s is to modify Z_0 rather than Z_s. As detailed in Sect. 2.2.2, the characteristic impedance Z_0 of a transmission line is highly dependent on the ratio of W_1 to W_2. However, since this ratio already serves as a tuning parameter for the second self-cavity coupling mechanism (see Sect. 8.3.2), it cannot be further adjusted for impedance engineering. While a $W_1 = W_2 = 3\,\mu$m geometry yields $Z_0 \approx 103\,\Omega$, a transmission line with $W_1 = 5\,\mu$m and $W_2 = 0.1\,\mu$m on the same substrate with $\epsilon_r = 10$ has a characteristic impedance of $Z_1 = 42\,\Omega$ [7]. In such a cavity sensing geometry, the transmission line is only locally narrowed down, as illustrated in Fig. A.10. As the transmission line on the sample side is equally narrowed down than on the sample side, the sample's impedance Z_s has to be compared to Z_1, not to Z_0 (see Fig. A.10).

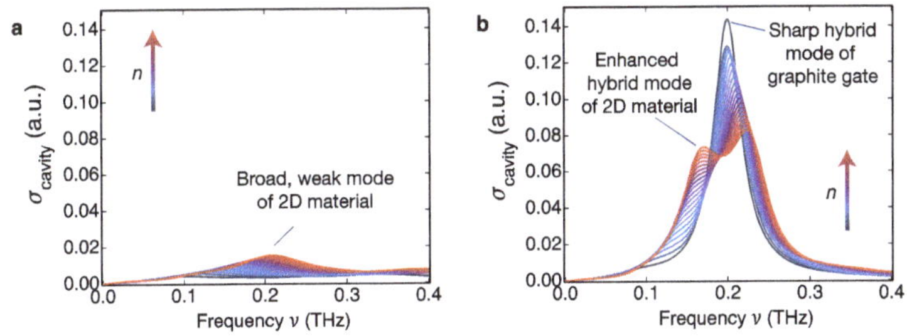

Fig. A.11 Detecting broad modes via ultrastrong coupling to a graphite gate. (**a**) Cavity conductivity spectrum of a weak mode in a 2D material with a large scattering rate, shown for various carrier densities. (**b**) Cavity spectrum of the same 2D material when coupled to the narrow plasmonic mode of a graphite gate

Thus, to yield $Z_0/Z_s \approx 2$ in a sensing cavity geometry, Z_s has to be engineered smaller, as achievable with a thick graphite gate. Additional procedures that help yielding $Z_0/Z_s \approx 2$ are:

- The usage of a substrate with lower ϵ_r, as this leads to an increasing characteristic impedance (see Sect. 2.2.2). In this thesis, sapphire substrates with $\epsilon_r = 10$ were used. However, if fused quartz substrates with $\epsilon_r \approx 3.7$ are used instead, the characteristic impedance of a given coplanar strip geometry can be increased by a factor of 1.6.
- Longer samples in y-direction do not change the self-cavity mode landscape, but decrease Z_s and thereby increase the contrast in the transmitted THz pulse.

In the cavity control geometry, where $W_1 < W_2$ locally, the condition of $Z_0/Z_s \approx 2$ is obtained more easily for low-conductive materials. The characteristic impedance of the transmission line increases for larger gap transmission lines and additionally, low amplitude or broad signals are amplified through ultrastrong coupling with a graphite gate (that yields good signal contrast in typical transmission line geometries), such that detecting a hybridized spectrum is generally possible for low-conductive samples (see Fig. A.11).

In contrast to samples with low-conductivity, samples with high conductivities (low Z_s) can exhibit $Z_0/Z_s \gg 2$. However, this condition is barely reachable in vdW heterostructures and is more easily obtained in grown or evaporated materials. In case of grown or evaporated superconducting thin films, the sample's width, thickness and length can easily be tailored (see Eq. A.35) to reach $Z_0/Z_s \approx 2$ [6].

A.9 On-chip THz Spectroscopy in DC Contact

While the 2D materials investigated both experimentally and theoretically in this thesis were all electrically isolated from the coplanar strip of the on-chip

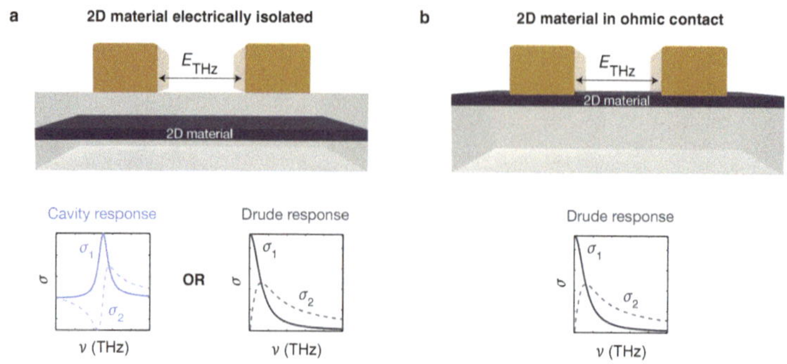

Fig. A.12 Electrically isolated vs electrically contacted 2D materials in on-chip THz spectroscopy. (**a**) 2D materials that are electrically isolated from the coplanar strip, as studied in this thesis, exhibit either a cavity response, for typically sized vdW heterostructures, or a Drude response, if the sample would be very large (see Section 6.6). (**b**) In contrast, a 2D material that is in direct electric contact to the coplanar strip always a Drude reponse, independent of the heterostructure size

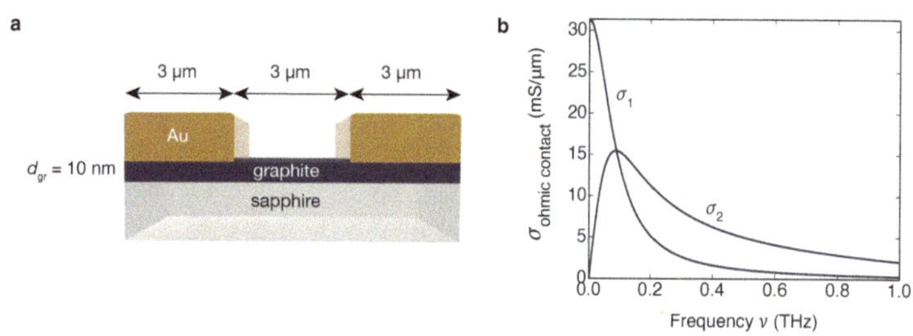

Fig. A.13 Full 3D electromagnetic simulation of a graphite heterostructure in ohmic contact to the circuitry. (**a**) Simulated heterostructure geometry. (**b**) Real and imaginary parts of the simulated near-field conductivity, that resembles a Drude response. Simulation parameters: $\Gamma = 0.5\,\text{THz}$, $\nu_{3D} = 118\,\text{THz}$, and the height of the gold traces is $d_{Au} = 300\,\text{nm}$

THz spectroscopy circuitry, one can also directly contact the 2D material with the coplanar strip (see Fig. A.12). While the cavity conductivity spectra of the electrically isolated 2D materials exhibited a plasmonic cavity response, where the cavity conductivity is zero at zero frequency, a Drude response can only be anticipated for very large samples (see Sect. 6.6). Thus, in the devices studied in this thesis, one can expect to sense *either* a cavity response *or* a Drude response (see Fig. A.12a). In contrast, if the 2D material would be placed in direct electric contact to the coplanar strip, a Drude response could be detected (see Fig. A.12b).

Figure A.13 shows the result of a full 3D electromagnetic simulation by Dr. Toru Matsuyama of such an ohmic contact geometry. As shown in Fig. A.13b, real and imaginary parts of the near-field conductivity show a Drude response. This is expected, as the electric field of the THz pulse can excite a current in the 2D material that can directly flow from metal trace to metal trace of the coplanar strip.

Interestingly, the filling factor C_{filling} (see Sect. 6.4.7) also has to be considered in such a geometry. The 2D conductivity of the graphite layer simulated in Fig. A.13 at zero frequency would be given by

$$\sigma_{\text{DC}} = \frac{\epsilon_0 (2\pi \nu_{3\text{D}})^2 d_{\text{gr}}}{2\pi \Gamma} \tag{A.42}$$

where $\nu_{3\text{D}} = 118\,\text{THz}$, $\Gamma = 0.08\,\text{THz}$ and $d_{\text{gr}} = 10\,\text{nm}$, such that

$$\sigma_{\text{DC}} = 96.83\,\text{mS}. \tag{A.43}$$

To match the simulated DC conductivity of $\sigma_{\text{ohmic contact}} = 33.3\,\text{mS}/\mu\text{m}$, a filling factor of $C_{\text{filling}} = 344\,\text{mm}^{-1}$ has to be multiplied. Such a high filling factor matches the expectation that the filling factor increases with decreasing distance of the 2D material from the coplanar strip (center), as also shown in Fig. 6.17.

The approach of placing a sample in direct ohmic contact to the transmission line broadens the possible applications of on-chip THz spectroscopy. For instance, if the quantum phases appearing in a Moiré superlattice sample, such as magic angle twisted bilayer graphene, shall be characterized, one could choose between a "cavity sensing" (Sect. A.7.1) or "direct electrical contact" geometry (Fig. A.12b). While in case of the "cavity sensing" geometry the cavity conductivity (see lower panels for σ_{cavity} in Fig. 10.4) would be probed, the 2D conductivity would be probed in a "direct electrical contact" geometry (see upper panels for $\sigma_{2\text{D}}$ in Fig. 10.4). When considering the filling factor discussed in Sect. 6.4.7 in measurements with either of the two geometries, both geometries enable the extraction of intrinsic material properties such as gap openings and magnitudes in superconducting or correlated insulating phases.

The advantage of the direct electrical contact geometry in this case is that only the coplanar strip gap region contributes to the signal. This would result in improved twist angle homogeneity as the detected sample area is reduced by over a factor of three. Further, the heterostructure width does not have to be considered when designing the device, as a Drude response is expected independent of the heterostructure width. However, a drawback of the direct electrical contact geometry, compared to the electrically isolated geometry explored in this thesis, is that the sensitive 2D material is not protected from its deleterious dielectric environment. This could lead to material degradation, strain, or inhomogeneity. While these concerns are negligible for robust thin films, they present substantial risks for delicate Moiré lattices like twisted bilayer graphene, which typically require both bottom and top electrostatic gates to tune the electronic properties of the material. Such a top electrostatic gate cannot be implemented in a direct contact geometry, making it essential to carefully weigh the trade-offs between the two on-chip THz spectroscopy sensing geometries. Furthermore, when an electrostatic (graphite) gate is required for measuring a 2D material, it must be electrically isolated from both the 2D material and the coplanar strips. As a result, finite size effects must always be considered when probing the THz electrodynamics of gate-tunable 2D quantum materials - an insight advanced by this thesis.

References

1. Novoselov KS, Geim AK, Morozov SV, Jiang DE, Zhang Y, Dubonos SV, Grigorieva IV, Firsov AA (2004) Electric field effect in atomically thin carbon films. Science 306:666–669. https://doi.org/10.1126/science.1102896
2. Yoon H, Yeung KY, Kim P, Ham D (2014) Plasmonics with two-dimensional conductors. Philos Trans Ro Soc A: Math Phys Eng Sci 372:20130104. https://doi.org/10.1098/rsta.2013.0104
3. Economou EN (1969) Surface plasmons in thin films. Phys Rev 182:539–554. https://doi.org/10.1103/PhysRev.182.539
4. Kipp G, Bretscher HM, Schulte B, Herrmann D, Kusyak K, Day MW, Kesavan S, Matsuyama T, Li X, Langner SM, Hagelstein J, Sturm F, Potts AM, Eckhardt CJ, Huang Y, Watanabe K, Taniguchi T, Rubio A, Kennes DM, Sentef MA, Baudin E, Meier G, Michael MH, McIver JW (2025) Cavity electrodynamics of van der Waals heterostructures. Nat Phys 21:1926–1933. https://doi.org/10.1038/s41567-025-03064-8
5. Pozar DM (2005) Microwave engineering, 3rd edn. Wiley, Hoboken
6. Potts AM (2023) On-chip terahertz time domain spectroscopy for sub-diffraction van der Waals heterostructures. PhD Thesis. University of California, Santa Barbara
7. https://zalophusdokdo.github.io. Coplanar strips calculator (2025)

FSC
www.fsc.org
MIX
Papier | Fördert gute Waldnutzung
FSC® C083411

Kolibri 360 GmbH
Ferdinand-Jühlke-Straße 7
99095 Erfurt, Deutschland
produktsicherheit@kolibri360.de